à ó

avec

comm

Military Politics

FROM BONAPARTE TO THE BOURBONS

Raymond Horricks

1994

et maintenant pour

Tony Clark

avec l'auteur's amitiés,

Ray

Sept., 1996

Military Politics

FROM BONAPARTE TO THE BOURBONS

The Life and Death of Michel Ney,
1769-1815

Raymond Horricks

With a New Introduction
by the Author

Transaction Publishers
New Brunswick (U.S.A.) and London (U.K.)

Library of Congress Catalog Number: 94-17380
ISBN: 1-56000-767-2
Printed in the United States of America

Library of Congress Cataloging-in-Publication Data

Horricks, Raymond, 1933-
    Military politics from Bonaparte to the Bourbons/Raymond Horricks.
    p.   cm.
    Includes bibliographical references and index.
    ISBN 1-56000-767-2
    1.   Ney,   Michel,   duc   d'Elchingen,   1769-1815.   2.
Marshals—France—Biography. 3. France. Armée—Biography. 4. Napoleon
I, Emperor of the French, 1769-1821—Relations with marshals. 5.
France—Military policy—History—19th century. I. Title.
DC198.N6H673 1994
944.05'092—dc20
[B]   94-17380
                                                                 CIP

*for Justine Gabrielle,
and for her mother*

'You can form no idea of Ney's brilliant courage—
equalled only in the Age of Chivalry.'

BERTHIER, *Chief-of-Staff to Napoleon, after the
Battle of Friedland*

'What is war? A barbarous profession whose art consists
in being stronger than the enemy at any given moment.'

NAPOLEON, *before Borodino*

'A Marshal of France never surrenders. One does not parley
under the fire of the enemy.'

MICHEL NEY, *during the retreat from Moscow*

'Unhappy man, you were reserved for French bullets!'

VICTOR HUGO

Contents

List of Illustrations

1. Ney with his staff, by Meissonier
2. Baptismal certificate of Michel Ney
3. General Kléber
4. General Moreau
5. General Hoche
6. Napoleon, by Vernet
7. Josephine de Beauharnais, after an engraving by Isabey
8. Aglaé Ney, by Gérard
9. Joseph Bonaparte, by Gérard
10. Marshal Soult
11. Talleyrand, by Godefroy
12. Marshal Masséna
13. Ney in the retreat from Moscow, by Yvon
14. The First Abdication, Fontainebleau, 1814, by Berne-Bellecour
15. Waterloo: After the battle
16. Arthur Wellesley, Duke of Wellington, by Goya
17. Louis XVIII, King of France, by Gros
18. Ney imprisoned in the Conciergerie
19. Joachim Murat, by Gérard
20. The execution, December 7, 1815
21. The grave at Père Lachaise
22. Detail from Père Lachaise

Introduction to the Transaction Edition

The personal genesis for this book is detailed in the prologue which follows; my enthusiasm as Ney's biographer grew as the research progressed. All the things I know or have decided about him are contained in the finished text. And several editions later I can find no reason to change my overall viewpoint. About the man, that is. However, with the writing of his life left behind, and now reinforced by a decade's hindsight, I would like to take the opportunity of making an addition to my held viewpoint, one which is both connected to, but at the same time exterior to the man. Namely that, without the social upheaval and consequent necessities of his time, very few of us could ever have been made aware of him. This is in no way to underestimate his bravery as an individual soldier, nor his later talents as a commanding officer. It is simply restating a cliché: 'the right man in the right place at the right time'. But then a cliché is seldom wrong. Indeed history almost invariably consists of them. They are the affirmations of proven truths.

Prior to the French Revolution, Michel Ney—in common with most of the Napoleonic marshals—should not have expected to rise above the rank of Senior NCO. A similar fate would have awaited him in the British Army right up to the exigencies of World War II; Field-Marshal, later Viscount, Slim being the exception to prove the rule here. This said, the French Revolution *did* intervene, there has been no more important social event in modern European affairs before or since, not even in Russia, and in terms of military politics, Ney, and much more so Napoleon Bonaparte, became the men of the hour.

The revolutionaries themselves were essentially of the middle class—who in turn made astute use of a largely discontented proletariat as their cutting arm. France's monarchy was by no means the only corrupt and inefficient one in Europe. Again though it was a matter of *certain* men being in *certain* places as the situation grew ripe. These revolutionaries were also gifted. Lawyers, editors, *et al.*: they too wanted a place in the sun. And having gained it, they were fully prepared to take on the rest of the world in order to keep it. For a time they succeeded brilliantly. Whatever its excesses in other directions, for administrative ability, shouldering a prodigious workload, and as a war cabinet, the Committee of Public Safety can hardly be faulted. Lazare Carnot almost on his own built whole Revolutionary armies practically from scratch. There was no obvious requirement for France to turn to a new, military-spearheaded form of leadership. But the fall of Robespierre has to be regarded as the watershed. Paul Barras and the Directors who followed proved as corrupt as the *ancien régime*; once more the situation grew ripe.

The advent of Napoleon Bonaparte in French politics lends itself to Thomas Carlyle's theory that history is actually the history of great men: simplistic, and even naive though this sometimes seems. Certainly in the case of France, her history has been marked by the appearance of such people just as the nation's fortunes have hit an exceptional low: Charles V, Joan of Arc in tandem with the future Charles VII, Louis XI, Henry IV, Cardinal Richelieu, Napoleon, and Charles de Gaulle. We do not have to like them, nevertheless they were restorers. Moreover they each promoted on merit. Under Napoleon, Michel Ney only had to do well.

Relations between the two were not always serene, of course. In fact, I am reminded of a *pensée* by Blaise Pascal at this point: 'We do not feel our bonds as long as we follow willingly him who leads; but as soon as we begin to resist and to draw away, then indeed we suffer'. Ney would experience some sufferings under Napoleon, but then he lived to discover the alternative as so much worse. As the biographer of his real life I have only a most tenuous right to speculate upon the aftermath if his execution, whatever its injustice, had not taken place. A term of imprisonment, followed by banishing back upon his estate at Coudreaux, perhaps? Some duck shooting and fishing there? Cards? With winters spent as a clubman in Paris? Hardly, I suggest. The French have a most expressive word for boredom—*l'ennui*—which in Ney's particular case could well have led on towards another symptom: *spleen*, especially when witnessing further events as ordained by the restored Bourbons; unless longevity had allowed him to note their own eventual, ignominious downfall in 1830, which other Bonapartist officers did survive to enjoy.

The political answer for the taming of any revolution, and decidedly when as socially important as France's of the late eighteenth century, is not to bring back more of what had been originally overthrown. Unfortunately, Europe's other leaders of that period were too blinkered to realize the fact. Subsequent readings incline me to believe that Britain's new foreign secretary, Lord Palmerston arrived at the right conclusion; but was forced to keep his private opinions under wraps—or limited to making the Napoleonic turncoat, Talleyrand as France's ambassador, wait for hours in his outer office!

History, I often hear proclaimed, is cyclical, with man at its epicentre being the constant, and only the details surrounding him changing; 'details' referring here to the advance of technology. I don't subscribe to this belief. Basic human nature—love, anger, generosity, greed—may not change, but whether we wish it or not, for better or worse, we are *honed* by the added acquaintance of discovery. It is after all our grand hope: to collect fresh experiences and benefit from them. What I will admit to though are cyclical coincidences, and, in conclusion, two such occur to me regarding Michel Ney. First, that he was born, but then also died under an absolute monarchy—while in between contributing to the most volcanic political process our European arena for events has yet seen. And second, he himself climaxed, by giving up a single life, a process of monstrous hypocrisy. For I can claim, at the risk of appearing prejudiced, that the Bourbons persuaded the rest of Europe's established rulers to band together against 'the upstart Bonaparte' because he symbolized 'the bloodthirstiness and tyranny' of the Revolution. But then, in coming back to power, they themselves resorted to exactly the same methods. However, in doing so, they turned Ney, as their principle victim, into twice the national hero he had proven himself, by his natural abilities, on the battlefield.

I rest my case. May the reader be the judge now.

Raymond Horricks
Isle of Wight, England
1994

PROLOGUE

A Journey in November

All the way across northern France it had rained and it was raining still.

After avoiding Paris and then driving due east, over the flatlands towards Champagne, it seemed as if the bad weather had become a permanent escort: nothing but black swollen skies and a steady sluice across the windscreen of the car. At Soissons another tremendous thunderstorm turned the streets into fast-flowing rivers, the squares into agitated lakes. By Reims enough was enough! Feeling fed-up even with being a Francophile, I pulled off the N31 and second-geared in and around the bucketing city to find a *hôtel*.

Later, over a solitary meal, and while night and the wet intermingled outside, I thought again about the extraordinary ends to which a natural curiosity can compel one.

Saint-Exupéry writes that:

> to come to man's estate it is not necessary to get oneself killed round Madrid, or to fly mail planes, or to struggle wearily in the snows out of respect for the dignity of life. The man who can see the miraculous in a poem, who can take pure joy from music, who can break his bread with comrades, opens his window to the same refreshing wind off the sea. He too learns a language of men.

'But,' he adds, 'too many men are left unawakened.'[1]

No doubt this is true; in which case the human predicament is very sad indeed. For how is the individual to be sure of *knowing*, of waking up to the right calls? All I do know is that the best things to happen in my own life have been as the result of a single-minded wish to discover more. And this book, if it has any merit, grew from one such wish.

I am not a professional historian. Nor do I consider myself a great expert on military affairs. On the other hand, and for a number of years, I have been fascinated by the career and times of Napoleon's Marshal Ney, *Le Brave des Braves*. He is an argued-about figure even today: much discussed as a general officer, and with blame frequently attached to some of the things he did. But hardly ever, anywhere, does one find him carefully discussed as a man. Why?

It is exactly this last problem, a relating of the human being with the soldier, which has fuelled my curiosity from the outset.

Writing the book, I admit, has been a kind of journey for me too. Perhaps by trying to lift the covers off a fellow-man one cannot help but go part of the way into oneself. At least though I believe I have developed an affinity with my subject, in the end liking more about him than I dislike, and seeing a unity where before there appeared much incoherence.

I date my aroused interest in Ney from a half-humorous incident in Paris when my daughter was very young. We were standing outside the Closerie des Lilas, at the intersection where the Boulevard du Montparnasse meets the Avenue de l'Observatoire. The Lilas, formerly a café where Ernest Hemingway wrote several of his best short stories, is now a dignified, expensive restaurant. Suddenly, as I thought about *Up In Michigan* and *The Battler* and *Big Two-Hearted River*, my daughter broke away. She darted into the restaurant by a side-entrance on the terrace, then up to an immaculate but by now astonished *maître d'hôtel*, announcing loudly: 'I want meat!' I nearly knocked over a table in the pursuit. Red-faced apologies were followed by our rapid retreat. Back outside to the last of the city's daylight and a gentle 'talking to'. . .

It was at this precise moment I first noticed the statue. Rude's statue of Ney, standing angled at the top of the Avenue de l'Observatoire, turned towards the famous fountain and its even rows of chestnut-trees. (Also—I found out later—near the place of his bizarre and unjust execution.)

A closer examination of the high, defiant bronze on its square stone pedestal reminded me how little I actually knew about the Marshal. For years he'd been a 'Big Name' in my mental catalogue of history; the next after Napoleon in every popular account of the *Grande Armée*. But that was the end of it. *Or rather a beginning.* Because, gazing up at the statue, I wanted to learn more.

Hemingway himself refers to the 'many days Ney had fought, personally, with the rearguard on the retreat from Moscow that Napoleon had ridden away from in the coach with Caulaincourt'. Although this comes after saying: 'What a fiasco he'd made of Waterloo. . .'.[2] While Brigadier Peter Young states: 'Had Soult, not Ney, commanded the left-wing of the Army of the North the story of Quatre Bras and Waterloo might have been very different'.[3] But surely such distinguished leads for the prosecution are omitting certain facts here. Ney remained directly under Napoleon's command at Waterloo (and nearly always under his eye). Also, he was up against Wellington, who fought with genius that day, yet who confessed later: 'It was the most desperate business I ever was in. I never took so much trouble about any battle, and never was so near being beat.'[4]

What was Ney really like? And what did his life contain before that was to have an influence on his performance at Waterloo? So many questions needed to be answered. All of them leading me to a broken journey on a miserable night and a small *hôtel* in Reims!

However the next morning came up dry if not yet clear—so that in turning the ignition-key my original enthusiasm also revived. Pools and puddles were everywhere; the city looked like Atlantis risen above its station. Already though a fresh wind licked at the water's edges, and overhead the clouds were racing and thinning towards Switzerland.

I decided wherever possible now to skip the major roads with their convoys of juggernauts. I was still heading east, but always some kilometres north of the busy Verdun-Metz route. . .

The weather would prepare at least two more ambushes in the course of that morning. One in the higher parts of Argonne, where a series of icy showers came at me almost diagonally just as the bends hairpinned and started to look dangerous. Then another beyond Étain, when hail the size of marbles bombed and bounced me to a standstill in a place called Fléville. ('Around here,' the barman making my coffee said, 'the winter gets so bad that the only time anyone ever opens a window it's to pass a coffin through.')

But the hail around Fléville was to be the last angry assault by an otherwise fleeing enemy. After leaving Briey and then turning left for Thionville, I began to pass people standing at bus-stops or riding *mobilettes* with a thin sun slanting across their faces. And soon I could feel my own face-muscles puckering up against it. Ahead, the sky was an odd kaleidoscope of ragged grey and broken white rapidly dissolving into blueness. The road wound on between tall trees: dark green conifers, with the occasional orange splash where a tree had died. Then it dipped briefly before rising again and snaking in towards Thionville.

Through the city and on. Prominent advertisements for a variety of strong fruit liqueurs had replaced the DUBONNET and BYRRH signs by this time. And driving along I recalled one more thing about Ney. Somewhere, someone had written that the Marshal, although never a close friend of Napoleon's, on occasions had dared to speak to him as no one else in France did.

I lowered the driver's-window slightly. The air was cold, in spite of the sun. But it carried inside a faint odour of the countryside. A compound of the earth itself and certain late herbs, acrid but pleasant.

The next signpost said: GERMANY 5k.

[1] *Wind, Sand and Stars*, New York, 1939.
[2] *A Moveable Feast*, New York, 1964.
[3] *Napoleon's Marshals*, London, 1973.
[4] Antony Brett-James: *The Hundred Days. Napoleon's Last Campaign from Eyewitness Accounts*, London, 1964.

CHAPTER 1

Early Life and Character

The first big surprise about Michel Ney is to discover that, at least to begin with, he wasn't really a Frenchman at all. Despite legal and some purely technical claims to the contrary, this most fervent French patriot was by family background, language, religion and even geography entirely German. He was born at Saarlouis under the deep-winter sign of Capricorn on 10 January 1769: obviously an exceptional year, because it also saw the births of Napoleon, the future Duke of Wellington, Lord Castlereagh and two of Ney's fellow-marshals, Jean Lannes and Nicholas-Jean de Dieu Soult.

Saarlouis today is a prosperous West German town of 50,000 self-made, hard-working, matter-of-fact people. It has coal and steel interests, a partly female taxi-service and a big modern area across the River Saar (Roden) where the station-*hôtel* has been turned into a hostel for Turkish immigrant workers. Certainly as one mixes with its people along the central Französische Strasse, or tackles a large helping of cream-and-chocolate cake on a Sunday afternoon, it is hard to imagine the town being anything else but essentially German. However its name is the giveaway. On account of its strategic position, during several periods of European history the place has been in French hands: most notably the Sun King's, Louis XIV, who ordered his chief military architect Vauban to surround it with 'impregnable fortifications'.

Much knocked-about during World War II, the new skyline of Saarlouis is dominated by a filigree clock-tower with an ornamental sun on top. The sun seems defiantly appropriate. In part by its reference to the town's one-time illustrious occupier. Also since this is an area where it rains a lot.

Not everything was destroyed in the war though. The spire and front-porch of the church where Ney was baptised are still standing; and so is the outer shell of the house where he was born. Perhaps of greater importance, although the original Town Hall was flattened, either by a miracle or just good German cellar-building the book which contains the actual registry of Ney's baptism has survived — allowing one to clear up at least two long standing arguments about the Marshal.

I owe these discoveries initially to the Saarlouis police, and in particular to

Heinz Nienas. I speak French but hardly any German; and approached the local *polizei* in a bid for further information. They responded with an officer who spoke reasonable English and excellent French. He not only guided me to Ney's true birthplace, but in addition gave me the name and address of the new *propriétaire* of the house, caused the Saarlouis museum to be opened up after its scheduled hours, showed me the Vauban-Insel with its 'other' statue of the Marshal and finally made an appointment for me the next day at the Town Hall: 'With Herr, or rather Monsieur Fontaine. He, quite definitely, will be able to answer your other questions if anyone can.' And so it proved.

The main arguments connected with Michel Ney's background and birth have revolved about the names of his parents and the identification of his religion. Everyone agrees that he came from humble origins; but several writers claim his father was named Nicholas and his mother Catherine or Catharine, and they dismiss the counter-claim that their names were Pierre and Marguerite as an error of confusion with Michel's elder brother and his sister. *The Memoirs of Marshal Ney*—compiled after his death—provide us with certain facts about the father but include no mention of his name. Meanwhile opinion would appear to be equally divided over whether Ney was a Protestant or a Catholic.

The case for 'Nicholas' and 'Catharine' rests upon a document, allegedly prepared for and signed by Madame Michel Ney on 15 June 1823, in which she renounces on her sons' behalf all rights to a modest legacy left by Dame Francisca Rothmund of Württemberg to 'her niece, Madame Catharine Ney, née Rossman, mother of the late *Monsieur le Maréchal, Prince de la Moscowa, Duc d'Elchingen*'. And this in turn is linked with an official record, filed at Friedrichshafen on the shores of Lake Constance, showing Catharine Rossman to be the wife of one Nicholas Ney, born there. How exactly such a document came to be accepted so easily is the real mystery.[1] If authentic, then perhaps Madame Ney made a genuine mistake. As for the Friedrichshafen record: it must involve a totally different couple. The name 'Ney' is common in these parts. I counted 43 Neys in the Saarlouis telephone-directory alone.

Anyway, it took only half an hour of Herr/Monsieur Fontaine's extremely valuable time at the Town Hall to demolish the whole theory. He produced the big, black, cloth-bound book for 1769—and there was the entry relevant to Michel's baptism. His father was Peter Ney, a master barrel-cooper of Saarlouis; his mother was Margarete, née Gräffin; and the godparents were Michel Winter (presumably as a family friend the source of the chosen Christian name) and Eva Renard, the daughter of Andreas Renard, another barrel-cooper.

Since Saarlouis was still being ruled from Paris at this time the entry in the register is made out in French. But the actual ceremony in St. Ludwig's, the church facing the Paradeplatz (now the Grosser Markt) was in Latin, plus a short address in the local tongue: in other words, very much in the conservative, entirely devout southern German Catholic tradition. And the priest who administered the sacrament was also German: Father Justin Bichelberger.

Afterwards family and friends returned for a brief celebration to the house along Bierstrasse where Michel had been born earlier in the month. An official handbook on the town lists the house as number 11, which is probably correct. When I visited Saarlouis in 1976 I noticed a small, blue-and-white plate above the doorway giving it as '13'; but the plate looked of more recent stamp and most likely got there due to a redistribution of numbers. I also found the interior of the house gutted and the roof a skeleton of beams—with a pile of builder's materials in the street outside. This was because the new owner, Herr Leopold Böhm from the suburb of Wallerfangen, had decided to restore the birthplace to its original specifications. However, even gutted, it's still possible to describe these. Three steps lead up to a stout wooden door set in a facade of yellow-grey stone blocks. The stone has been dressed but is otherwise free of any carving. The windows on either side are tall with shutters, and above these are three smaller windows, set dormer-style into the slanting, tiled roof. Inside, the house was obviously divided around a straight, central staircase. There is evidence of four good-sized rooms downstairs and four bedrooms with sloping ceilings. In addition there were cellars and a yard behind, where Peter Ney could have worked from time to time—although as a master-cooper he had a shop and larger working premises elsewhere.

He must have been an interesting man, Ney's father. In fact, probably both parents were; and they were truly perceptive about different aspects of the young Michel. It was the father who recognised that this second son was brighter, more forward than his brother Pierre, and consequently decided he should aim higher than just assisting in the family business. But it was Margarete Ney who saw something deeper, more complex in her son's nature. She must have been unusually sensitive as well as loving, for she recognised the fire and spirit of adventure in him; likewise his tendency to associate loyalty with a definite cause. As a result, she was less taken aback and far less hurt than her husband when Michel suddenly embarked upon his chosen career.

Peter Ney, despite his German background, felt strongly sympathetic towards France. He had fought with the army of Louis XV against Frederick the Great and was now a reasonably contented subject of Louis XVI. As a sergeant he had been one of the few on the French side to emerge with any credit from the disastrous battle of Rossbach, earning the written praise of his captain, and he hoped 'never to see the day when Prussia got a hold over the Saar'. It's not surprising, therefore, that these pro-French views, regularly bandied about the family dinner-table, made a great impact on the developing thoughts of his sons. Likewise his 'old sweat' tales of army life. *And yet*, paradoxically, the elder Ney had become extremely prejudiced against joining the army as a career. It wasn't just the defeat at Rossbach that turned him to making and trading in casks. As he explained: 'It's the same problem with every European army. Brains and courage are not enough. You have to belong to the nobility in order to gain advancement.' And he told his sons never to entertain the idea.

Various historians have described the Marshal as being 'crude', 'boorish', 'ill-educated' and 'scarcely literate'; even incredibly, as 'a hot-headed *Gascon*'! All of which is decidedly untrue. He grew up to be bilingual, fluent in German and in French, an accomplished author of reports and when the occasion demanded a convincing speaker. In private he was regarded as a lively talker, a good companion at mealtimes and always—with one celebrated exception—gracious in the presence of ladies. He was also more widely-read than is commonly supposed; not so much in the field of classic literature, but as time went on as a keen student of the theories underlying the practice of fighting a war. No writer who in later years visited the Ney house in Paris ever found him lacking or searching for *le mot juste*, while his receptions were frequented and enjoyed by painters and especially musicians, for he was an enthusiastic—although purely amateur—flute-player. Only the politicians and courtiers disliked going to his house. Because if they did he pointedly ignored them.

For the start of his education, which as with most sensible human beings became a continuing process, he definitely had to thank his father. After noting the boy's quick and eager disposition, Peter Ney—now moderately successful in his trade—began to put money aside and when Michel was six he was sent to the best school in the area: the Collège des Augustins. Their discipline was severe, and if they caught him organising sham battles—as was his wont—then they would send him home with a note of disapproval guaranteed to bring a paternal wallop. But their teaching methods were sound and as the years passed Michel began to earn good marks. So much so that at thirteen, Peter Ney was able to obtain for his son a position in the office of the town's leading lawyer, Monsieur Valette. It seemed an auspicious beginning to what might eventually land him a job with the French government, perhaps even 'high office'.

But Michel's heart was not in it. At an impressionable age he had soaked up, sponge-like, all his father's anecdotes about life with the regiment and in the heat of battle. Moreover he knew every foot of Vauban's casements surrounding Saarlouis and had persuaded the guards on the island in the middle of the river to let him examine the fortifications there too.[2] Now, sitting on a tall stool in the lawyer's office, between copying out deeds and marriage-contracts in what was 'a remarkably fair hand', he found his attention wandering outside the window to the Paradeplatz. To the troops drilling; but even more to those marching off towards service in the east.

Monsieur Valette, a kindly man, was also a clever one and not slow to notice these things. He thought highly of the young Ney, but considered he disliked the work too much to make a good lawyer. 'Every afternoon your handwriting deteriorates because your mind is already intent upon the next fencing-lesson,' he told him. However, he not only thought highly of Michel, he quite genuinely liked him; and he was a person of influence. He used this influence to obtain for Michel (now aged fifteen) a job with the *Procureur du Roi*. It was an office-job again: concerned with the indictment of prisoners, serving summonses on tax-defaulters and generally writing out any other document the *Procureur* required. But at least it was government service, and consequently a step nearer the army which Valette knew the youngster really wanted.

One step at a time was not enough though. By the end of the year Michel had persuaded his father to let him resign the job and take another at Appenweiler: running the office for a mine-owner. Having abandoned the idea of his son ever becoming a lawyer, Peter Ney hoped Michel had at last found his real vocation and would progress to be an ironmaster; a prospect which gained credence when the following year he was offered the position of overseer at the Saleck ironworks. Not bad at the age of seventeen, his father thought. . . .

What the elder Ney couldn't possibly know, however, was that Michel would tear through his correspondence in the office in order to spend the rest of his day in the works. He was mad keen to learn the smelting process, whereby chunks of ore were turned into soft iron: destined for the manufacture of all kinds of things. And one section of the works fascinated him especially. For at Saleck a normal part of the output consisted of the tools of war, including cannon.

In all he spent nearly three years as the overseer at Saleck, during which time he grew to manhood.

There is general agreement that he grew up to be tall; and to be very strong: but again one finds widely differing viewpoints about his true height, physical appearance, colouring, etc. To take the height question first. Here the views vary between six feet one (claimed to be an official measurement taken after his execution, but with no proof that the measurer did not begin at the big toe instead of the heel) and at least two histories which state he was five feet eight. The latter still say he was tall according to the standards of the day. They point out that Napoleon at five six and a bit was considered the French average then. So Ney at five eight, plus one and a half inches of boot-heel, plus his broad-shouldered, barrel-chested physique must have looked imposing.

However, if his marshal's dress-uniform in the Musée de l'Armée at Les Invalides is genuine (and I thought only the shoes and stockings had been added from a department store), then he must have been taller. I'd put his height at five eleven: which would certainly have made him stand out among his contemporaries—and even more so on the big Normandy or German stallions he favoured.

Actually a visit to Les Invalides is useful for clearing up the other physical points as well, because the museum contains the famous oil-painting of Ney by Baron Gérard and a head modelled in plaster after the bronze by Houdon. Both artists were rich and successful without being flatterers; and by and large they agree on the shape of the Marshal's head and features. He had an unusually high forehead, which the combing forward and across of his wavy hair only partly disguises. This gives the impression of a long face, even though his other features are very much in proportion. Also the lower face gradually juts forward, so giving the chin its defiant look. The eyes are large but somewhat heavy-lidded (at least in repose), while the nose is longish and slightly concave: more so with Houdon than Gérard, who starts it *under* the brow.

As regards colouring Gérard is again the best authority, and is supported by Louis David,[3] who otherwise depicts the Marshal with a somewhat irritable expression. As a child Michel was obviously extremely fair. But the notion that he grew up to have bright red or carroty hair is quite false. His hair was a rich chestnut colour, almost auburn when the light caught it. The mistake probably derives from his nickname with the troops: *Le Rougeaud*. And yet this isn't a word normally applied to hair. The French call 'a red-haired man' simply *un rouge*. *Rougeaud* means a person who has a red or ruddy complexion. In which case as a nickname it was particularly apt, for Gérard gives the Marshal a very high colouring, and one that must have intensified quite dramatically during a battle when his blood was up. In fact looking at the portrait, I came to the conclusion that if Ney had escaped execution in 1815 and settled into a desk-job and a life of ease, most likely blood-pressure would have been the sad story of his advancing years.

Happily the painters and the writers do agree about one thing: the colour of his eyes. These were a brilliant blue, in turn emphasised by their largeness, and apparently they flashed with an almost electric quality when he was commanding in battle . . . or as danger-signals if he was annoyed.

On the other hand, if a person's eyes are acceptable (seriously) as 'the mirrors of the soul', as two-way thoroughfares and a key to personality, then Ney's painters must have heaved sighs of relief at being able to put on canvas just what they saw and nothing else. For if there have been arguments among the writers over his background and physical appearance, these are as nothing compared with the ones which have raged about his character and subsequent actions. 'Hot-headed' is a singularly mild form of criticism. Others directed against him have been 'a turncoat', 'a traitor', 'a brave soldier, but a muddled thinker who never got his priorities right' and of course that most terrible accusation from St.-Helena, by the memoirist trying to set the record straight in his own favour: *Ney was the principal cause of France's defeat at Waterloo.*

Well, I can only write what I believe, based upon all I have read or managed to find out. If it reads rather strangely to the Marshal's admirers as well as to those who continue to denigrate him, then I admit to its being an entirely individual interpretation. Nevertheless I see no other way. Curiosity has led to conviction. And this means the development and application of one's viewpoint to the facts and events throughout the remainder of his life.

'Hot-headed' will do to begin with. Michel Ney was certainly a quick-tempered man, but there is absolutely no evidence to suggest that he was a bad-tempered one. Moreover his anger, when it blew up, was invariably directed against one of two things. Either what he considered to be the incompetence of another officer, or against some act or slight which touched upon his honour. He was never a vindictive person, and I have come across only one instance (in Galicia in 1809) when he took illogical, irresponsible reprisals against his enemy. The truth seems to be that he cooled down almost as quickly as he blew up, and this—balanced with his qualities of forgiveness, warmth and generosity—goes a long way towards explaining why he was loved by his troops and liked by his opponents. Although a tough and often demanding leader, he would show his men precisely what they must do *by setting an example*; and he always acted quite naturally in their midst. As regards his honour though he could be very prickly indeed, especially after he became fully involved with the Empire under Napoleon: and this undoubtedly led to a great deal of conflict with politicians and other well-known figures at court. For to Ney it wasn't just a question of maintaining his personal honour and the glory he had won by fighting. In addition, he felt compelled to maintain the honour attached to the rank of marshal, to his family's position in society and above all to the France he had become a part of and believed he represented.

'Hot-headed' would also do to cover certain other superficial aspects of Ney the person, his sudden enthusiasms, for instance, and the occasions when he acted rashly or impulsively. If the latter occurred in battle, and despite everything he succeeded, then it would be attributed to his personal bravery. If not, then he was said to have 'blundered badly'. (As a Capricorn he must confound the astrologers! For the will-power to succeed in what he does best is there; also the inner desire for glory. But in his directness, his bouts of single-mindedness and the flights of enthusiasm he is much more typical of people the astrologers associate with Aries or Gemini.)

Where 'hot-headed' fails totally, however, is in describing the deeper, interior part of his nature. This was the area which his mother first recognised and learned to understand. A dual area of apparent contradictions until one discovers that, although sharply divided, the two sides of it are also twinned, allowing the man to be both unpredictable and consistent, of alternating moods and yet always himself. Furthermore it was the part of him which supplied the motives (and therefore supplies us with the explanations) for countless actions and many of the things he said during his life.

What is most important about a man is not necessarily the easiest part of him to define; and in dealing with a considerable individual one is hesitant to slip into psychoanalytical jargon. For convenience though there has to be some way of referring to these twinned, contrasting facets of the inner Ney, and so I have decided to call them 'the romance' and 'the real'. No doubt 'romantic' and 'realist' scan better: and I will certainly make a point of using them in subsequent stages of the book. But initially the romance and the real are more *fitting* descriptions of the manner in which, ideally, I believe he would have preferred to live out his life, drawing upon his natural instincts in order to fulfil an honest public destiny. To a limited extent he succeeded. But in the end he fell—because his contemporaries refused either to see, or if they did, to accept, the essential dichotomy in his make-up.

The romance. Michel Ney possessed so strongly the romantic spirit that one wonders whether in his learning French the *Chanson de Roland* became his favourite reading matter. Or perhaps he simply heard the legends locally as a child. After all, even if a long way from Roncevaux, the Saar had formerly belonged to the old Frankish empire of Charlemagne. It took great pride in this fact; and in several of his letters and speeches as a general and marshal it is interesting to note how Ney keeps referring back to Charlemagne. What we can be reasonably sure of, anyway, is that Ney grew up to have a firm moral sense, embracing justice, a soldier's devotion to duty and his personal integrity; but at the same time with a code of behaviour closely akin to the Age of Chivalry. And it is this 'code' which opens the door upon his romantic self, for it was something he applied equally to his life in war and at peace; completely without reserve.

As an officer Ney was one of the fiercest fighters in military history. And like Napoleon he fought to win. He was a brave, forceful and inspiring commander. But he was never a bloodthirsty man. When his adopted nation required it he went to war with the desire to be a hero: courageous and unrelenting in pursuit of victory, steadfast when the going got rough and magnanimous towards his opponents. He was a man of deeds and fewer words; but the deeds had to be *right*, and it wouldn't have occurred to him—for example—to ill-treat or feed prisoners any differently to his own men. As for plunder, he regarded it as one does a contagious disease. Above all though he loved the grand gesture in war. Whether there are further (subconscious) echoes of Roland here, of defending the pass at Roncevaux and proudly refusing to blow the great horn until the last minute, it's difficult to say. But one is continually coming upon instances of him seizing a rifle and bayonet and dashing in amongst his hard-press infantry; or galloping across at his peril to turn a dangerous charge; or again shaking hands with and inviting to dinner the general who has just surrendered to him. Such moments, one senses, are what he felt fighting ought to consist of instead of what by experience he knew it to be.

Obviously the romance part of Ney served him far better in war than it did in peace. It enabled him to develop into a kind of nineteenth century paladin, a shining example of courage and fair-play when so much going on around him was brutal and squalid. Away from the wars, unfortunately, it was a different story. The nineteenth century has been described as a romantic one, and no doubt it was by comparison with our own. But what purported to be the romantic 'ideal' was confined to literature, to a lesser extent the art of love-making and to isolated individuals. In the field of politics and especially at court it scarcely existed: although it was paid a good deal of lip-service. There all was self-seeking, intrigue and corruption. As it had been in the century before. As it still is today.

No wonder Ney periodically expressed his disgust with life in Paris. It wasn't, as has been argued, that he completely misunderstood politics. On the contrary, if at times he appears somewhat naïve, there are other occasions when he seems to know what is going on rather too well. No, the truth is that he disliked politicians so much he would only deal with them in an extreme situation.[4] He couldn't eliminate them (as I suspect he really wished), and therefore he preferred to avoid them, concentrating instead upon the ideal of *France*: of the land and the nation's continuing existence and greatness despite the transient sordidness of its politics. Peter Ney's dinner-table indoctrination had worked with remarkable effectiveness. His second son became far more than a Francophile and a marshal in the service of his adopted country. In the end he grew so passionately attached to the ideal of 'France one and indivisible' that he was prepared to change sides—twice—and eventually abandoned his life for it.

The real. Michel Ney would have been lucky to survive one battle let alone over a hundred if the romance in his nature had not been balanced by another, more practical side. In reality he became a keen student of warfare: of its techniques, its tactics and its strategy. Moreover he was interested in every smallest detail. The idea that he couldn't be bothered with anything not promising to add to his store of glory is again false. He wanted to be the master of his craft, and would go to great lengths to unravel the problems of logistics, use of terrain, even how much an infantryman could carry in his pack and still move quickly. I've already mentioned that he was widely-read in military history; and as time went on (aided and abetted by Baron Jomini) he delved deeply into the actual theories of war. The result was a soldier of tremendous fighting qualities who could also be an extremely resourceful one.

This became increasingly evident as the Empire's fortunes declined and the marshals were forced to exchange glory for a war of national defence. However, where I find *the real* most interesting in Ney is after his romantic inclinations had taken a bad knock and he was forced to lean upon their reverse side quite heavily. It could then display an amazing strength and resilience, buttressing his whole personality through the difficult days until events brought about an upturn in his fortunes. I am thinking in particular of the year after his virtual 'disgrace' in the Peninsular War, and of the manner in which he coped during the retreat from Moscow. But also I am reminded here of the dignity and calm he showed in the final months of his life: when the tragedy of his fate was compounded by its inevitability.

At Waterloo the romantic and the realist had gone down to defeat together; but in the aftermath to the battle his equilibrium was gradually restored, so that he died with his permanent ideals intact and a clear awareness that his sentence was unjust. Time, he accepted, could be relied upon to sort out his reputation as a French patriot and hero.

If Ney's dual nature was often misunderstood by his contemporaries, this is not to say that his case was without parallel. The most obvious one to be drawn is with a man with whom he had some slight acquaintance, the writer Stendhal. As Henri Beyle Stendhal was also an ardent Bonapartist, and the two met briefly in Russia when Beyle was serving as a senior quartermaster with the *Grande Armée*. (There are several flattering references to Ney in his letters and private diaries.) Unlike the Marshal he was always an administrator, never a fighting soldier. Nevertheless he possessed the same romantic/realist axis within himself, giving rise to similarly alternating moods and attitudes. Not being a man of war the romantic in him was expressed in a series of hopeless love-affairs; and later of course in his writings. Again like Ney he was generally a person of spontaneous enthusiasms and optimism. On the other hand his books are fascinating to explore because the realist is there as well in the acute observation, character analysis and attention to detail of his novels, similarly in the practical concerns of his letters, whether over money matters or the planning of his travel. Finally he makes the prediction that he too will not be appreciated until long after his death.[5]

Other parallels have been drawn with certain fellow-officers, notably Jean Lannes who became a good friend and was the first of Napoleon's marshals to die on the battlefield (after Essling, 1809). Both Ney and Lannes were indomitable fighting generals and over an extended period of time Napoleon liked to use one or other of them when he decided to attack his enemies head-on.

Even so, I think the soldier Michel Ney resembles rather more closely is of an earlier period and well-known in Great Britain: namely Prince Rupert of English Civil War fame. He too was a brilliant cavalryman who learned how to handle infantry. He too has been accused of rashness; and at Marston

Moor before York in 1644 he made a terrible error of judgement which had the gravest impact upon his own cause—comparable with Ney's mistake on the final day at Waterloo. Also like Ney he was an eager student of war; although because he lived longer he could say he never made the same mistake twice. Much later, as a professional under Charles II, he was destined for an equally distinguished career at sea.

In other ways the resemblance with Ney is uncanny. In fact at this point I would like to incorporate a quotation from C. V. Wedgewood's definitive history of the English Civil War. Delete the period references and her character-sketch of Rupert might well be the French Marshal's epitaph:

> His naturally frank and generous disposition had been overlaid with austerity and reserve during his three years' close imprisonment in an Austrian fortress, a long frustration which had also sharpened his naturally hasty temper. But at the Viennese Court after his release he had impressed the Venetian envoy by the *nobility of his ideas*; the observation was just and penetrating, for the Prince was distinguished among the many ambitious and frivolous men about the King by an elevated and candid spirit, free from conceit and selfishness. He had wanted to be a sailor and was more deeply interested in the mechanical part of his profession—artillery, fortification and siege warfare—than in the cavalry fighting for which he had so immediately established a reputation. The speed with which he took decisions and acted in the field, and his impatience with the King's older and slower advisers at the Council table reflected the quickness of his intellect, for he could be thorough and patient when occasion required, as for instance in the training of his men. (Later he was to acquire great skill as an engraver, an art which demands unlimited patience and concentration.)
>
> With so many qualities, he lacked the easy social gifts, and was apt to show his uneasiness with courtiers and politicians by a sardonic wit and a contemptuous manner. But he inspired in those who knew him best a deep and lasting loyalty, and he easily gained and held the affection and respect of his troops, for he was enduring in hardship, fearless in danger, quick to recognise merit, and as indifferent to praise himself as he was generous in giving credit to others.[6]

Only one phrase in the final sentence could not be applied. Michel Ney was indeed generous in giving credit to others. But he also liked praise himself. It was his Achilles heel—for it left him vulnerable to exploitation.

In November 1788 and without warning Michel threw up his job at Saleck and returned to the family-home in Saarlouis. His brother Pierre had already defied their father and joined a regiment of infantry. The second son could contain himself no longer. When he announced his intentions Margarete Ney expressed little surprise. She wept for his safety, but knew it was pointless to object. His sister hoped he would wear a fine uniform. 'Don't worry,' he told them, 'when I come back it will be as a general!' Then he kissed them both. In contrast Peter Ney fumed—and resorted to all his previous arguments against a career in the army. Not that it did any good. His only consolation was that

Michel had opted to travel west: away from the better pay incentives offered by the armies of Austria and Prussia.

At the beginning of December the future marshal arrived in Metz, having walked the equivalent of 55 kilometres. When he entered the big, grim cavalry-barracks there he was just six weeks short of his twentieth birthday and had taken the first positive steps towards becoming a real Frenchman.[7]

[1] I am particularly suspicious because 'Nicholas' and 'Catharine' were the names of his so-called parents given by Peter Stuart Ney, the North Carolina schoolteacher, who when much in his cups claimed to be Marshal Ney and that he'd escaped death near the Luxembourg Palace after the firing-squad had been 'fixed' and a bladder of pig's blood substituted for his own.

[2] The Vauban-Insel or 'isle', where the Saar curls around outside the town's Die Kasematten which forms its present main-entrance, has been preserved more or less intact. Some flagged paths have been laid out, electric-lamps introduced and the interior of the single, central building—which anticipated the modern 'blockhouse' —turned into a fashionable restaurant. Otherwise all the typical Vauban breast-works, sloping down and outwards into the river, have been preserved; similarly the many deep gun-emplacements. The only new and dramatic addition is the modern statue of Ney: sited on the grass mound which covers the roof of the blockhouse-restaurant. Buff-coloured with blue and red touches, made of concrete, its wrap-around shoulder-to-ground cloak, high semi-circle of collar and backward tilt of the head suggest a very self-contained warrior who is as fearless and resolute in defence as the Rude statue makes him out to be in attack. The sabre is huge; leaving one with the distinct impression that it would be surrendered only in the moment of death.

[3] *Collection Raoul et Jean Brunon*, Château de l'Emperi.

[4] e.g. with Fouché after Waterloo.

[5] For a reason not fully explained Stendhal disliked Madame Ney. He observed her in Italy, he claims, and later had 'a dogfight of an argument concerning Marshal Ney. His widow has no dignity. She has brought grief into disrepute in this country.' But he adds nothing to give his opinion substance, and in an earlier diary-entry (24 August 1804) is obviously confused about her Christian names.

[6] C. V. Wedgewood: *The King's War, 1641-1647*, London, 1958.

[7] The barracks at Metz is still partly in use—although daily shaken by a thunder-trail of juggernauts.

Other Ney 'landmarks' vary in the extent to which they can be recognised. At Valmy, for instance, where as a lieutenant he had his baptism of fire, the village can hardly have changed much in size since the battle was fought. The famous ridge where the French braved the German cannonade remains unaltered. There is the same open-topped centre with the tiny, thick wood on the right and on the left the windmill (Kellermann's command-post) which Goethe refers to. At Longwy, on the other hand, where Ney saw more action in the Revolutionary wars, it's hard to imagine that the place ever had a past. It looks like an across-the-border extension of the Grand Duchy of Luxembourg's coal and steel belt. The smell of iron is in the air. A maze of railway lines surrounds works and plant with Giacometti-thin complexes, spirals of orange-coloured smoke and what I think are blast-furnaces with doors leading directly to hell. A French Sheffield.

CHAPTER 2
A Soldier with the
Revolutionary Armies
1791-1800

Before visiting Saarlouis several rare and valuable books had passed through my hands. First, the three volumes of Bonnal's *La Vie Militaire du Maréchal Ney*: published in Paris between 1906 and 1914. I regret very much that General Bonnal didn't live to complete the work; nevertheless his research, especially among the *Archives de Guerre* in France, has been my principal source of information for the chapter which follows and for many of the facts in later ones. Secondly, *The Memoirs of Marshal Ney* (two volumes published in London in 1833). These are not Ney's own memoirs but were compiled by General Foy and the Marshal's second son, the Duke of Elchingen, who—his publishers claimed—'affixed his signature to every sheet sent to the press'. They are largely based on the Ney family's private papers and transcribe a number of interesting documents and letters. Next, *Les Mémoires d'Une Contemporaine* by Ida St.-Elme (Paris, 1896). This extraordinary person, half intellectual, half camp-follower, was for a time the mistress of General Moreau and even (so she claims) of Ney himself. What is more certain is that she was present at some of the most important battles fought by the armies of the Revolution and the Empire, proving herself a lively 'war reporter' well before the description came into being. For the use of all three works I am deeply grateful. Likewise for the details about promotions, decorations, monetary rewards *et al* contained in Georges Six's *Dictionnaire Biographique des Généraux et Amiraux Français de la Révolution et de l'Empire* (Paris, 1934).

Hussar Ney was an eager, efficient soldier and popular with his comrades. He also became, quite quickly, a superb horseman. He had the right build for it: long in the leg, strong at the waist. And once having developed a good seat, throughout his career it remained a personal boast that he could mount any decent charger and immediately make it respond to his command. All of which was in marked contrast to Napoleon's efforts on horseback. The Emperor, trained as an artillery officer, rode from necessity (on ceremonial occasions or when commanding his armies) and in later life for exercise. But he was notoriously lacking in style. His plump figure rolled all around the saddle, resulting in frequent falls; these last despite the efforts by Jardin *père*, His Majesty's head-groom, to break in his horses:

They were taught to bear pain without even wincing, being struck repeatedly over the head and the ears with a whip. Pistols and maroon-rockets were let off near their ears; drums were beaten and flags were waved close to their eyes, while heavy obstacles—sometimes live sheep and pigs—were flung under their hoofs.[1]

(To be fair though his seat was rendered additionally uncomfortable by the onset of haemorrhoids.)

Michel, with no such problems, rapidly came to the notice of his superiors as the most intrepid *sabreur* in his troop: a fierce front-rider once the charge was sounded. He had enlisted in the *Régiment Colonel-Général des Hussards* (later to become the 5th Hussars). This meant that he dressed in a dark blue tunic frogged with yellow braid, covered by a long red *pelisse* and on top of that a light blue coat. The young Duke of Chartres was Honorary Commander of the regiment, so his family's coat-of-arms had to be embroidered on each hussar's wide red sash. And to crown the uniform there was a huge, more or less cylindrical black hat known as a 'shako', complete with its red pendant and white *aigrette* tuft. Both his sabre and scabbard gleamed with brass finishings; while his horse's saddle rested upon a white sheepskin saddle-cloth, again bordered with red.

He seemed determined to look and live the part. After the fashion of the hussars he wore his rich, chestnut-coloured hair in a plaited *queue,* as well as growing sideburns and a luxuriant moustache. Sometimes he found the drilling and battle-manoeuvres around Metz repetitive, but he enjoyed the 'old sweat' talk at the end of the day, usually sparking it off himself with a piece of calculated banter. To keep fit he relied on primitive gymnastics together with a sensible diet.

He ate good bread with fresh vegetables to make up for the poorish meat, and to drink there was beer, rough red wine (the best possible laxative) or—when the regiment had pleased some visiting VIP—brandy or *marc.* It was also the done thing for rankers to chew tobacco; although in subsequent years, after coming into contact with Parisian etiquette, Ney exchanged this habit for the more sophisticated ritual of snuff-taking.

Less pleasing to those officers keenest on discipline was his remarkable progress at fencing, which nearly brought him in front of a court martial. The regiment's fencing-master had been severely wounded by his opposite number in the Chasseurs, their local rivals at Metz, and the gifted Michel was selected by his comrades to redeem their honour. But it needed to be arranged with the utmost secrecy, for duelling was strictly forbidden in the French army. All was set, in the gymnasium, and with lookouts posted. But their attention must have been distracted towards the fencing, for hardly had the match begun when in marched Ney's colonel, grasped him by the *queue* and jerked him backwards into the rim of spectators. 'You are under arrest! Take him away!' he said. The colonel then threatened to have him shot, only relenting after a petition signed by more than half the regiment was handed to him. Despite this, following Ney's release the duel was arranged again—and to his comrades' satisfaction he succeeded in wounding the boastful *chasseur.*

Altogether he remained an ordinary hussar for two years and almost one month. It could have been a lot longer, because promotion in the army of France's Louis XVI was inevitably the result of either privilege or patronage. However the social earthquake known in our history books as the French Revolution now occurred: a series of upheavals destined to change the face of the nation's army just as much and as rapidly as they altered its structure of government.

The basic situation and the conditions inside France which brought about the Revolution had been sending up warning tremors through three reigns at least. An inequality whereby the nobles and clergy paid no taxes while the middle classes and farmers paid *all* of the taxes was bound, in the end, to lead to disaster. The so-called *ancien régime*, although still buttressed by a king's government, merely proved itself oppressive, corrupt and increasingly inefficient. Voltaire had mocked it, Diderot and the Encyclopaedists castigated it and Montesquieu (who favoured a constitutional monarchy) and Jean-Jacques Rousseau (in his more radical *Social Contract*) put forward suggestions for improvement or, even better, an acceptable alternative. Sadly the ill-advised Louis XVI ('an amiable blockhead' André Maurois calls him) failed to understand the mood of his people. As per usual he did too little too late. Consequently, once the real disturbances came what followed was swift, at times terrifying and yet—insofar as the French were concerned— irreversible. Lawyers, doctors and other articulate members of the Third Estate gave the Revolution its characteristic voice. 'Dismiss these proud lackeys who form your escort,' a young deputy from Arras, Maximilien Robespierre, urged King Louis; while to the clergy his words were more like an order: 'Spend your vile superfluity on food for the starving!'[2] In the meantime their new-found allies, the workers and peasants, provided the Revolution with its edge of violence. While the French army, traditionally the keeper of their monarch's person, stood back and watched, its officers remained loyal but more often than not found themselves impotent, disobeyed, even placed under arrest by their own men. France's ordinary soldiers, as distinct from the Bourbons' Swiss and German mercenaries, had listened to the new ideas and become equally imbued with a revolutionary spirit. . .

1789. Having been forced to call an Estates-General in January—the nearest thing to a parliament in France—King Louis was unable to prevent this body declaring itself a National Assembly in June. On 14 July the ancient fortress-prison known as La Bastille was stormed by 'a leaderless multitude of persons belonging to every class'[3] and its surrender proclaimed a national victory when soldiers of the garrison refused to fire on the populace. Almost two years of haggling followed between the interested parties on both sides. But from June 1791, when the King's attempted flight resulted in his arrest at Varennes, all effective government passed into the hands of the revolutionaries. As with most insurrectionist groups they revealed themselves to be half good, half bad. In the short-term the new National Assembly, soon to be renamed 'Legislative Assembly', freed the middle classes from the worst of their frustrations and put a brake on the excesses of the mob. But there were destined to be more ugly scenes, various internal struggles for power, the execution of King and Queen, then further bloodshed under Danton, Marat and Robespierre before France found stability again. Moreover—and here Michel Ney comes back into focus—the Revolution by its very nature would plunge the country into a generation of warfare with its alarmed (because still largely reactionary) neighbours. And with what sort of an army could France hope to defend herself?

For once the great powers of Europe were able to set aside their own power politics and unite against a single nation whose social changes they feared. Austria, Great Britain, Holland, Prussia, Russia. Each was ruled in one form or another by an hereditary monarch: upon whom what was happening in Paris and throughout France had the same effect as a tocsin-bell being rung. Unless overwhelmed and *eradicated at source*, then the French Revolution seemed to be merely a prelude to their own separate downfalls, and none would admit that their behaviour was in any way at fault. They were confirmed in such beliefs by the hordes of French aristocrats, the *émigrés*, who now descended upon them bearing lurid tales—some true, some false—of infanticide and public executioners dipping their bread in the victims' blood. Before long all the professional troops in Europe were converging upon France. Their aim, or rather the aims of their official paymasters: *complete suppression of the Revolution and a restoration of the Bourbon monarchy.*

Facing them at the outset, and if nothing else fully supporting its revolutionary government, was one of the most remarkable, ragtag and bobtail armies in the entire annals of war. In the main it consisted of previously-serving Frenchmen together with the rump of Lafayette's National Guard. Nor did it lack volunteers. But with these its apparent attributes ended. It was short of weapons, uniforms, forage for the horses (when there were horses), proper billets and even shoes. The sight, on the parade-ground, of a conventionally-dressed and armed fusilier standing beside a peasant carrying a pitchfork and whose only uniform was the tricolour was not unfamiliar. In Paris the talented and energetic Carnot, now Minister for War, strove without ceasing to remedy this situation; but the nation's finances were in dire straights and Europe's merchant bankers refused him help. Furthermore, after being supplied by Carnot the soldiers didn't always respond to the necessities of military discipline. They too were caught up with their own interpretations of the Revolution.

Here, for instance, are two extracts from a letter by the national volunteer Thiébault, later to become a general:

1) On 1 October 1792, when I heard that the Prussians were marching on Paris, I joined as a volunteer grenadier in the first battalion of the Butte-des-Moulins regiment and left for the front. I marched for several days with the battalion but marching wearied me and so I lagged behind . . .

2) I should mention that service in the battalion of the Butte-des-Moulins was in no sense obligatory. Two hundred young men had already left this battalion before it reached Belgium. Among them were M. Odiot the goldsmith, who only proceeded as far as Châlons and who was a lieutenant in the Grenadiers. Bertaux, the engraver; Grasset, at present conductor of the orchestra of the Bouffes (Italian Opera) and a captain in

the Grenadiers, as well as Messrs. Lafargue, Leqoc, Devismes, de Vigearde and others, who left the battalion soon after I did. Moreover, there had never been any question of fighting except for the express purpose of driving the enemy out of France and for my part I was so anxious to be considered not as a soldier but as a volunteer, in the fullest sense of the word, that I consistently refused to receive either pay or rations for myself and provided my own arms and equipment. In addition, when I left Paris, I had reserved for myself the post which I occupied under M. Dufresne de Saint-Léon, when the business was temporarily wound up, and my right to return to it was I believe actually guaranteed by decree . . .[4]

Above all though the Revolutionary army suffered from a chronic shortage of good officers. Its former senior commanders, being of aristocratic origin, had fled with the *émigrés*—all of them prepared to accept commissions of service under France's enemies. To complicate matters, in Paris an Assembly now dominated by the Girondists (moderate republicans) and characterised by its excess of ideological zeal decided that the soldiers could elect their own officers. *Extract from the regulations governing national volunteers*:

The administration of battalions, Article 2. The officers, non-commissioned officers and volunteer members of the council of administration shall be elected by the entire battalion according to the number of votes by each individual. They shall be elected for one year after which they shall be subject to re-election. By the same process an officer, a non-commissioned officer of each grade and four volunteers shall be chosen as substitutes for those members of the Council who may be sick or absent.[5]

Soldiers also took it upon themselves to complain about their officers directly to the government:

. . . in the midst of my deep distress, I cannot refrain from making a number of comments on this disaster. Why was the General unaware that the enemy had crossed the river? Why had not the posts along the river been reinforced as the water was going down? Why was a garrison of 6,000 men (at most) considered adequate to guard 15 leagues of country? Why were we abandoned by the rest of the force at a distance of only 6 leagues from headquarters? Why were isolated battalions left exposed to attack? . . . I appeal to your experience, Citizen Minister: cannot the person or persons responsible for our advanced positions at least be blamed for criminal negligence? I beg you to communicate this letter to the National Convention. I am sending a copy to the Citizen Commissioners.

Signed and certified by Beauge-Grenadier, Boulinois-Grenadier.[6] With such barrack-room lawyers as these breathing down their necks it's hardly surprising that the lists of officer-candidates rapidly dwindled. And more so when the Convention started shooting commanders convicted of inefficiency. 'When a battalion quits its post, the cause must be ascribed to the cowardice of the officers and to their failure to maintain discipline and to inculcate in their soldiers the love of glory . . .' Signed: Saint-Just, Gillet and Guyton, the 29th Prairial, in the Year II of the French Republic, one and indivisible.[7]

In the end though the First Republic was saved from foreign conquest by three interrelated developments. 1) The Army, despite its inadequacies, fought with a truly patriotic conviction. The men would fight to the death; and as time went on and training improved they became more than a match for the mercenaries and conscripts of Austria and Prussia. 2) Carnot's absolute insistence on the soldiers obeying orders and accepting an iron code of military discipline. Gunner Bricard records how, in Thermidor, Year II, of six little drummer-boys convicted for stealing and assaulting an old lady, two were shot and the remaining four forced to assist at their execution.[8] And finally 3) alongside the army being shaped into a cohesive whole, it was made possible for soldiers of talent like Michel Ney to assume responsible positions.

Ney's rise to fame under the direction of his new masters can only be described as meteoric.

His first promotion was on 1 January 1791: to the modest rank of *brigadier fourrier* (corporal quartermaster). It aroused in him feelings of considerable pride, as he explained in letters to his mother and Monsieur Valette. By this time too he was a young man with firm Republican ideas; and the hours not devoted to training were turned into urgent, and often heated debate with his colleagues: either analysing the latest bulletins from Paris or discussing the experiments in democracy across the Atlantic. However, such beliefs as he held received a shock, and then a flash of enlightenment in November of that year with the news of his mother's sudden death. When he heard, Margarete Ney was already buried. 'And I promised her I would come back a general . . .' he told his friends in the billet. Out of consideration for his grief they avoided smiling, but really, whoever expected the most junior kind of NCO to become a general? Ney in turn plunged them into a discussion about God. Previously he had considered himself a Christian, but without thinking very deeply on the subject. 'We talk of liberty, equality, fraternity—here around us,' he began: 'But isn't there also another, an after-life?' His colleagues were inclined to dismiss the idea, or to link it back with the more worldly aspects of religion. 'The Church offers a promise of heaven but has permitted a hell in France,' one reminded him. Michel was unimpressed though. 'It cannot be that I shall never see her again,' he said firmly; and it is interesting to record that, with war about to engulf him, his Republicanism never involved atheism nor led him to join the (then) fashionable anti-clericals. He would become a Mason, but that was much later and had rather more mysterious connotations.

On 1 February 1792 Ney was made a sergeant-of-horse (*maréchal des logis*). At just twenty-three he held the rank which under the former regime he would have acquired after a lifetime of service. Three months later came his appointment to sergeant-major. In April France declared war on Austria and Prussia; and on 14 July the anniversary now known as Bastille Day, Ney was promoted to *adjudant sous-officier* (something like an RSM and the highest non-commissioned rank). It meant he spent a lot of his time in the orderly-tent, handling company orders, guard duties, even applications for leave (and his reluctant studies in the law-office came in useful). Not that this paperwork would occupy him for long. He was with the hurriedly-assembled Army of the North, and his regiment (re-styled the 5th Hussars) stood directly in the path of the Duke of Brunswick, the commander who promised his Prussians 'a holiday march on Paris'. He issued a proclamation: 'Austria and Prussia would restore Louis XVI to his rightful place . . . any Frenchman taken with arms would be hanged . . . if the king should be harmed then Paris would be destroyed!' It only provoked French anger. In the capital a mob attacked the Tuileries: the Swiss guards were murdered and the King put back under close-arrest. Meanwhile big, ugly Danton with his pock-marked butcher's face became Minister of Justice. 'We must dare, and dare again, and ever dare,' he roared at the Parisians. 'If we do, then France is saved!'

But to dare in itself was not enough. In Belgium French troops broke and fled at the mere sight of the enemy. In the east two key fortresses, Longwy and Verdun, agreed to capitulate without even token resistance. It looked as if Brunswick would have his 'holiday march'. He had occupied Saarlouis over a month before (which caused Pierre Ney, earlier demobbed, to re-enlist in Danton's Volunteers). Only the weather—three weeks of almost continuous rain—seemed to be holding up the Austrian and Prussian advance. What the French badly needed was a boost to morale, but how to obtain this? All their remaining forces were concentrated in Argonne under the command of General Charles François Dumouriez. If they broke as well then the Revolution was finished. 'Comrades, we are fighting for the freedom of the nation,' Dumouriez told the troops. 'What the soldiers of the despots are afraid of is cold steel!' Ney and the 5th Hussars had been posted away from Metz (just in time to avoid its blockade) and up to Carignan near the Belgian frontier, where they found billets in Lafayette's old camp. Now they received orders to pull back into the narrow valleys of western Argonne. It appeared Dumouriez had decided to make his stand at Valmy, between Verdun and Reims. Coinciding with these orders Michel Ney was promoted to *sous-lieutenant*: the rank in which, almost immediately, he would have his baptism of fire.

Valmy proved to be the turning-point in France's fortunes. On 20 September in drizzling rain, 131,000 Austrians, Prussians and French *émigrés* came up against Dumouriez' army of 82,000. Most of the latter's battalion commanders had been chosen by their men: Suchet, Jourdan, Victor, Oudinot and Marceau, but not Lieutenant-General Kellermann, Dumouriez' deputy and a former officer under the Bourbons. After failing to manoeuvre round the French, Brunswick announced to the King of Prussia he would 'soften them with gunfire'.[9] His belief being that the high proportion of raw recruits on the French side could never withstand a prolonged and fierce cannonade. But in this he was mistaken. The men of the Revolution not only withstood his fire; before very long they were returning it.

Kellermann's report to the War Office describes what happened:

Finding it difficult to break through, the enemy prolonged their line on my right under the protection of an intense artillery bombardment. To meet them I drew up all my troops in battle order. My position was not a pleasant one, but I was reluctant to believe that a large proportion of their forces had passed through the gap at Grandpré and so I prepared for the battle which continued from before seven in the morning until after seven at night. They never dared to come at us in spite of their very superior numbers, and the day was spent in a short range artillery duel lasting nearly fourteen hours, which cost us the lives of many brave fellows. I hear that the enemy's losses were also very heavy, especially in cavalry and artillery.

He goes on:

I cannot say enough in praise of the courage and devotion of my officers —generals, commanders of units and subordinates—and of the troops under their command. At one point I saw whole ranks of soldiers destroyed by the explosion of three ammunition wagons caused by a shell, but the men did not flinch or lose their alignment. Some of the cavalry and in particular the riflemen were frequently exposed to a murderous fire, but they showed a fine example of courage and steadiness.[10]

The Prussians' observer that day was no less a person than Wolfgang Goethe. He confirms the ferocity of fire which converged upon the exposed windmill of Valmy (Kellermann's command-post) and writes with exhilaration of the ammunition wagons going up: 'We thought with joy of the damage the explosion must have caused in the ranks of the enemy!' In typical phrases he likens the noise of the cannonballs to 'something between the humming of a top, the splash of water and the cry of a bird,'; and he also refers to the heat ('I felt as though I was in the middle of a furnace') and to 'the thunder of the guns, the snoring, whistling and chattering voices of the bullets as they cleave the air'. But he is absolutely truthful over the growing consternation felt on the Prussian side as the day wore on and the French failed to crack:

. . . they (the French) remained unmoved. Kellermann had improved his position. Meantime our troops had been withdrawn from the zone of fire and it seemed as if nothing at all had been achieved. In the morning we had been talking of spitting and eating the French. I myself had been drawn into this dangerous adventure by the confidence in our splendid Army and in the Duke of Brunswick. Now everyone was thinking again. People avoided each other's eyes and the only words one heard were oaths and imprecations. In the twilight we assembled and were sitting in a circle. We had not even been able to light a fire as we usually did. Only a few men spoke and their reflections seemed illogical or frivolous. At last they pressed me to say what I had thought of the events of the day, as my terse comments had often interested or amused our little company. So I simply said: 'From this place and from this day forth commences a

new era in the world's history and you can all say that you were present
at its birth.'[11]

Michel Ney was only one of several young officers appointed at the last
minute who distinguished themselves at Valmy. But his 'holding hard' under
fire brought him to the attention of his superiors; and when Kellermann was
detached to chase Brunswick out of France—via Longwy—he found himself
kept back to accompany Dumouriez' renewed invasion of Belgium.
Dumouriez promoted him to (temporary) full *lieutenant* and in this capacity,
in November, he took part in the extraordinary French victory at Jemappes.

'Extraordinary' because the French, refusing to consider themselves the
underdogs, took on and vanquished what were regarded as the pick of
Austria's Imperial troops. Ney was always in the thick of it and his sabre-arm,
rising twice as fast as anyone else's, caught the eye of (provisional) Brigadier-
General Bernadotte. 'Lieutenant Ney will go far,' he noted; '*If* he isn't killed
first. . . .' Another brigadier-general (*maréchal de camp*) present at Jemappes
was Dampierre. His account leaves little to the imagination. 'We took the first
line of trenches with the bayonet, but the enemy struggled desperately to
hold the other redoubts. Cannons firing grape-shot, musketry, cavalry
charges, every form of attack was employed but all failed before the invincible
valour of the French. . .'

The 5th Hussars helped a regiment of Vivarais infantry sustain three
murderous charges by the Coburg dragoons, but were then free to charge,
jump over and turn upon the Hungarian defenders of the last system of
trenches. Dampierre goes on: 'The troops were raked on several occasions by
grape-shot at pistol-range. That was the enemy's last effort. After our charge
they fled in the greatest disorder.'[12]

A Prussian officer drew his own conclusions from this French victory:

> The French *émigrés* have deceived our good King and all the foreigners
> in the most infamous fashion. They had assured us that the counter-
> revolution would occur as soon as we showed our faces. They had also
> told us that the French troops of the line were a collection of riff-raff
> and that the National Guards would take to their heels at the first shot
> fired. Not a word of this was true. The *émigrés* have contributed nothing
> to our forces and the French troops resembled in no way the picture they
> had drawn of them. We found among them plenty of fine men and a well-
> mounted force of cavalry. Their discipline is as good as that of our troops
> and we have seen them performing evolutions which have compelled the
> admiration of our generals.[13]

Two hundred years of Austrian domination of the Belgian provinces were
now coming to an end. At Jemappes and before the great concentration of
guns at Mons the French troops' courageous assaults proved unstoppable.
The battalions from Bender and Würtzburg, the Coburg dragoons and the
famed hussars of Blankenstein, even the dogged Hungarians: all were put to
flight. Eight days later Michel Ney was one of the first French soldiers into
Brussels. He was still stationed there when news came through that Danton

and the Jacobin clubs—with Robespierre and Saint-Just in the ascendant—
had taken advantage of the army's victories to execute the king. 'For myself, I
see no middle course: this man must either reign or perish!' Saint-Just
announced to the new Convention and so swayed a majority of the voters.

Since before Jemappes Ney had been given extra duties as provisional *aide-
de-camp* to General François Joseph Drouot, *dit* Lamarche and on 3 February
1793 he was confirmed in the rank of *lieutenant*. It coincided with things
taking a turn for the worse. Early in March the Austrians began a drive to
recapture Brussels. Lamarche (thanks to a spirited charged by Ney's troop)
halted them at Gossoncourt and took many prisoners. But in other places
garrisons were surrendering: at Juliers, Altenhoven, Aix-la-Chapelle and
Liège; while reports came in of Royalist revolts in Brittany and the Vendée.
More serious still, from Louvain Dumouriez sent a letter to the Convention
openly denouncing its style of government. Danton and Lacroix arrived by
the fastest carriage to regain his support. Having failed in this they ordered
his arrest, which Dumouriez ignored. He went out to suffer a defeat at
Neerwinden and then to take a thousand of his troops over to the Austrians.
Ney was horrified and refused to join him. 'France', he insisted, 'means more
than the sum of any personalities ruling in Paris.'

In April Carnot changed the 5th Hussars into the 4th and three months
later Robespierre's Committee of Public Safety suspended Lamarche. The
general was lucky; in the capital heads were falling off 'like roofing-slates'.
However, upon leaving he gave the cavalryman his first written testimonial:

> I, General Lamarche, commander-in-chief of the Army of the Ardennes,
> hereby certify that Lieutenant Ney of the Fourth Regiment of Hussars
> has been employed by me as *aide-de-camp* from 19 October 1792 to 4 July
> 1793, and that he has fulfilled the duties of his position with all the
> intelligence, intrepidity, activity, and courage for which such a post gives
> occasion, and that in all circumstances in which he has been employed,
> even in the midst of danger, he has displayed a discernment and a tactical
> insight that is seldom found.

The Hussars themselves gave him a second one in October, this time signed
in the democratic fashion of the day:

> We, officers, non-commissioned officers, and hussars of the Fourth
> Regiment, certify that Citizen Michel Ney, *lieutenant* in De Boye's
> company of the said Regiment, has in every grade in which we have
> known him, acted as a brave soldier and a true Republican, and has at
> all times served with zeal for the interests and welfare of the Republic,
> and that it is only according to what we know of his way of thinking and
> his conduct that we have signed these presents for his use and for the
> information of all whom it may concern.

In December he became *aide-de-camp* to a new general, Colaud, and then on
12 April 1794 he was promoted to captain and sent back to the Hussars—
expressly to command a party of 500 horse under General Jean-Baptiste

Kléber. That summer Kléber decided to form a special 'Flying Squadron' of dragoons and *chasseurs*: its duties being 'to guard the supply columns and protect the flanks of the army against surprise attacks!' Upon the advice of Pajol, his ADC, Ney was appointed to lead it; but when Pajol rode over to him with the news he declined to accept, saying he 'was not yet equal to such responsibilities'. Whereupon Kléber lost his temper. He sent Pajol back to inform 'Captain Ney that even in these days of democracy generals still have a certain power. You will accept this post!' He also created him a (temporary) *adjudant général chef d'escadrons* or major (31 July).

The French, in this sector renamed the Army of the Sambre-et-Meuse, were now moving further into Belgium. One day—'towards Louvain'—Ney with only thirty dragoons tackled and scattered a large force of Austrian cavalry who were attacking French infantry units. Kléber happened to ride up from another direction in time to see the skirmish and that night he wrote to Gillet, Representative of the People and the Government's official observer in the region: 'Citizen. Captain Ney, who is doing duty as adjutant-general, has performed prodigies of valour. With thirty of the 7th Dragoons and a few *chasseurs* as orderlies, he charged two hundred of Blankenstein's Hussars, and threw them into the greatest disorder.' Gillet in turn was 'delighted' to further promote Ney to the rank of *adjudant général chef de bataillon* (major of the staff) in recognition of 'his military talents and patriotism'—a decision which was confirmed by Carnot on behalf of the Committee of Public Safety (9 September 1794).

Delight though was hardly what Gillet felt in the weeks to come, when reports reached him that the young officer had adopted 'a very individual way' of treating prisoners. Before their separate downfalls in the April and July of this year, Danton and Robespierre had laid down rigid rules relating to persons captured by the Army. All former aristocrats, *émigrés*, priests or any suspected Royalist sympathisers were to be handed over to the Government 'immediately': which in effect meant to the *guillotine*. So far these orders had not been revoked and yet Major Ney—Gillet's spies informed him—quite blatantly avoided recognising former *émigrés*, letting them mingle and lose their identities among the foreign soldiers he captured. The crunch came when two priests were allowed to escape. Even Kléber was worried about the possible repercussions; what would Gillet do? 'Damn Ney!' he fumed. 'He should have been more careful. There's no excuse. He's an intelligent man. . .' Suddenly though he became aware that the severe Representative of the People was eyeing him with an unaccustomed amusement. 'Yes, General, your friend is intelligent,' he replied. 'And he knows how to spare the blood of his countrymen!' With Robespierre's passing, he went on, there was a different mood in Paris. Things were taking their time, unfortunately, and no doubt there would be some further executions; but use of 'The Terror' as a political instrument of survival had gone forever. 'We must remain vigilant, the fight to keep our Republic will hardly become easier. But let us forget about these two priests. For France's soldiers to display a common humanity is no bad thing.'

After this Ney's career continued on the upswing, but always as a result of boldness and vigour in the field. More glamorous and therefore newsworthy deeds were being performed in the south and soon in Italy by Napoleon Bonaparte, a former officer-cadet at Brienne, whose tactics were said to be as novel as they were brilliant. In the north, however, the war against the *émigrés* and the Coalition continued to follow a grim, unrelenting pattern. At times it seemed more like a war of attrition, except that across the open lands of Belgium, Holland and western Germany it yielded up its reluctant victories to those who operated with the greater mobility.

In August 1794 Ney's Flying Squadron, out raiding supply trains, was betrayed by a deserter and cut off from its base at Diest by a large force of Austrian cavalry. The leader's reaction was instantaneous—and typical of his early style. 'Straight ahead, men!' It meant charging up a slope, but Ney led from the front and timed his gallop perfectly, striking into the Austrians with maximum impact. As well as fighting their way through they also captured the Austrians' general, 'a present' which Kléber celebrated the receipt of by promoting his intrepid subordinate to full colonel. Next it was Bernadotte's turn to enthuse about Ney. 'I owe great praise to this man,' he wrote to Kléber after driving the Austrians back upon the Rhine. 'He has assisted me with the intelligence you know so well . . . and I must say, to speak the exact truth, that he counts for much in the success we have obtained.' Kléber dutifully passed on this report to the Government.

However, on 22 December the cavalryman suffered his first bad wound. Bernadotte had bottled up the Austrian rearguard in Mayence (Mainz) and laid siege to the place. On the 22nd Ney organised a 'feint' against the enemy's earthworks with several companies of infantry; then—as the Austrians' attention was drawn to this attack—he rushed out of the French positions with his cavalry and endeavoured to take the defenders from behind. Unfortunately he had not bargained for a cavalry-trap at this point: a wide, deep ditch to the rear of the earthworks. Ney made it across, but the other horses refused to jump and in slashing his way to freedom Ney received a bullet in the upper arm. At first it was feared the main bones were shattered; he had lockjaw, grew delirious and there was serious talk of amputation. In the end the surgeons saved his arm, but recovery of its use proved slow and Kléber ordered convalescence at home. He was seconded in the decision by Merlin de Thionville, Gillet's replacement as Representative of the People. At Mainz Thionville had been so impressed by Ney's exploits that he wanted to create him brigadier-general on the actual battlefield. Now he wrote: 'My brave friend. Go and complete your cure in the Saar, at your birthplace. I have despatched an order to Surgeon First Class Bonaventure to send one of his pupils with you. Return soon, Citizen, and lend us your powerful aid against the enemies of your country. Health and Fraternity—*Merlin.*'

So the colonel of twenty-six and an even younger doctor travelled by coach along the snow-covered roads to Saarlouis, or Saar-Libre as the new Republic insisted on calling it. Pierre Ney was still away with the Volunteers, but Marguerite acted as the doctor's assistant, overcoming her initial shock at the depth of the hole in Michel's arm.

Meanwhile Peter Ney and Monsieur Valette vied with one another in singing his praises. Peter, totally won over from his previous position of disliking his son's military career, now delighted in reading and re-reading the 'get well' letters that arrived from comrades in the field. Nor could he understand Michel's stubborn refusal of Thionville's second offer to promote him. 'This isn't mere modesty,' his son explained. 'I want to be a general; but I have to feel ready. It's a question of experience.' From a soldier with undoubted ambitions who was also a confident front-line fighter (the perfect candidate for *la gloire* as Napoleon later judged him) this reveals an extremely conscientious attitude. A firm grasp of the realities of war had transformed his youthful dreaming into the self-knowledge which equips a man for command. On the other hand it did nothing to placate his masters in Paris. The Government, now a mixture of cynical politicians like Tallien and Paul Barras with some former 'idealists' making huge black-market profits, decided that Ney lacked *civisme*. Again Kléber came to the rescue: assuring everyone that 'Ney has commanded his troops with distinction. In all his operations he shows courage and good thinking. . . .' He also referred to 'his bullet wound which interrupts his active service until the re-establishment of his health.'

When Ney returned to active service General Jourdan was in overall command along the Rhine and the Moselle, but he took part in several further battles under Kléber before his protector was replaced by General Grenier. The warfare seemed sad and empty though. Often they were in strategic retreat: underpinning the actions of Jourdan, as at the bridge of Neuwied. But Ney was in the successful combat at Altenkirchen (4 June 1796), and a month later took the town and citadel of Forcheim. On 15 August yet another reshuffled Government (a second Directory, manipulated by Barras) insisted that he accept promotion to brigadier-general. 'Experienced enough?' Kléber remarked drily when the news came through. It was the last occasion the two met. After repeated disagreements with Jourdan (who was ill) and Pichegru (whose loyalty he suspected; correctly as it turned out), Kléber decided to accept his replacement by Grenier and withdrew into private life, not fighting again until Napoleon's Egyptian expedition, where he was assassinated by a religious fanatic in 1800.

In the short term though it was Michel Ney who experienced the more surprising change of fortune. Having become a general of brigade—to everyone's approval—and within months of his twenty-seventh birthday, he celebrated both these events by getting himself taken prisoner! It was the one and only time it happened to him and occurred immediately following his winning a small battle at Kirchberg (19 April 1797). He moved on to Giessen, where the next day his cavalry and a battery of horse-artillery attacked the Austrians' line of retreat. Suddenly a regiment of heavy dragoons appeared from behind to cut him off. Ney whirled, prepared to fight them, and at that moment his horse went down, entangling him in the stirrups. The retreat went on and the Austrians took him with them.

But if fate had dealt Michel a peculiar blow then at least he wasn't forgotten. Within twenty-four hours both armies were informed that an armistice had been signed by General Bonaparte, representing the Directory, and Austria's Archduke Charles. This gave General Lazare Hoche, briefly back on the Rhine after putting down the Royalists in La Vendée and Brittany, the opportunity to make Ney's repatriation a number one priority.

'My dear General,' he wrote from Giessen on 21 April (although dating his letter *2nd Floréal, Fifth Year of the Republic*):

> You must know me well enough . . . to understand how afflicted I am at the terrible event that has happened to you. I have sufficient confidence in the spirit of mutual courtesy with which the Austrian Generals act, that they will treat you as we have dealt with such of their colleagues as we have made prisoners in Italy. I am asking Monsieur Elnitz to release you on parole, and I am looking forward impatiently to the moment when I shall again grasp your hand. Let me know if I can send you anything helpful. *Adieu*, my dear Ney. Believe in my sincere and constant friendship.

Two days later he forwarded on Ney's broad tricolour sash, silk and fringed with gold, newly arrived from Paris. 'I do not pretend,' he said in an accompanying note, '. . . to recompense either your success or your merit. I ask you only to accept it as a poor testimony of my personal esteem and my unalterable friendship. Let me have news of your health.'

It took the energetic Hoche less than a fortnight to secure his release, exchanging him for a senior Austrian. Ney must have felt his luck was in; especially since the certificate of exchange was accompanied by a note from the Government stating that 'his capture would in no way damage his reputation or his prospects'. Moreover he returned to a period of comparative calm. In Italy Napoleon Bonaparte, now widely recognised as Barras' *protégé* (even to the extent of marrying his ex-mistress, Josephine de Beauharnais) was turning the armistice into the Treaty of Campo Formio. He drove a hard bargain, forcing the Austrians to give France the southern Netherlands, Venice, the Venetian fleet, the Ionian Isles and security of tenure along the Rhine. Genoa became a republic, while French troops were allowed to occupy the Papal States. This left just England supporting the *émigrés*, but 'the Roast Beefs' (it was observed) fought mainly to protect their trade interests and in any case were only effective at sea. France could afford to relax therefore, enabling the nation to take stock of itself. A window had been opened on the Revolution—and through it the old idealism, fervour and intrigue were free to mingle with pride of achievement, some elegance and a taste for high-living.

It was also a time for the more politically-minded generals to further their ambitions away from the battlefield. General Bonaparte, just back from Campo Formio, seemed to be here, there and everywhere: one moment proposing an 'Army of England' to be based on the Channel ports, the next organising his expedition to Egypt, which would involve botanists, cartographers and medical scientists as well as Kléber, Desaix and the other fighting-men. He was to be encountered in all the best Paris salons, charming, bursting with energy, making useful contacts and recruiting adherents from every walk of life. Even Hoche was sucked into the maelstrom of Directory politics—and he died in his bed on 18 September 1797: poisoned, it was whispered, by jealous rivals.

Ney would have none of this. He treated politicians with a fatal disdain; largely on account of their impractical meddling in the army's affairs. Nor was he prepared to be a mere creature of society. And so he spent the remainder of the year and the whole of 1796, either in billets near Amiens or at La Petite Malgrange, a modest country house and small farm near Nancy, where he installed his father and sister Marguerite. It had cost him eighty-thousand francs in those inflation-ridden days, and he was proud of the fact that his purchasing money came from accumulated salary and gratuities, not from looting or graft. What is certain is that he read a great deal at this time: every military manual he could lay his hands on—because, as he stressed, 'I am still inexperienced in several departments of warfare, and particularly where the infantry is concerned'.

It has been claimed too that this was the period when he joined the Freemasons.[14] The story runs that he entered the Lodge of the Nine Sisters in Paris (where he later became a Master Mason) following an introduction by Kléber, who it is claimed explained to Ney something of the brotherhood of Masonry and its religious and charitable significance. Kléber is also supposed to have said: 'Some day you may find yourself in distress. Remember the distress signal. It may serve you in good stead.'[15] Personally I discount this theory. Kléber might well have convinced him of the values of Freemasonry when they served together along the Rhine. But there is too much evidence that Ney did not visit the capital until May 1801 and my friends who are Masons tell me it would have been inconceivable, given the elaborate initiation ceremony, for him to have been enrolled by proxy. I think it far more likely that he joined the lodge *after* his first encounter with Napoleon, when he made a number of changes in his life-style, either at the dictates of fashion or as a result of the shift in his allegiance.

Hostilities recommenced early in 1799. England's foreign trade profits plunged disastrously into the red once the French were manufacturing again and began to win over European markets. British Government ministers consequently went all out to persuade their former partners in the Coalition to break Bonaparte's Treaty. To the Russians they proposed a joint expeditionary force, while to the Austrians they agreed to pay generous subsidies. Not that this amounted to very much. Within a year the Anglo-Russian troops landed in Holland were in headlong flight towards their ships, disgraced and ridiculed after a series of incompetent performances against General Guillaume Brune. And within two years Michel Ney had gained a national reputation thanks to some startling successes over the Austrians.

His first victory came almost immediately and was the result of an unbelievable gambit: something more akin to the fall of Troy or Harun-al-Rashid in *The Arabian Nights* than to a stolid war between trained opponents in early nineteenth century Europe. Ney's cavalry units were placed at the service of Bernadotte, and their principal objective was Mannheim, a strongly-defended city on the far side of the Rhine. By rights it belonged to France under the Treaty of Campo Formio, but the Elector Palatine had refused to hand it over. Ney, who disliked the use of a formal siege train (he considered them mediaeval), proposed that they tackle the job 'by stealth and cunning' and Bernadotte, although dubious, agreed to let him try.

Accordingly he placed two companies of infantry and three guns on the French side of the river: in full view of the enemy patrols who imagined they were a token body. The remaining infantry and the cavalry he kept well out of sight. Then, disguised as a peasant selling vegetables, Ney crossed the river, walked alone down the road and entered the city unchallenged. Speaking fluent German he was able to move about the streets at will, estimating the size of the garrison ('a large force, but slovenly') and the condition of the defences ('good, thick walls, difficult to breach with cannon'). The main difficulty was a big, deep moat, together with an ancient drawbridge which when up sealed off the city completely. However, on the way in the general had encountered a young woman, obviously in the last stages of pregnancy, who said the officer in command of the drawbridge had given orders for it to be lowered, even at night, if her labour pains came on and she needed the midwife and doctor. With this in mind Ney hurried back to his own camp, selected a hundred and fifty of his best men and after dark had them transported across the river in skiffs. There they waited in the shadows near the drawbridge. Luck was on their side. An hour later the pregnant woman appeared, the drawbridge was lowered and in one swift surge Ney's 'commandos' were inside. They kicked up so much noise, and Ney himself issued such dire threats that the Austrian commander thought they were a full brigade and agreed to surrender. Mannheim was now a French possession.

Beyond lay Philipsburg, and here Ney adopted an entirely different strategy. He received information which led him to believe that its commander would surrender the town in exchange for fifty thousand francs. He suggested therefore that they pay this sum in order to save time and lives, and Bernadotte was in enthusiastic agreement:

I have just received your letter. I approve of all you have done. In war, when one is not strong, one must be a bit canny. It is very trying that I cannot spare more troops with which to invest Philipsburg. Promise 500,000 francs: promise 600,000 and even more if necessary. I pledge my word of honour to have them paid down the same day that the fortress is handed over to us, or at the latest within twenty-four hours. We shall pay it all by levying contributions. Be open-handed in supplying your emissaries with money. Try to get into correspondence with the most influential officers. The man who is not brave, my dear Ney, will nearly always allow himself to be corrupted with gold; to profit by his weakness is an art one must master, and it is not easy to find another opportunity if the first time fails.

Unfortunately on this occasion it did fail. The secret negotiations broke down and Ney was compelled to effect a hated blockade. Perhaps he was still feeling depressed about it when the Minister for War's letter arrived, informing him that by a decree of the Directory he had been promoted to *général de division* (lieutenant-general). At any rate it put him in a refusing mood again:

I have received, Citizen Minister, your letter of the eighth of this month, (April 1799) *he wrote back*, in which was enclosed the degree of the Executive Directory promoting me to the rank of General of Division. The Directory, in elevating me to these new functions, has probably considered only the favourable reports about me. I would submit to the honour of this promotion if my talents were sufficient to justify the Government's kind actions. I hope my refusal will be interpreted as proof of the disinterested civic spirit which guides me in my functions as General of Brigade, and I beg that you will assure the Government that my conduct will never have any other object than to deserve, more and more, its esteem.

To Bernadotte, whose report on the capture of Mannheim had brought about this promotion, he wrote:

I have the honour to request you, *mon général*, to forward to the Minister for War, by the first available courier, my refusal of the rank of General of Division. This offer of promotion flatters me, and at the same time inspires me with a very lively sense of gratitude for him who, by his favourable reports, has induced the Government to confer this new rank on me. I hope that while you may not approve of my conduct in this matter, you will continue to manifest your good will towards me, and I shall always respond to it with the most sincere and lasting affection.

Neither of these letters produced the result he'd intended. On 4 May the Minister replied that 'The Directory insists upon your acceptance'. He assured him '. . . the Government sees in your modesty a valuable pledge that you will continue to serve the Republic with your accustomed distinction. In consequence of which I herewith again forward the decree of your appointment.' Bernadotte was more blunt, and in effect told him not to be a damned fool:

Look around you, my dear Ney, and say candidly whether your conscience does not tell you to lay aside a modesty which is out of place and even dangerous when carried to excess. We must have ardent souls and hearts as inaccessible to fear as to seduction to be able to lead the armies of France. Who more than yourself is gifted with these qualities? It would be an act of weakness, then, to shrink from the career that is opening up before you. Adieu, my dear Ney. You will, I know, listen to everything from one who is attached to you by ties of the warmest friendship and the most perfect esteem.

(Bernadotte himself would soon be taking up other duties, including a short period at the War Ministry; but the 'ties of friendship' he refers to had an enduring strength, surviving both Napoleon's anger and the period when they fought on opposite sides.)

Further refusal was clearly useless, and it is worth noting that from this point on Michel accepted each new advancement as and when it was bestowed. But it had been a hard thing to digest, so that as late as November one finds him still grumbling about it in a letter to General Lecourbe. Patriotism, he insists, should always come before ambition. 'My sole ambition is to carry out my duties. I shall never be low enough only to serve men. I think all the time of my country with my whole solicitude, and I shall always make the sacrifices for it which circumstances demand.' The plain fact was that his brave and novel capture of Mannheim had given him a reputation. He had captured the imagination of Paris as well. It wasn't just Government ministers who discussed the affair at Mannheim, but people in wine-shops and restaurants, at the fashionable *salons*, even those pilot-fish who were calculating the possibilities of a Bonapartist *coup d'état* (Napoleon himself was still away in Egypt). Previously they had relished and helped spread stories about Jean Lannes, Joachim Murat and other daredevils fighting under the little Corsican in sunnier climes. Now, they realised, the Army of the Rhine held a man with similar abilities.

The ladies were attracted to him without even knowing what he looked like—although I doubt whether any went quite so far as Ida St.-Elme. A fascinating woman (of Dutch and Russian origins), she had already left a wealthy husband to become the unofficial 'wife' of General Moreau. Obsessed by military men, she often contrived to follow them into battle—disguised in male attire—and later still would devote herself to Napoleon's cause. (He flirted with her in Milan, if one can believe her *mémoires*.) Now though it was suddenly the popular figure with the Rhine army who held her interest. Still only an impressionable twenty-one she decided to write to Ney. 'At least', she told herself, 'I shall let him know what a high opinion I have of him.' *Dear General,* her letter was addressed:

I obey the dictates of my heart without waiting for vain excuses. I am not familiar with the art of disguising my feelings. Besides, there is something in the bottom of my heart which tells me if what I am doing offends against the conventional rules, it may find favour in the sight of a man of such noble honesty as yours. Only once have I seen you with my eyes and yet your picture is engraved in my soul. Always with you in thought, I have trembled whenever you were in peril, I have rejoiced at all your triumphs, and I have enthusiastically applauded every account of your fine deeds. My position in the world is splendid: there are women who envy me. I would give it all up in a moment to become a partner with you in danger. Respect and gratitude have bound me to General Moreau. To confess it in a letter such as this, isn't it running the risk of making myself contemptible to you? But I feel quite unable to choke back the irresistible cry of my heart. In making this avowal of the sentiment which destroys my peace, I have no other purpose than of letting you know that far way from you is a woman to whom your fame is no less dear than it is to yourself.[16]

Ney already had a mistress by this time, a German lady whose name has eluded all his biographers but with whom he certainly spent more than one leave at La Petite Malgrange. In her highly-charged state Ida St.-Elme sealed this passionate 'fan' letter in an envelope addressed to Moreau, while Ney received a perplexing one filled with domestic humdrum. Moreau, not knowing for whom the letter was intended—*although obviously not himself*—promptly dumped the wayward lady. Ney didn't know what to think, but he remembered her name as someone he might look up if ever he visited the capital.

Otherwise his attentions were fully directed towards the war again. Transferred to Masséna's armies of the Danube and Switzerland (4 May 1799), he was first given command of the light cavalry, then of an advanced division under Oudinot (23 May). 'Division' proved an optimistic misnomer. He had only two thousand men, no artillery and six hussars and a corporal for his cavalry! Yet with these at Winterthur on 27 May he faced a whole wing of the Austrian army under the renowned Archduke Charles. The weather was intensely cold, with out-of-season snow squalls reducing visibility to about fifty yards. A General Tharreau had promised support, but this never came and Ney (outnumbered almost five to one) was forced back behind Winterthur itself. This was his first real experience of a retreat and he seems to have handled it most creditably. The French losses were heavy, but there was no disorder and in the moments when they were hardest pressed they turned and defended themselves with the bayonet. Ney fought as one of them, was cheered for his efforts and red in the face picked up the nickname which was to stick, *Le Rougeaud*. He also collected three wounds: from a musket-ball in the leg (which troubled him on and off for the rest of his life), a bayonet thrust into his foot and a pistol-shot through the hand. The next day, still bleeding badly from the leg wound, he reluctantly relinquished his command to Colonel Gazin. As he explained in a report to Masséna, his losses were over six hundred killed or wounded, with another hundred captured, and without support it had been impossible to hold on. Masséna later replied that 'he should not feel too downhearted'; his gallant containment of the Austrians had left the main French forces free to defeat the Russians under Suvorov and Korsakov near Zürich.

Ney requested permission to go to Colmar to recover from his wounds, and it must have been there—calamity apparently piling upon calamity—that he received news of Pierre's death in Italy. His brother had served as a Lieutenant under another future marshal, Macdonald, and was killed in the preliminaries to a disastrous campaign which would result in Joubert's Army of Italy being routed and Joubert himself left dead on the field. 'I feel myself wounded both inside and without . . .' he wrote to his father by way of consolation. And he expressed alarm: because the Austrians would soon be free to redouble their efforts along the Rhine.

Reporting back for duty he was from 25 September provisionally in command along the river: due to the incompetence of General Leonard Muller, who sat biting his nails miles behind the front, afraid to make any decisions. Ney drove the Austrians away from Beilbrun in spectacular fashion. Again he was outnumbered, but this time he had three cannon and used them with such skill that the enemy retreated leaving several hundred dead and the French had not suffered a single casualty. He nearly died himself shortly afterwards though. He was defending Mannheim when his horse was shot from under him and the cannon-ball reopened his leg wound. Then as he stood up a bullet caught him full in the chest, hurling him to the ground. By sheer good luck the bullet had come a long way; its force was spent and merely gave him a bad bruising.

Once the Directory had confirmed his replacement of Muller Michel wrote a round-robin letter to each of his subordinate-commanders:

> The Executive Directory has forced me, dear comrade, to accept provisionally the post of commander-in-chief of the Army in place of General Muller, who is recalled to Paris. You know how insufficient are my military talents for this position, especially in such critical circumstances. I shall perhaps be the victim of my devotion, but I cannot avoid taking this step. I rely on your care for the safety of the troops confided to your charge, and on your special goodwill towards myself . . .

Humility had replaced refusals. With each copy of the letter he enclosed specific proposals for going over to the offensive. Then he moved his HQ to Hagenau, purged the army of several more inefficient officers and divided it into two groups: one to hold the river-line, the other to execute lightning blows against the Archduke. He was just completing this reorganisation when the news came through of '19 *Brumaire*, Year VIII' (10 November 1799).

It reached Ney via the new semaphore station at Strasburg and the message was sent on by General Colaud:

> My dear General. The director of the telegraph has just communicated to me the following two despatches: First despatch, dated 18 Brumaire . . . *The Legislature has been transferred to St.-Cloud. Bonaparte has been appointed commander of Paris. All is quiet and peaceful*; Second despatch, dated noon the 19th . . . *The Directory has resigned. General Moreau is in command at the Palace. Everything* . . .

And here it ended. 'Everything WHAT?' Ney complained to his officers. 'This is only half a message!'

What he did not know, but soon learned, was that by midnight on 10 November Napoleon had made himself virtually the dictator of France. In England Edmund Burke had already prophesied it. 'The liberty which France has gained will fall a victim to the first great soldier who contrives to draw the eyes of all Frenchmen upon himself,' he wrote and now it was true. Returning from the Middle East Napoleon found Barras and the other politicians quarrelling amongst themselves, doing little to promote France's trade and industry and nothing to support her embattled armies. 'When the house is crumbling,' he said to Marmont, his *aide-de-camp*, '. . . it is time to busy oneself with the garden? A change here is indispensable.'[17]

Accordingly he consulted with a group of willing conspirators: the Abbé Sieyès, Fouché, Roederer, Bruix, Cambacérès and Talleyrand. And at St.-Cloud on 10 November an impassioned speech by Lucien Bonaparte was followed by some suitable sabre-rattling from Murat, Leclerc and a handful of soldiers who formed the nucleus of a praetorian guard. Lucien 'appealed to the military to free the council from the menaces of some deputies—men, he declared, who were not only armed with daggers, but were in the pay of England; and this happy reference to daggers and to English gold saved the situation.'[18] Murat and the assembled guards 'took up the cry on behalf of

Napoleon . . . and then, drums sounding an advance and troops pouring into the hall, the legislators were driven out pell-mell, their cries of *Vive la république* notwithstanding.'[19]

The Directory 'just expired'. That night in the Orangerie at St.-Cloud:

> denuded of its windowpanes, a few smoking lamps and four candles in sconces above the empty tribune shed a sepulchral light over the overturned benches, tattered hangings, bonnets shorn of plumes, torn cloaks, togas, badges—all vestiges of the parliamentary rout. People came and went, talking aloud. It was the usual audience of great political events: elegantly dressed women exposing themselves to the maximum, foppish men, intriguers, generals, flunkeys, prophets of the future.[20]

They discussed the day's events, 'congratulated one another, and saluted the dawn of a new day and the happy era about to commence.'[21] More especially they were discussing the immediate outcome, which was that in future France would be ruled by three Consuls: General Bonaparte, Sieyès and Roger Ducos. (The last two were soon replaced by Cambacérès and Lebrun.) No one had any doubts where the real power lay though: very effectively in the hands of Napoleon.

These changes caused Ney no particular worries. On the whole he preferred the idea of a general governing France to the senseless squabbling by Barras and his fellow-politicians. Also the new First Consul was a man of undeniable talents who displayed a no-nonsense, no-time-to-waste attitude from the outset. One early result of his rule came with the posting out of General Moreau as Ney's commander-in-chief. Not that the saturnine-faced Moreau, a suspicious Breton, regarded the command as a sign of great favour. 'The First Consul has chosen his associates from those fawning men who went with him to Italy,' he complained to Ney when they met in Basel. 'After all, he is Italian. And he seems to feel that I may be an obstacle in his path.' Even so, Napoleon had entrusted him with what was now the largest of France's armies—and in April 1800 he marched into Germany, Ney's division leading. (If Moreau had discovered that Ney was the object of Ida St.-Elme's passion then he never mentioned it. Personal relations between the two men remained good. In any case Moreau had recently married the ambitious Mlle Hullot, a creole and former friend of Josephine. Presumably he didn't care to throw up awkward spectres of his troublesome mistress.)

The French advance continued on a broad front. Moreau preferred this traditional method to Ney's divided, 'anchor and strike' technique, allowing his units to spread out over a hundred miles. At one point Ney drove in the enemy's rearguard, taking two thousand prisoners, and in July he easily defeated General Neu at Ingolstadt. '*Nez-à-nez*' as the wags in his division called it. Then the campaign somehow got bogged down, and in the autumn, with peace talks already under way, Moreau sent him on leave. However, hardly had he reached La Petite Malgrange when a letter caught up with him from HQ. 'The Austrians have rejected General Bonaparte's terms,' it stated. 'Hostilities will be resumed on 18 November . . .' And Moreau had scrawled across the bottom: 'Travel day and night. Your presence was never more necessary.'

Early on 2 December 1800 the French and Austrian armies collided and
began to fight what would escalate into one of the most significant battles of
the Revolutionary wars. Certainly from the French point of view it was their
most valuable victory in the northern sector since the cannonade at Valmy;
while the Austrians' appalling defeat dealt a final blow to England's
subsidised Second Coalition, led to the Peace of Amiens and condemned the
émigrés to another fourteen years in the political wilderness.

The battle took its name from the crossroads of Hohenlinden, a village
midway along the paved carriage-way that left the River Inn at Muhldorf and
ran due west into Munich. But really this is just a convenient identification.
Although Hohenlinden saw some of the fiercest fighting, the whole affair was
one of great mobility, covering a wide area and often swinging back and forth
through the neighbouring forests. It also took place in atrocious weather: with
freezing fog at night, then rain and snow squalls bringing a thaw during the
day. Underfoot the mud was sometimes knee-deep; which made the
movement of cannon and horses well-nigh impossible and so threw the main
weight of the conflict back upon the infantry.

The situation was as follows. Moreau had sixty thousand men, Austria's
Archduke John a little in excess of seventy thousand. The latter had hoped to
cross the Inn and circle around behind the French so that a complete cordon
would be thrown around their rear, cutting the supply-lines which were
already sticking in the mud. But this did not take into account any movement
on Moreau's part, and to his everlasting reputation he kept his hungry army
so poised that—at any given moment—he could swivel at least two French
brigades to face every Austrian one. Ney's division was stationed at the vital
crossroads, both a die-hard centre and the army's pivot, with Grouchy ready
to move at a few minutes notice on his left and Richepanse free to swerve
over to the other side.

Moreau guessed the Austrians couldn't be far away, and so during the night
of 1 December he set up a line of watch-fires in front of the French positions,
thus tempting the Archduke John to make the opening gambit. At the
crossroads meanwhile Ney and his men found themselves sandwiched on a
narrow ridge with the River Ibsen at their backs. Moreau visited them one
hour before dawn, just in time to witness the Austrians' supposedly surprise
attack. They were dressed in white, difficult to pick out against the slushy
snow. But Ney had posted sentries in the trees beyond the line of fires and
their empty bellies and the cold had kept them awake. 'Shall I move in
Grouchy to help hold them?' Moreau shouted to Ney. 'Why? I'm just about
to launch a counterattack!' came the reply. And he did: capturing a gun and
over four hundred Austrians. It gave Moreau the freedom to begin a
remarkable encircling movement. Grouchy he still kept fairly close to Ney,
but Richepanse was ordered into the forest on their far right. Seeing
something of these movements the Austrian reconnaissance officers decided
that Moreau was retreating upon the larger Ebersberg Forest. At noon,
therefore, they launched another murderous attack upon Ney's positions, and
this time Ney was more than grateful for Grouchy's spirited support. They
defended desperately until nightfall: when the Archduke John grew
convinced he would beat them with one more attack the following morning.

3 December. After a second freezing, starving night, Ney and Grouchy, two trained cavalrymen sliding about in the muck of winter, withstood this final Austrian assault. And all the while Richepanse was edging his way around behind the enemy. Movement through the snowdrifts in between the trees proved difficult, and at the last moment he was in turn attacked by a smaller Austrian column, which almost cut his force in two. But he succeeded in driving past them, overran the Archduke's baggage-train and reserve artillery and then from the cover of the trees started pouring volley after volley into the Austrian rear. It coincided with Ney's noticing the first signs of hesitation along their front. Without more ado he had his cannon dragged up and then ordered a charge with the bayonet, Grouchy breaking away to fall upon the Austrian right flank. Within minutes the Archduke's once-proud army had dissolved into chaos and headlong flight. Except that it hardly knew where to fly: coming under a hail of bullets whichever way it turned. The carnage was terrible. When Richepanse eventually called off the chase at Mattenpott the Austrians altogether had lost eight thousand men dead and another twelve thousand captured.

'The vigorous efforts of my division,' Ney reported to Moreau the next day, '—assisted by the combined movements of those near it, placed in our hands more than eighty guns, with an immense quantity of ammunition wagons, many standards and about ten thousand prisoners, among whom are several generals and many officers of rank.' He concluded by requesting special gallantry awards and promotions for his adjutant Ruffin, Major Passinges and Citizen-Soldiers Brayer, Schwiter, Daiker, Randon and Perrier.

Hohenlinden increased his own reputation immeasurably. Its strategy and a part of its tactics were Moreau's. Napoleon also benefited, for it set the seal on his success as First Consul and by Christmas the Austrians were forced to accept his armistice terms. But Ney was the principal fighting general involved, and the battle revealed his toughness when in extreme difficulties. This was no longer just an exciting cavalry officer; a man who could take towns by clever scheming. Here at last was a commander of quality—and an inspiration to all who served under him.

That Moreau and Napoleon appreciated his qualities there can be no doubt. Still jealous rivals, each now set out to win Ney's allegiance and subordination by different methods. At Christmas, at Burghausen, he received from Moreau four orders upon the Treasury. Two were for six thousand francs to be paid to each of his brigadiers, a third authorised payment of three thousand frances to his chief-of-staff and the final one ordered that Ney himself be paid ten thousand francs. Napoleon's approach was much more subtle. He could have simply increased these gratuities. But he had scrutinised the written reports from Hohenlinden very carefully, and heard one or two more at first hand. He began to form his own, quite distinct opinions about General Ney. As a result, when the spring came and Michel was on leave and helping his father farm La Petite Malgrange, he sent him a letter. It seemed rather late to be praising him for his services along the Rhine and the Danube. However, this was not the real purpose of the letter. At the end it 'invited' him to visit Paris. Coming from the First Consul this was more or less a summons . . .

[1] Constant: *Mémoires*, Paris 1830-31.

[2] Etienne Dumont: *Recollections of Mirabeau and the first two Legislative Assemblies*, Paris, 1832. Quoted by Georges Pernoud and Sabine Flaissier in *The French Revolution*, trans. Richard Graves, London, 1960.

[3] *Histoire de la Révolution de France de 1789, par Deux Amis de la Liberté*, Paris, 1790-1803.

[4] Chassin and Hennet: *The National Volunteers during the Revolution*, Paris, 1906. Quoted by Pernoud and Flaissier in *The French Revolution*.

[5] Ibid.

[6] Ibid.

[7] Ibid.

[8] Ibid.

[9] Comte R de Damas. *Mémoires, 1787-1814*, Paris, 1912.

[10] *Gazette Nationale*, Paris, 1792; also *Le Moniteur Universel*, No. 268, 24 September 1792.

[11] Wolfgang Goethe: *Campagne in Frankreich, 1792*, Paris, 1891.

[12] *The National Volunteers during the Revolution*.

[13] *The French Revolution*, ed. Pernoud and Flaissier.

[14] By Dr J. E. Smoot in *Marshal Ney, Before and After Execution*, quoted by Legette Blythe in *Marshal Ney: A Dual Life*, London, 1937.

[15] Ibid.

[16] This must have been at Valmy, where she had gone to console a Captain Marescot, later a friend of Ney's in his Kléber period.

[17] Auguste Frédéric Louis Viesse de Marmont: *Mémoires du Duc de Raguse de 1792 à 1832*, Paris, 1857.

[18] H Richardson: *A Dictionary of Napoleon*, London, 1920.

[19] Ibid.

[20] Commandant Henry Lachouque: *The Anatomy of Glory*, trans. Anne S. K. Brown, London, 1961.

[21] Ibid.

CHAPTER 3

"Accept, General Consul, the Imperial Crown..."

Michel Ney was destined to fall under the equivalent in human terms of a spell. It would bring him great glory, several sharp rebukes and ultimately death and a controversial place in the history of warfare. Moreover, despite some notable, but evidently superficial interruptions, it wasn't truly broken until 18 June 1815 on the field of Waterloo. In its headier years (1802-1812, if one leaves out the worst incidents of the Peninsular War) the romantic in him achieved full fruition, with the realist limited to his military techniques. Then after Borodino profound disillusion set in and the realist took over. But at Waterloo the twinned facets of his personality came together again in a quite dramatic balance, enabling him to perform during the battle with a flawed, even frenzied splendour: mixing up bravery, the occasional grand gesture and at least one rash mistake with tenacity, endurance and a piece of brilliant improvisation that brought Wellington to the very brink of defeat. 'A damned near run thing!' was how the Iron Duke later described the battle; and with good reason.

Ney can be forgiven for falling under Napoleon's spell. Nearly everyone else in France did. The man undoubtedly possessed genius. He'd shown the practical side of it in the course of the siege at Toulon, and later in other ways on the battlefields of Italy. Now he was showing it as a civilian administrator. Above all though he had the genius to lead. Whenever he put pen to paper, either in his contributions to the enormous drafting of the *Code Civile des Français* or in hasty letters to his *préfets* and officers, the go-getting First Consul moved men's hearts. And it was just the same with his oratory; instead of windy rhetoric he used the right words and a telling glance to achieve the desired effect. His personal charisma was such that even his intimates went in awe of him. He knew when to charm and when to command, while his expressions of ill-humour were generally devastating. As only one instance of a remarkable ability to impose his own strong will upon others there is the story dating from 1796 which involved three future marshals of the Empire: Masséna, Augereau and Sérurier:

> They were all older men and older soldiers than this little Corsican, and when he came into the room they did not so much as deign to uncover.

36

General Bonaparte took off his hat, obliging them to do so too. No sooner had they doffed their plumed hats than he clapped his own on his head again and stared them in the eye. They lowered their gaze. Then he gave them his orders: polite, but curt and unequivocal. Once outside, the veterans looked at one another uneasily. Masséna, who knew more about people than the other two put together, spoke for all: *That little scoundrel only just missed frightening me!*[1]

After this, the moment Napoleon told them to do anything, they jumped.

Naturally so powerful a character was loved, looked up to or feared for a wide variety of different reasons. In Michel Ney's case, even if somewhat complex, these reasons are not too difficult to find since they stem directly from what we know of his make-up and thinking. He didn't love Napoleon in the manner of Bertrand, Drouot, Duroc and later on the Imperial Guard; but he looked up to him and remained an adherent to his cause for a very long time. He was by nature an excellent subordinate officer provided he could admire the man who was leading him. He was impulsive, and occasionally unruly, but he needed to ride alongside a sun-god's chariot, in fact positively enjoyed doing so, and as a result Napoleon found him much easier to handle and if necessary discipline than the jealous marshals (Murat and Bernadotte) or the greedy ones (Masséna and Soult). If he had a latent vice it was the desire for glory, which Napoleon recognised before Michel himself did and consequently aroused as a means of exploiting his talents. Add to this Josephine's careful, strategic binding of Ney to the First Consul in a purely social way and one has explanation enough. The ex-Hussar from Saarlouis would never be cynically disloyal.

Still more important, however, was his basic allegiance to France; and here Napoleon emerged as the sole beneficiary. It has been pointed out by theologians down the centuries that a convert will frequently manifest more devotion than a person actually born into his religion, and something similar happened between Ney and his adopted country. The ties were both cerebral and emotional, evidently started off by his father's views as well as the political situation in the Saar during his formative years. But in effect Michel fell into a love affair with France which grew stronger rather than diminishing as time passed. And it was sufficiently strong when the vital moment came to override his considerable scruples about being a good Republican. For a decade at least he had fought to save the French Republic he genuinely believed in: only to keep hearing of its being as storm-tossed at home due to the incompetence of successive waves of politicians as it was abroad facing the high tides of two European Coalitions. In their wake Napoleon Bonaparte brought efficiency, energy and all the signs of a new prosperity. Consequently, although his rule was autocratic, as a man he began to symbolise France for Ney; or as Stendhal would have put it, the love affair 'crystallised around him'. With Napoleon going from success to success so France's fortunes rose up with him, and Michel the dedicated Francophile became ever increasingly bound up with his leader's objectives. Waterloo, as I mentioned at the beginning of this chapter, would finally break the spell—but that was fifteen years on and by then it was too late . . .

In the meantime May 1801 and their first meeting, one discovers, did not go at all well. Upon reflection this isn't really so surprising. Such intricate relationships take time to develop. Napoleon was seeking to isolate Moreau, the man he still regarded as a threat to his present position; and to do so he had to win over the soldiers of the northern sector behind an accredited leader. Ney was the obvious choice because of his popularity and high reputation. Accordingly the First Consul set out to disperse the other Rhine Army commanders. He retired Lecourbe on half pay, offered Colaud a promotion that kept him away from Paris and encouraged Richepanse to join Leclerc's expedition to the West Indies against the Negro rebel Toussaint l'Ouverture. Moreau himself he called to Paris—where he presented him with a pair of pistols on which the names of his victories were studded with diamonds, also suggesting he took some overdue leave on his estate at Gosblois. Only then did he feel free to receive Ney in an effort to gain his support. However, he wasn't looking forward to the encounter. He'd heard a lot about the general's undiluted Republicanism and for this reason alone considered he might well persist in being Moreau's man.

Nor did Michel make the journey to the capital with any kind of enthusiasm. He had always associated unseen Paris with self-seekers and political intrigue; while the new First Consul was as yet an unknown quantity, surrounded—or so Moreau claimed—by a group of opportunist officers from the Army of Italy. (He couldn't possibly know that Napoleon advanced only those he either trusted or needed.) But to refuse the invitation outright would have meant very probably the end of his career. So he followed his father's advice: which was to go and just take things as he found them.

What he found at the outset was a city even busier and more crowded than usual. Napoleon had inherited 'in the exchequer exactly 167,000 francs in cash, and debts amounting to 474 million. The country was flooded with almost worthless paper money.'[2] His first act as Consul therefore had been to raise fourteen million francs (from French and Genoese bankers and by means of a national lottery). This tided him over while he instituted a new, more effective system of tax collection based upon income, wine, playing cards, carriages, salt and eventually tobacco, which he made a state monopoly. Only then, but with the clever support of his fellow-consul Lebrun (who turned out to be something of a financial wizard), was he able to restore the value of the currency, cut the rate of inflation, encourage investment and pump state-aid into the expansion of trade and industry. Within a year though his methods were working and working well. Greatly encouraged, he decided to hand the whole mechanism over to Lebrun and concentrate upon the affairs of government which interested him most: justice, social legislation (including state education) and, as he himself described it, 'generally reshaping France'.

Naturally this included Paris, so that by the time of Ney's arrival there the old, fascinating, but dirty and unhealthy capital ('a legacy from the Middle Ages') was beginning to be changed out of all recognition. Although personally thrifty, for reasons of national prestige the First Consul expressed determination to make Paris 'the most beautiful capital in the world . . . within ten years' time of three million inhabitants'. It meant Chaptal, his

Minister of the Interior but with special responsibilities for public works, soon became the most harassed member of the administrative team. The Minister's own first choice for improvement was the hospital scheme and, as he reminded Napoleon: 'One cannot improvise population; from where do you obtain enough good drinking water?' The First Consul demanded to know. Recovering from his surprise Chaptal offered him two alternatives—artesian wells or the bringing of fresh water from the River Ourcq. 'I adopt the latter plan,' Napoleon replied without a moment's hesitation. 'Go back to your office and order 500 men to set to work tomorrow at La Villette to dig the canal.' Then he sat down and wrote out a draft on the Treasury 'equal to half a million pounds sterling'.[3]

It was merely the start of a whole series of innovations for improving the city. That same year Napoleon decided the Louvre should become 'a people's treasure-house of world culture' and ordered the building of galleries to connect it with the Tuileries, 'making a façade to the Rue de Rivoli' and ensuring that 'only workmen of the city are employed'. New bridges were to be constructed over the Seine and the quays down alongside the river greatly extended. Meanwhile on the left bank the Panthéon or 'Temple of Fame' was being embellished (Rousseau's body having followed the remains of Voltaire and Mirabeau into it) and a group of scientific colleagues from the Egyptian expedition received permission to develop another pet project, the Jardin des Plantes. Schools, squares, public parks and a new Chamber of Commerce came next on the list, with a major reconstruction of the Cathedral of St.-Denis and other churches despoiled by the revolutionaries 'mentally planned' for when the French became reconciled with the Holy See again.

As a result the Paris on which Michel Ney first clapped eyes was occupied by hordes of agitated workmen, either swinging picks and hammers or swarming like monkeys over the wooden scaffolds covering the new and partially-restored buildings. Interspersed with these navvies were the skilled masons, moving at a delicate, deliberate pace upon the stone exteriors and apparently immune to the din made below in the streets as further groups of workmen tore up old drains and installed the new water pipes. Napoleon preserved everything from the French Renaissance (Francis I, Catherine de' Medici, Henrys III and IV) and everything important from the Middle Ages (whether Louis IX's churches or Philip the Fair's towers); but towards the ancient, decaying medieval plaster-and-wood houses he was absolutely ruthless if they got in the way of his road-widening or bridge-building. Whole blocks of them went under to make room for his huge storehouses for food (in and around Les Halles), and the already horrendous noise was increased once the peasant-farmers were tempted to drive their big carts into the city centre to sell their produce. On the other hand Ney couldn't help but notice that the streets were not only being widened: they were also clean. Under the Directory, so he'd been told, the capital had boasted a kind of sleazy ostentation. Now it featured grandiose designs, tremendous activity and a growing self-respect. 'La Gloire' as instituted by the First Consul was not just a military prerogative!

Although having corresponded with various important people in the government Michel knew none of them personally and so spent the night of his arrival in Paris in lodgings. His interview with the First Consul was arranged for the following morning and both men were firm believers in punctuality. At the Tuileries he was at once greeted by Louis-Alexandre Berthier, Napoleon's Chief-of-Staff in Italy and since the *coup d'état* at St.-Cloud the Minister for War. Together they then crossed the marbled entrance-hall towards the main stairs. (Napoleon both lived and worked in Louis XVI's old suite of eight rooms on the first floor.) No doubt the visitor felt slightly uneasy at the sight of the Consular footmen toing and froing in pale blue livery decorated with silver lace. Didn't their very subservience give a monarchal look to the place? But his attention was quickly diverted to the collection of busts and statues Lucien Bonaparte had assembled for display in the Great Gallery. It's unlikely that he attached any particular significance to the choice of historical heroes: to the good Republicans, Cicero, Cato, Brutus and George Washington, ranged alongside empire-builders and empire-supporters such as Alexander, Julius Caesar, Gustavus Adolphus, Turenne, Condé, Marlborough and Frederick the Great. Mirabeau's inclusion probably gained his approval. However his wide set eyebrows shot up at the dead presence of several second-rate Revolutionary commanders, Dampierre, Marceau and Joubert. If these, then why not Hoche, or Kléber or even Moreau? Was the distinguished general of Italy afraid of any contemporary talents which seemed fit to rival his own? This possibly caused the tightening of Ney's lips and the severe expression on his face as Berthier ushered him into the all-important presence.

They had entered a room incredibly, almost unbearably hot. Despite the spring sunshine outside there was a huge log-fire burning in this, the larger of two apartments which Napoleon used for offices. And as he rattled off the last of his dictation beads of perspiration gathered over his secretary Bourrienne's upper-lip. The First Consul sat before the ornate roll-top writing table that Louis XVI had used so seldom, a table inlaid with mosaics representing musical instruments and loaded with gilt-bronze ornaments and piles of documents. His plump thighs were crossed, and so were his well-kept hands upon the red frock-coat embroidered with gold braid that his valet Constant had brushed with such care earlier in the morning. As they stood waiting he rose and moved towards them, instinctively producing his famous smile. He signalled for Berthier to cut out the formal introductions.

But the smile disguised the fact that Napoleon's initial impressions of General Ney were extremely unfavourable. Even after doffing his hat the visitor overtopped his own stocky, round-chested figure by more than four inches. Also there was in his candid blue eyes a steadfast look which the First Consul mistook for defiance, a bold affront to his show of friendliness. Worst of all though was the man's arrival wearing a tricolour sash, and with his hair still done in a pigtail! These 'badges of original Republicanism' represented a clear disapproval of everything going on under the Consulate, or so Napoleon believed and he promptly abandoned the idea of winning Ney over. He concealed his real displeasure, but at the same time made no allowance for the other's rigid features being partly a brave face to cover his feeling awkward and shy. 'Welcome, Citizen General, to Paris and to the Tuileries . . .' he had begun. Now, to fill the conversational void, he plunged into a technical discussion of military affairs along the Rhine. The two bent over maps and so were able to avoid each other's eyes. One thing did impress Napoleon as the minutes went by. Of his many recent visitors Ney was the first who did not seize the opportunity to ask him for some gift or position. And from the answers to his questions he judged him the type of general one would find most valuable in an emergency. But he still marked him down as Moreau's creature, and instead of seeking to keep him in Paris decided to let him go back to La Petite Malgrange 'and rusticate'.

In the end the difficulties between them were to be relieved (and perhaps, as has been claimed, actually resolved) by the calculated intervention of Josephine. This was not the only occasion when her tact and femininity helped save the day, but in Ney's case the role she played as intermediary became especially important. She made her surprise entry with the usual swish and rustle of Lyon silk just before the interview was due to finish, and blushingly apologised for disturbing them. Then, sensing the tension, she brought the whole of her considerable ingenuity to bear upon relaxing the gauche, ill-at-ease general from the east. With her hair gathered up in a mass of ringlets, the exact amount of delightful bosom showing and her teeth (her one bad feature) hidden behind a pretty, tight-lipped, well-practised smile, she fully lived up to her reputation as the uncrowned Queen of Paris. And she drew from Ney a spontaneous outflow of his own natural politeness towards women. Noting her success Napoleon himself felt able to relax. He didn't change his mind about letting Ney go. But at least they parted on a more cordial note.

As Michel withdrew so Josephine accompanied him downstairs. (Naturally enough since her private apartments were on the ground floor.) He enquired after his *cabriolet*. It had not yet arrived, and she used this for her excuse to show him the Tuileries gardens. Orderly, well-tended gardens, she explained as they strolled between the profusion of blooms and hedges, were for her the greatest joy in life after the love she felt for Napoleon. Most of all though gardens with roses. 'Rose,' she informed him had been her original Christian name, and the one she was called by from childhood—until Napoleon had decided to lengthen her second name, Josèphe, to Josephine.

They were just turning the corner to another box-bordered path when suddenly they came upon three remarkably attractive young women. Exclaiming with surprise and delight at what was possibly pre-arranged, Josephine hastened to make the necessary introductions. First to Hortense, her own daughter by a previous marriage to Alexandre de Beauharnais, executed during the Terror; then to Hortense's friends, the Auguié sisters, Adèle and Aglaé. Did the general hesitate and appear to take rather longer in bowing over the hand of the latter? Certainly this became the subject for discussion among the ladies once Ney's cab was announced and he departed. Hortense, a born matchmaker and recorder of this scene, commented upon his striking physique, while Josephine alluded to his bravery and other qualities as an officer. But Aglaé, easily distinguished as the tallest of the group, merely tossed her head of dark-brown curls. 'Your friend, Madame Bonaparte, looks somewhat old-fashioned,' she said. 'He ought to cut his pigtail and shave off those abominable whiskers!'

Michel remained at La Petite Malgrange until the beginning of December. Not being political and with France still at peace he did not regard this as a period of exile. He helped harvest a good cash crop of oats, another of barley, and adding the profit to his salary and the Hohenlinden gratuities he found himself fairly comfortably off. It coincided with his noting how quickly and efficiently Napoleon had re-established monetary values. Just one of a number of things which now brought him around to the viewpoint that the First Consul's rule was providing the nation with a better form of government than he could ever remember.

Also at this time in Paris Josephine was working upon her personal scheme to bind Ney to her husband's cause by a combination of promotion and marriage. She evidently broached both of these subjects at Malmaison, during the day-and-a-half's break which concluded each *décade*, the Republican ten-day week. Napoleon was always at his most approachable then. 'True to his prejudices, he never admitted her into political matters, yet in other directions he often made use of her good sense and judgement.'[4] On this particular occasion he listened to her ideas with interest and growing admiration. Certainly he was prepared to have another go at winning Ney over. 'The Moreau Club', he suspected, were hatching plots and it was essential that General Ney did not join them. Accordingly he promised to offer him a favoured situation; if there wasn't one vacant, well he would have to invent something! As for the marital plans, although he approved of these, he left the details strictly to Josephine. She would find the best way of handling it all. Especially in the matter of talking round young Mademoiselle Auguié . . .

Aglaé Louise Auguié. She was not yet twenty-one years of age in contrast to Michel's thirty-two, and she came from a totally different background. Her father, Pierre César Auguié, had been *receveur-général des finances* under Louis XVI. Her mother (*née* Genest) was the daughter of a former first secretary at the Foreign Office and after her marriage became lady-in-waiting to Marie Antoinette. As such she had braved the fury of the Parisians when they massacred the Swiss Guard and broke into the Tuileries. On that August day in 1792 her bold stance enabled the Queen to slip away and find temporary refuge with her husband and children in a stenographer's box inside the Riding School where the National Assembly was sitting. But one lady-in-waiting could do nothing against the determination of the Committee of Public Safety. By October 1795 Marie Antoinette was tried and condemned; and when Madame Auguié heard of her execution she lost her reason, killing herself by jumping from a high window. The weight of this tragedy upon Aglaé is easy to imagine. She was barely twelve. Only the week before her father had been thrown into prison and now awaited trial. After a privileged, comfortable childhood it appeared as if the whole world had gone mad! In a sense, of course, it had: to one who was young, sensitive and who had never previously experienced violence. And yet these events were of the utmost importance to her upbringing and future, for they led to both she and her sister Adèle passing into the care of their formidable aunt, Jeanne Louise Henrietta Campan.

Madame Campan too had served Marie Antoinette, as First Lady of the

Bedchamber. Before which at fifteen she had been appointed reader to the three daughters of Louis XV, subsequently developing into an expert on manners and the graces of life. She was extremely cultured: speaking good English, Italian and possessing a thorough knowledge of classical literature and music. Also she was a great favourite at court, the King giving her as dowry an annuity of 5000 *livres* upon her marriage to M. Campan, the chief secretary to his Cabinet.[5] In spite of so many close associations with the *ancien régime* (and in particular as a known friend of Marie Antoinette) she managed to elude the agents of the Terror. However, she was left impoverished, and under the relaxed rule of the Directory decided to open a school for young ladies:

> Not only was Mme Campan a governess born, not made, but with government and society settling down to a normal system . . . her venture was opportune and succeeded beyond expectation. As a link with the old regime she was looked upon as the arbiter of taste and etiquette, and her school became the fashion not only for children of Republican families but for those of returned *émigrés*.[6]

Meanwhile Monsieur Auguié had been released from prison. Feeling broken in spirit if not physically, he readily agreed to trust his daughter's further board and education to the *Institution Nationale de Saint-Germain* as his sister-in-law's academy was named.

This school consequently must be regarded as the principal formative influence upon Aglaé's adolescent years. It was first of all sited in a house at St.-Germain-en-Laye which Madame Campan judged too small, so she leased the greater part of the old Hôtel de Rohan in the Rue de l'Unité and quite soon her pupils included a number of girls who would marry men famous under the Empire. At least one future queen, two princesses and several duchesses passed through the establishment. All later expressed the greatest admiration for its methods.[7]

Aglaé for her part found Madame Campan strict, but just; and gradually she transferred to her aunt the deep and sincere affections she had once bestowed upon her own mother. 'Love and respect intermingled.' The school side of things she discovered hard going though. There were lessons every day, including Sunday. Eugène de Beauharnais was not exaggerating when he recalled visiting his sister there. 'Afterwards one could almost smell, see, and hear Madame Campan's seminary on a Sunday afternoon: the polished beeswax parquet, the drawing-boards, the practising of scales . . .'[8] And yet, as Aglaé herself admitted, the head of the school was far from being a conventional teacher. For instance, both of the Auguié girls had to take lessons in domestic economy: previously unheard of in the French curriculum. Other classes were devoted to literature, music and the social arts, especially the art of pleasing. But it was Madame Campan's belief that studies should be made more interesting, and she broke the accepted routine of school by teaching Racine, Corneille and Molière not from books but by having her pupils produce and act the plays as public entertainments. Aglaé responded to such methods and through them gained in learning, charm, deportment and not least in conversation. She never forgot her aunt's quaint, but effective rules regarding a choice of subjects for the latter:

> Any choice, *she said*, was not by the tastes and inclinations of the guests, but strangely enough by their numbers. If there were twelve guests at table, travel and literature were to be discussed; if eight, then art, science and new inventions. When six were present, politics and philosophy might be essayed; if four, affairs of sentiment and romantic adventures were allowable. Two guests, then—she says each talks of himself—a *tête-à-tête* belongs to the egoist![9]

At the same time Aglaé grew up to be extremely pretty. She was tall for a member of her sex at this period, five feet five or six inches, and had a slender but well-formed figure. Her hair was dark brown, almost black and naturally curling; while she had large, dark, melting, almond-shaped eyes set in an oval face where the nose was slightly too long, but straight and redeemed by a very good mouth. Moreover her complexion had little need of artificial aids (a fact confirmed in the later portrait by Baron Gérard). It was soft, pink and delicate, with no hint of olive or sallowness. In manner she was gay and lively, which together with her conversational skill gave her overall personality a most attractive animation. She had done well to overcome the tragic events of several years before. On the other hand she remained of a somewhat nervous disposition and was unusually susceptible to taunts or slights. Michel Ney would never do anything consciously to hurt her; but others would and did.

Her best friend at the school, Hortense de Beauharnais, was if anything even more excitable and highly-strung, and perhaps to begin with their mutual misfortunes were responsible for drawing the girls together. (Each had one parent killed by the Revolution; with the other escaping the same fate by a hair's breadth.) When Hortense entered Madame Campan's she was in very difficult circumstances, so Aglaé cannot be accused of befriending her in order to pursue her own social ambitions.[10] Josephine was not yet the companion and hostess for Paul Barras, and there is a telltale footnote to her biography which describes how Madame Campan agreed to take Hortense at half-price and provided her with a second-hand uniform. After her marriage to Napoleon, however, the situation changed dramatically and Hortense's new importance swept Aglaé and her sister along with it. It wasn't long before General Bonaparte himself visited the school. He came as guest-of-honour to one of their entertainments and was 'so pleased with Madame Campan's methods . . . and the refined manners of the pupils, that he at once said he must send his *ignorant* little sister Caroline there.'[11] (Later he is reputed to have told Madame Campan: 'If ever I were to create a women's Republic I would make you First Consul!'; and when he founded an academy at Écouen for the education of daughters and sisters of members of the Legion of Honour, he appointed her his superintendent.)

Hortense in a letter to her brother Eugène describes the excitement she shared with Aglaé after the *coup* of Brumaire:

> General Murat, a true knight-errant, sent us four grenadiers of the

Guard, of which he was the commander, to tell us what had taken place at St.-Cloud, and the appointment of (General) Bonaparte to the Consulate. Imagine the effect of four grenadiers knocking on the doors of a convent in the middle of the night! Everyone got a terrible shock and Madame Campan blamed this military method of sending news. Caroline (B) read it as a proof of love . . .[12]

But the letter which more truly reveals her closeness with Aglaé is one dated August 1799 and sent to Eugène when he was serving in Egypt. '*Maman* (Josephine) has brought La Malmaison, which is near St.-Germain. I go there nearly every week. She is living . . . very quietly and seeing nobody except Madame Campan and her nieces Mlles Auguié, who often go with me . . .'[13] In effect Aglaé became like an unofficial lady-in-waiting to Josephine. Probably without even realising the fact since the Consul's wife always treated her in a friendly and informal manner. However, it did mean she was readily available whenever Josephine chose to extol the various merits of General Ney—which happened quite often after the meeting in the Tuileries gardens.

By 1 December 1801 Napoleon had decided to create Ney 'Inspector-General' of all France's cavalry. The appointment was intended to single him out as a favoured officer, and Berthier wrote to La Petite Malgrange stating that these new duties were to commence on the first of the year; would he please come to Paris in order to make the necessary arrangements? But Ney did not see it this way. The job struck him as being routine at a time when he felt in the mood for adventure. On 4 December therefore, he wrote back requesting permission to join the expedition to the West Indies. Napoleon was understandably piqued. He did not yet know that Michel now viewed him with a marked degree of admiration. Consequently, he took the refusal of his offer as being not only ungrateful but proof that Ney preferred a kind of exile to serving him in any direct or personal capacity. 'Very well! Let him perish in the fever-swamps of San Domingo,' he snapped at Berthier—and he stabbed his signature on a document authorising Ney to command Leclerc's cavalry. The main expedition had already sailed from Brest. General Ney would have to follow it.

Michel travelled to the capital and received his official appointment on the morning of 18 December. But that same evening Josephine contrived to let him see Aglaé again—whether at the theatre or some reception we do not know—and suddenly the desire to go to the West Indies vanished from Ney's mind. Obviously his first glimpse of the young lady had been at work subconsciously, because he now realised and accepted that he was deeply in love. His thoughts crystallised about Aglaé just as passionately as they had upon France, leading from admiration to the idea of 'What a pleasure it would be to kiss her, in fact to marry her!' Using Stendhal's own phrase, he began to adorn the woman he loved with a thousand perfections. The following day he hurried to the War Ministry. 'Urgent family matters,' he informed Berthier, 'might require my staying on in France.' And on 1 January 1802 he sent in a formal request: to give up the West Indian command and be appointed Inspector of cavalry after all! Napoleon was amazed, considerably perplexed, but nevertheless delighted at what he regarded as a change of heart towards himself. Immediately, therefore, he dictated his assent.

Josephine's next step was to take Michel into her confidence and arrange for a proper meeting with Aglaé. Since his release from prison Monsieur Auguié had been living in semi-retirement at Grignon, near Versailles. The family owned a small *château*, which the Consulate had restored to him, and in February Josephine persuaded him to invite the General there, assuming that love could be relied upon to do the rest.

But it wasn't that easy. At Grignon Michel proved to be as shy and withdrawn as he had at the first meeting with Napoleon. Aglaé found the conversation between them heavy going, which reduced her own vivacity to a point where she seemed cool and studied. Ney was largely to blame in another way through. He had turned up at the *château* still very much the rough, tough Republican general. Josephine had caused Aglaé to be already half in love with him as a man, but she could not stand his actual appearance. And afterwards she told Josephine so.

Michel returned to Paris with grave doubts. Not of his genuine love for Aglaé, but because having conceived hope he had then encountered what appeared to be her indifference. 'She has all the perfections,' he reminded himself: 'But does she love me? Where is the evidence?' He decided the visit had been a waste of time. . .

Perhaps Ida St.-Elme caught him on the rebound. Anyway, this was the moment when she claims to have become his mistress. Her account is written in a decidedly racy style, and one suspects exaggeration, but she states that he arrived at her home in the Rue de Babylone and poured out the whole, sad story of his misplaced infatuation with Mlle Auguié. She concealed her annoyance and invited him back the next day, when she told him of her efforts to succeed as an actress. (Even with the Minister of the Interior as a patron her debut at the Comédie Française had been disastrous.) 'What a pity you are an actress!' Ney exclaimed, bursting out laughing. 'Why, I would rather see you as a canteen-woman!' 'That would suit me perfectly,' she replied, 'for it would allow me to see you all the time!' Ney laughed again, but then turned serious: no doubt recalling the various Mother Courages he had seen along the Rhine. 'No. That life wouldn't be suitable for you either, for with canteen-women not even ugliness guarantees the preservation of their virtue.' He was now thoroughly at ease, she continues, and after a further exchange of anecdotes they progressed beyond all formalities towards the bedroom.

Had Ida St.-Elme lived in the twentieth century she would have endeavoured to sleep with film-actors and popular singers rather than generals. She definitely preferred images of success to the basic man. But she did have certain qualities of her own, and afterwards, when Ney met her on or near the European battlefields, he always greeted her with friendship and a show of camaraderie. She in turn leaves us with a marvellous tribute to him. 'If Ney had been an ordinary human being, one would almost have noticed a touch of ugliness in his face,' she writes. 'But with his noble figure; with his overall attitude and that look of his which expressed the whole man; by seeing so much glory one thought that one saw beauty. After a few words had been exchanged all our embarrassment had gone. We were completely at ease with one another, just as if we had been friends for twenty years.'

If the events she describes are true then Michel must have turned to her for consolation in the way that many frustrated lovers are apt to do. And yet the affair—if it existed—had to be fleeting, for Josephine was not prepared to accept defeat. She undoubtedly lectured Aglaé and she then had a few, feminine words with Michel. By March another stay was arranged at Grignon, and this time Ney arrived clean-shaven, with his hair cut short into fashionable waves, dressed in a smart new, non-combatant uniform and carrying a beautiful gilt-and-enamelled snuff box manufactured in Limoges. (His teeth, one should add, had been brushed clean of all their tobacco stains.) Aglaé was bowled over by the transformation, and so were her sisters and father. General Ney would make a most suitable match Monsieur Auguié decided.

Soon afterwards Michel sent Aglaé some jewellery. In an accompanying note he expressed his regrets for its small value. 'I cannot,' he wrote, 'offer you pearls and diamonds because according to my beliefs the sword should be used to win glory, not wealth.' By May he was sending appeals to Josephine to assist his formal proposal of marriage. She responded with a glowing testimonial to Monsieur Auguié, posted via Michel with the following note:

> I send you, General, the letter you have asked of me for Citizen Auguié. May I beg that you read it? I have not said in it all the good things that I know and think of you. I want to let this worthy family have the satisfaction of discovering for themselves all your good qualities; but I repeat to you the assurance of the interest which Bonaparte and myself take in this marriage, and of the satisfaction he feels in thinking that he will thus secure the happiness of two people for whom he has a special good will and esteem. I share with him both these feelings. La Pagerie Bonaparte—Malmaison, 30 May 1802.[14]

The wedding took place at Grignon on 5 August. There was a civil ceremony mid-morning in the *mairie*. Then the family were joined by a large number of guests (Consulate representatives, old army colleagues of Ney's and the Auguiés' friends and estate-workers) for another ceremony in the private chapel of the *château*—religion having become acceptable again to the French under Napoleon's guidance. Both the chapel and the rooms of the *château* had been decorated by Jean-Baptiste Isabey, sent by Josephine in preference to her husband's own favourite artist, the firebrand revolutionary Jacques-Louis David. Isabey was an expert in painting miniatures who decided to give the chapel 'the look of a little forest glade'. By the time he'd finished it looked more like the main conservatory in a botanical gardens, with banks of greenery around the walls, rare potted plants massed on either side of the altar and ivy and garlands of cut-flowers dangling from the beams and chandeliers. The foliage completely concealed the musicians of the military band in the gallery, who played all the louder to make up for it; including the marching song written by Hortense de Beauharnais for the Army of Egypt, *Partant pour la Syrie.* However, few seemed to mind the tendrils getting in their hair and eyes and Isabey was quite prepared to serve as his own best critic. 'Ah, but it is beautiful,' he lisped. 'It is nature perfected. *Anyone* could be married painlessly in all this beauty!'[15]

Two couples stood before the altar. Aglaé dressed in plain white, her veil encircled by yellow roses, held the arm of Michel, in full dress uniform, who with his other hand clutched his wedding-present from Napoleon: a magnificent jewelled sabre brought back from the Middle East. Beside them were two elderly peasants who also lived at Grignon and were celebrating their golden-wedding anniversary. Michel had thought it right and proper to let them renew their nuptial benediction at the same time. 'It will be a good omen for the happiness of our own married life,' he said. 'Also it will remind me of my humble birth.'[16]

The ceremony ended, Madame Campan took over. There was first of all a play, with Hortense (now married to Louis Bonaparte) and Aglaé's other sister Antoinette taking the leading parts. Then a fête, which included folk-dancing and a return of the—by this time somewhat inebriated—military band. Finally, as darkness fell, they were served supper and carried their refilled glasses to an open-air ball: with music provided by a more sedate ensemble. Again Madame Campan was in charge, and she saw to it that for the opening quadrille Aglaé danced with the golden-wedding peasant and Michel with his wife. (He danced, seemingly, with a muscular agility to make up for his lack of technique.)

Isabey, annoyed at having been up-staged by Madame Campan in the afternoon, was determined to gather the last plaudits though. Looking conspicuous by his presence—he had donned a Parisian's idea of peasant costume—at his given signal troops of boys in the trees lit hundreds of candles commemorating Ney's battles. Mannheim, Winterthur, Mainz, Würzburg, Altenkirchen and so on, ending with Hohenlinden. As one by one the candles flickered into spellings Aglaé marvelled at the number he had fought in and still survived. She had no way of knowing that these were merely a prelude to the more famous battles of his career.

Of all the twenty-six marshals created by Napoleon Michel Ney probably enjoyed the happiest married life. He was not naturally a philanderer and once having fallen in love he stayed that way. If any misdemeanours occurred in his campaign-tent then they were never publicised or used to drag down Aglaé's self-respect. He was generous towards her, tender and her only causes for reproach were the long periods he spent away with the army and the fact that he wasn't a very regular letter writer, more a scribbler of impromptu notes. At the same time though he could be fiery and stinging if there was any evidence that someone had hurt or insulted her.

Aglaé in turn, even if the affair had begun less passionately for her, was a model of good behaviour and her love for Michel gained strength over the years. Her lively disposition and extreme sensitivity were balanced by her grace, conscientiousness and loyalty. She proved an excellent domestic manager, an ideal hostess and in the course of time presented her husband with four strapping sons, all devoted to their father. She suffered terrible anxieties when Ney was away fighting, but in the end she learned to keep herself busy at home or on lengthy visits to her friends, especially Hortense whose marital problems were soon to become acute.

Ney's first departure came a lot earlier than he'd expected. Hardly had they settled into La Petite Malgrange when Napoleon wrote on 29 September ordering him to proceed immediately to Switzerland—his task being 'to re-establish order' in the friendly, but now precarious Helvetian Republic. It was certainly an important mission, involving diplomacy as well as soldiering, and the appointment signified that the First Consul not only favoured but trusted him completely. On the other hand there was also practical reasoning behind it: Ney being the only general close to Napoleon who could speak fluent German. Talleyrand, the Minister for Foreign Affairs, opposed his going. He disliked military men becoming involved in what he considered his own prerogative and, as he argued, the Swiss problem was very knotty indeed. But Napoleon was adamant and in the end events proved him right.

Switzerland, 'though small and politically insignificant, had become a country whose strategical importance was early recognised both by France and by her enemies'.[17] The principles of the French Revolution had easily penetrated the Swiss cantons, sweeping aside the old feudal federation with its oligarchic privileges and leading to the creation of a new, Helvetian Republic with civil equality and religious toleration. Following which, for several years the country remained firmly allied to France, no matter who was governing in Paris. However, in 1802 the Republic was faced by civil war. Provoked and subsidised by the British government, counter-revolutionary forces in the outer, more mountainous cantons had risen against the central government in Berne. The march on Berne was led by a General Bachmann, formerly an officer under Louis XVI and now in receipt of a British pension. But the real leader of the mountain cantons was Aloys von Reding of Schwyz, a genuine patriot who hoped to re-establish the old Swiss Confederation as a first step towards making the nation independent of France. To complicate matters several of the cantons were at loggerheads with their neighbours. The people of Vaud disliked Berne, but they disliked the people of the Valais even more. Meanwhile 'the French party was at war with the autonomists; democrats strove with oligarchs; federalists with unionists; Jacobins with Girondins. Even the *coup d'état* was naturalised on Swiss soil: effected now in this interest; now in that. . . .'[18] The situation was chaotic and the First Consul, plainly worried, decided upon a policy of 'either a Switzerland friendly to France, or no Switzerland at all!' When the Republican government appealed for his help he promptly drew up and despatched the well-known proclamation offering his mediation as the one hope of restoring peace in Switzerland.

Dictated at St.-Cloud, and headed *Bonaparte, First Consul of the French Republic to the Cantons of the Helvetian Republic,* he proposed, among other things, 'a form of government which would satisfy both Federalists and Confederates'. But in private he also meant 'without giving too much power to either; to give peace and contentment to Switzerland, and yet to leave her wholly dependent upon France!'[19] He declared that 'all government and parliamentary bodies formed since the counter-revolution should be dissolved', all armed forces in the cantons 'should disarm and disband' and the Senate of Berne as well as the inhabitants of each canton 'were to send three deputies to Paris in order to make known the means of restoring order and tranquillity, and of conciliating all the parties'.[20] For 'mediation' therefore read 'ultimatum' and Ney had the ticklish job of seeing it through. He would be given troops, of course, as many as he needed, but it was Napoleon's intention that these should be used only in the last resort. He desired the Swiss as allies. It might mean bullying them a little, but he genuinely hoped to avoid bloodshed and a French occupation.[21]

After waving goodbye to a tearful Aglaé Michel arrived at Geneva to learn that his allocated troops were in a state of readiness around the frontiers. Also that Colonel Jean Rapp had preceded him into the country distributing copies of Napoleon's proclamation and informing the cantons of 'General Ney's proposed entry. If necessary at the head of a strong army, or as Minister-Plenipotentiary if they showed themselves disposed to live in mutual harmony.' It appeared the proclamation was having its intended effect, because by mid-October the Republican Senators were able to return to Berne. Meanwhile Ney had received his own appointment and preliminary instructions in the form of a long letter from Talleyrand:

Général. I am directed by the First Consul to inform you that he has been pleased to appoint you Minister-Plenipotentiary from this Republic to the Helvetian Republic. You will therefore proceed to Berne, where you will receive the further instructions he has directed me to send you, and you will there fulfil the duties of your mission.

A few days since, Helvetia was in agitation; the flame of civil war burst forth in every part of it; but the proclamation of the First Consul has given ideas of order and peace to all its inhabitants. The citizens of that country, struck with the wisdom of the advice given them by the First Consul, have lost no time in following it. The principal object of your mission is to maintain and direct them in this just and prudent deference. From your title of Minister-Plenipotentiary, your former office, and your talents you will derive means of influence which you will employ, more particularly in preventing any marked opposition to the government. The constant principle of your conduct lies in the execution of the clauses of the First Consul's proclamation . . .

Do not cease to impress upon the minds of the citizens of Helvetia that the First Consul has most particularly in view the repose, happiness and greatness of Helvetia; that the Helvetian Republic can be neither rich, nor happy, except by its union with France. All that you may say to the persons with whom you communicate must tend to prove that the First Consul will suffer nothing against the repose and power of Helvetia; that he considers it a duty to renew the friendly relations which have at all times united Helvetia to France . . .

I am happy, Citizen General, that the choice of the First Consul has fallen upon you to direct the legation of the republic in Helvetia, for it gives me an opportunity of corresponding with you, and of acquainting the First Consul with the proofs of prudence and zeal which you will give in the course of your mission . . .[22]

The next day he set off for Berne, accompanied by two ADCs and leaving his troops under the command of General Seras. Talleyrand's second letter was waiting for him, containing specific instructions that he must avoid all friction between the returned Senate and the local government of Berne. 'It is of the greatest importance that the orders of the First Consul be carried out in accordance with his just and impartial views. He has proved that he will not show preference to any one faction.' He was also to find men who would co-operate with France's policy: 'Those men . . . who have shown a loyal preference for the voice of conciliation to the voice of war.' And finally Talleyrand could not resist another sally at Ney's own chosen profession of arms:

> Above all you should avoid giving the impression of the military man issuing orders. Everything that might make you appear in Swiss eyes as the General in command of forces placed upon their frontiers should be carefully avoided. You are the Minister of a friendly power which wishes to give good advice and operate only through its wisdom . . .[23]

The Foreign Minister contradicts himself here, of course. After Colonel Rapp's visitations the Swiss were left in no doubt that Napoleon meant business.

In fact, events had already overtaken his instructions. In Berne Ney discovered Aloys von Reding was still in armed revolt, so was Bachmann and that they were planning to declare their independence in a strongly-fortified Zürich. Within hours, entirely on his own initiative, he had ordered Seras to move towards Zürich, breaking up any insurgent groups he encountered on the way: mainly peasants armed with pitchforks, scythes and spiked clubs, but 'to treat them gently'. And he dashed off the following note to Napoleon:

> I propose to occupy the Helvetian territory with twelve battalions of infantry, six squadrons of cavalry and twelve guns, drawn from the bodies of troops assembled at Chiavenna, Como, in the Valais, at Geneva, Pontarlier and Hüningen. The operation will destroy forever the hopes of the insurgents and protect the election of the deputies, whom the cantons are to send to Paris in accordance with your proclamation.[24]

Seras marched his division at the rate of nearly forty miles each day, an incredible feat. By 28 October they were outside Zürich and Ney's official report to Berthier describes what happened next:

> On the morning of 29 October, as the insurgents went to place their sentries outside the city, so our troops entered Zürich through three different gates. Colonel Meyer was sent to protest against French intervention in their political quarrels; General Seras ordered him to conform to the proclamation which the First Consul addressed to the inhabitants. About six hundred regular troops, very well turned out, equipped and armed and paid by the insurrectionist chief twenty *sols* a day, stood in battle formation in the square together with a more

considerable number of peasants armed with iron bars, scythes, and all manner of weapons.

The music played at the head of the French column, inspired all these insurgents with a genuine respect. The peasants shouted *Vive le grand Bonaparte! Vive la France!* General Seras had the first-named disarmed, and sent everybody home with words of peace. The legitimate authorities have been installed. The unfortunate victims of the insurrection who filled the prisons, have been released . . .

We found Zürich well-provisioned with war materials. The whole operation reflects the greatest credit on the wise dispositions of General Seras and has, I trust, put an end to the pretensions of the Diet of Schwyz (*von Reding's group*) without a shot being fired. The Diet has been dissolved on 28 October. General Bachmann, Commander-in-Chief of the insurrectionists, was at Zürich; he took no part and has asked for a safe-conduct to Munich.[25]

It was a clever (and bloodless) victory because Ney had used his own judgement and timed his show of strength to perfection. But to make absolutely sure he wrote to Berthier again on 3 November, proposing: 'We should get hold of the chiefs of the insurrection . . . to counter their muffled conspiracy. A decision of the First Consul in this respect is vital. Apart from here (around Zürich) the rest of Switzerland is quiet.'[26] Napoleon agreed, whereupon Ney ordered Seras to arrest Reding together with a dozen or so troublesome representatives of the old, feudal regime: mainly noblemen who were behind the insurrection in order to regain their lost privileges. He had them shut up in the fortress of Aarburg and guarded by French troops.

From this point onwards he was able to become the man of peace and to be, as even Talleyrand admitted, the ideal diplomat. Towards his prisoners he extended the utmost courtesy, sending an ADC to ensure that 'the orders which I have given that the detained men be treated with humanity and the respect due to adversity are obeyed. You will see these men, receive from them in writing all complaints they have to make and consult with the French Commandant on the means which could ease their position without endangering security.'[27] Soon afterwards he visited Aarburg himself and held several lengthy, bilingual conversations with Reding, the two men making a good impression on one another. Ney could understand and came to admire Reding's severe, but deeply held patriotic beliefs, so much so that when his second son was born, he added the name Louis for 'Aloys' to his list of Christian names. The Swiss in turn was impressed by his captor's frankness, his honest desire to see the country settled in a way that would benefit all concerned—and by his promise of an early release.

At the same time Ney proceeded along the lines of finding and sending suitable delegates to the conference in France. Again he succeeded admirably. He extolled the virtues and good intentions of the First Consul on every conceivable occasion, until in the end even the self-centred Confederates were prepared to trust him. Meanwhile France's troops behaved with scrupulous impartiality; which considerably undermined Britain's frantic efforts to create more insurrection by smearing the French as 'hated, irresponsible conquerors'. Finally it was the Swiss themselves who chased out Britain's guinea-scattering agents, William Wickham and the Duke of Argyll—while Ney pulled off his biggest diplomatic *coup* by persuading Colonel von Mulinen, former Chief Magistrate in Berne and an acknowledged leader of the old, Confederate nobility to join the delegates at St.-Cloud.

By mid-February 1803 not only was internal peace fully secured but Napoleon and the delegates had signed the Act of Mediation: in effect the basis of modern Switzerland. After consultations conducted at St.-Cloud in an atmosphere of great friendliness, and with Napoleon displaying towards the Swiss a most benevolent respect, what emerged by general consent was a brand-new constitution, with in-built strengths to cope with any future problems which might arise. Napoleon acted upon the experience of the years 1798-1802, which made it 'abundantly clear that the Swiss—the German, French, and Italian peoples combined by a freak of nature or of circumstance —were not going to settle down in acceptance of a unified Republic.'[28] Consequently his Act of Mediation was 'a distinct improvement' upon the Helvetian Republic:

> It recognised the sovereignty of the cantons, adding to the original thirteen six new cantons representing the allied and subject lands, such as Vaud, Ticino and Grisons. Into the new cantons the principle of representative democracy was introduced; the old ones were divided into rural cantons with their primitive *Landsgemeinden* and urban cantons under burgher aristocracies. Upon the sovereign cantons, new and old, was superimposed a central government with a Federal Diet.[29]

Under the constitution each canton retained control of its own laws, justice, education and finance; but to the Diet passed all responsibility for foreign affairs, a common currency, defence and the customs-posts. Within the Diet, cantons were regarded as being equal, despite the larger ones being represented by two deputies, the smaller by one apiece. Executive power meanwhile was vested in a *Landamann* or chief magistrate assisted by a council of four, the *Landamann* to be elected annually—although by popular consent the position would first of all be filled by Monsieur d'Affry, an ex-officer in the Swiss Guards and a man whose abilities Ney had praised in his despatches to Talleyrand. In return for French mediation the delegates offered Napoleon an alliance based on trade and defence, and promised to supply France's army with four regiments. Nor did they ever give him cause to doubt their sincerity. Even during the Empire's later, darker days there was not the slightest question of their reverting to mediaeval tradition and selling themselves as mercenaries to the highest bidder. The Germans and Austrians would desert France for gain, the Italians through fear. But Swiss loyalty and goodwill remained as prominent as their own high mountains.

Ney accompanied Landamann d'Affry to the new capital, Freiburg; and won further praise for his tactful assistance to the Federal Council. Once its machinery of government was running smoothly, however, he obtained leave of absence, hurrying back to join Aglaé at Grignon. Their first child was born on 8 May: with Ney present at the birth. And as some indication of how clearly Michel now supported the First Consul they named him Napoleon Joseph. Napoleon himself—'very pleased with General Ney's successes in Switzerland'—and Josephine stood as godparents at the baby's christening. Following which the First Consul agreed that once they were strong enough Aglaé and her son could join Michel in Freiburg.

On 30 June 1803 Ney was France's official representative at the celebrations in honour of the first Diet being opened. There were bilingual services in Freiburg Cathedral and big displays of military pomp in the square outside. When it came to the speeches Landamann d'Affry was full of praise for Napoleon, describing him as 'the protector of peace, freedom and prosperity'. Then the legislators turned to Ney for his reply: a long, formal speech, written in French and delivered in clear, ringing tones:

> The day of 18 *Brumaire, he told them,* from which France dates the revival of her prosperity, was also the moment when the people of Switzerland began to have a ray of hope for a change in their affairs towards greater stability and an order more appropriate to their customs. Their expectations have been justified, and if this did not happen at an earlier date it was because the deep wounds of the French Revolution could not be healed at once. Great changes in the state had become indispensable, and these fully occupied the precious moments that Bonaparte consecrated to the welfare of the people whose interests he had so gloriously defended. His successes have brought about the tranquillity you now enjoy, and which it is up to you to perpetuate. The First Consul's mediation in Swiss affairs represents a legislative masterpiece and has been inspired by the benevolence that characterises this extraordinary genius . . .[30]

There was much more in the same vein, all of it sounding far removed from the unyielding, suspicious Republican general who had once been the essential conscience of France's Rhine army. Moreover it included one passage which curiously enough is underlined in his draft. 'You are all of you convinced, gentlemen, that the prosperity which Switzerland enjoyed before the unfortunate epoch of revolutionary fluctuations was due largely to the numberless good deeds of the French Monarchy . . .' Was this merely inserted to praise the First Consul's effort by comparison, or did it hint that various rumours now going the rounds in Paris had already reached the Minister-Plenipotentiary in Freiburg? Anyway, he finished on a note which most certainly did express his personal beliefs. The Swiss, he told them, must never feel demeaned in accepting the mediation of the First Consul and of France, their most powerful neighbour. 'For power,' he said, 'has no need to corrupt. It is only the lesser men who will stoop to injustice . . . enslavement . . . and tyranny.' These words went down well with the Fathers of Switzerland, and as Ney took his seat again they broke into a great round of applause.

In contrast the months which led on from this euphoric day were spent hammering out the proposed alliances, with the Swiss reverting to their normal hard-headedness whenever the subject of money was raised. The haggling over customs-tariffs seemed endless, likewise who bore the costs of supporting the Federation's local militias. Also the Swiss deputies insisted on restricting the services of their troops to the continent of Europe, and only with the greatest difficulty were persuaded to compromise upon a stipulation that no service in Asia or America would be required. These sessions called for all of Ney's powers as a diplomat, and on one occasion he sighed to his aides 'how much easier it would have been to settle this matter with powder and cold steel!' Eventually though the work was completed and the treaties prepared for signing.

Happily Aglaé and the baby were there to enliven his leisure hours and towards Christmas they were joined by Monsieur Auguié and Adèle. The latter naturally acquainted him with all the gossip from Paris. Hortense and her husband Louis Bonaparte were said to be quarrelling; and Napoleon's other relatives were still urging him to divorce Josephine. However, they added, emphasising what they considered of far greater importance, the rumours circulating that Napoleon wanted to reintroduce a monarchy had suddenly increased. Two years before such information would have provoked a furious outburst from the staunch Republican, but on this occasion he accepted the prospect with equanimity. So convinced had he grown of Napoleon being France's Man of Destiny that he didn't even reflect upon the possibly divisive repercussions within the nation if such an event came to pass. Instead he duly sent off the final copies of his treaties to Paris for Napoleon's approval; and soon received the following reply from Talleyrand:

> The First Consul, before whom I laid the articles of the treaty of alliance, and of the military capitulation agreed upon between you and the Helvetian commission, has directed me to express his satisfaction at the zeal you have displayed in following up and closing this negotiation. You are hereby authorised to sign the two treaties, and even to yield, if you deem it necessary, to the subsequent demands made relative to the recruits.[31]

It meant that his ambassadorship to Switzerland was over. Already Napoleon required his services elsewhere. On 17 January 1804 he was recalled from Freiburg and assigned to the revived Army of England. King George III's government had deliberately violated the Treaty of Amiens the previous May, and ironically Ney's much-publicised activities inside Switzerland were a contributory factor to the London decision. In the House of Commons, Charles James Fox made the greatest speech of his career against involving Europe in another unnecessary war. But 'German' George's bought-Tories had made up their minds. Even the disapproval of England's former Coalition partners could not sway them. In effect, as France's envoy Andréossy sadly wrote, *they feared the peace because it made them feel isolated and allowed their old rivals to prosper.* 'It is not such and such a fact but the totality of facts comprising the First Consul's *gloire* and the greatness of France that frightens the English.'[32]

Before leaving the Republic though Michel paid a friendly call on Aloys von Reding, now restored to an honoured position in Swiss society, and something of a folk-hero in the mountain cantons, akin to William Tell. Also he received a most flattering letter from Landamann d'Affry, thanking him for 'the good you have done us' and expressing Switzerland's gratitude:

> The cantons have expressed a wish that you would accept a feeble pledge of their attachment . . . and seeing the preparations for your departure, I have requested Monsieur Maillardoz to present it to you at Paris. It is a token of remembrance and nothing more; but we should esteem ourselves happy, if, by calling to your recollection a nation whom you have so essentially obliged, it should prove the means of your not forgetting the sentiments which every member of that nation will forever feel towards you.[33]

Five days later, back in Paris, General Ney received from Monsieur Maillardoz 'in the name of the Helvetian Republic' a gold snuffbox studded with diamonds.[34]

Today the facts make strange reading, anticipating many modern novels involving government espionage and paid 'hit'-men; nevertheless England's Tory ministry probably did more than anyone else to turn Napoleon from being First Consul for life into the Emperor of the French. It wasn't their declaration of war and subsequent attacks upon French merchant-shipping which helped to realise his monarchal ambitions. The people of France were by no means eager for another period of fighting, and some of them even blamed Napoleon for the situation, viewing with considerable alarm his assembly of a strong invasion force at Boulogne. But when London started running the Bourbons' agents into France and financing their unsuccessful, yet notorious attempts on his life then the French nation was quick to react in his favour. 'They want to kill Bonaparte. Very well, we must defend him and make him immortal,' is how Councillor Regnault, by no means a sycophant, saw it; although he took care to add: 'It is important to establish that it is the people, not God, who give crowns,' a reminder that should Napoleon be assassinated the French were still not prepared to welcome the Bourbons back. Perhaps though the truest words were those spoken by the leader of the would-be assassins, Georges Cadoudal, while in prison awaiting execution: 'We have done more than we thought. We came to give Paris back her King, and instead we have given France an Emperor.'

When Ney took up his new command on 4 March 1804 the infiltration of killers was considered so serious that three-quarters of the Paris police had joined in the hunt for them. No one in the capital could possibly forget that horrific Christmas Eve in 1800 when the First Consul, together with Josephine, Hortense and Caroline, narrowly escaped being blown up on their way to the opera. Only the fast—and some said drunken—driving of his coachman César had saved the Consul, but the explosion was so violent that nine innocent people died and twenty-six more were seriously injured.[35] Intensive investigations followed and the police succeeded in arresting the small-fry: Limoëlan and the aptly-named François Carbon. Saint-Réjant however, who lit the fuse, escaped to America, while Georges Cadoudal was already safe in England—where immediately he set up a training camp for conspirators and guerrillas at Romsey. Cadoudal himself was an extremely dangerous character. 'A squat red-haired Breton peasant of immense strength—Goliath to his friends—with a bull neck, broken nose, red sideburns, and one grey eye bigger than the other.'[36] Unmarried, 'dedicated body and soul to the Bourbons', he took his orders from Louis XVI's brother, the Count of Artois, their liaison being the Polignacs. But English government money financed his Romsey camp, finding its way to him via 'Fighting' William Windham, a close friend of Pitt's and otherwise known to the public for his vigorous defence of bear-baiting. A further one million francs was reserved for the actual assassination attempts.[37]

By 1804 Cadoudal was back in France at the head of no less than sixty trained assassins. Réal, Napoleon's Chief-of-Police, discovered this after his men had the good luck to arrest Bouvet de Lozier, the conspirators' Number Two, on 14 February. The overall plan was for the killing of Napoleon to coincide with the re-entry into France via Alsace of a Bourbon prince, the Duke of Enghien: a fact confirmed by Talleyrand's spy in Munich, Captain Rosey, who disguised as a discontented émigré received 10,000 pounds from the English agent there, one Francis Drake, to stir up rebellions in Besançon and Strasbourg.[38] Further to this Lozier admitted, General Charles Pichegru had returned from exile in Guiana to join the plot: his specific task being to win over as many senior army officers as possible. And General Moreau was also said to be implicated. (As a result he was arrested the next day—15 February—and held pending further enquiries.)

Ney had visited Moreau only a fortnight before. But if his commander at Hohenlinden entertained any hopes of winning him away from Napoleon then he was to be disappointed. 'I hear you go to the Tuileries now,' he began; 'So you have become the courtier!' 'Indeed I do,' Michel replied, stung by the accusation: 'Just as I would if you were First Consul!' 'How he has deceived us ...' Moreau went on. But Ney had had enough. 'I shall always be grateful to him for the rapid and wonderful way he administers public affairs,' he said and abruptly switched to a discussion of the intended invasion of England. 'It's madness,' Moreau told him. 'All you have are gun-boats and troop-carriers. How can you possibly hope to elude the British Navy with gun-boats?' Ney was undeterred though, pointing out that the Prince of Orange once eluded an entire fleet under the Earl of Dartmouth. Evidently he had been studying the merits of small ships in combination with winds and tides.[39]

The invasion continued to occupy his thoughts on the journey to Montreuil, the most southerly of the encampments along the Channel coast which made up the 200,000 strong 'Camp of Boulogne'. He had been given the new VI Corps (comprising three infantry divisions and a brigade of cavalry) and upon arrival found a majority of these troops quartered in wooden huts near the village of Étaples and along the banks of the River Canche, just opposite the modern resort of Le Touquet. They, and the gunboats and transports sheltering in the river estuary were in turn protected by an elaborate system of land batteries. His own HQ was seven miles up-river. Meanwhile the equivalent commands to the north were being taken up by Generals Bernadotte, Lannes, Davout, Marmont and Soult; with Joachim Murat leading the Reserve Cavalry.

Without delay Michel threw himself into the job of turning VI Corps into a cohesive and effective fighting force. He trained and drilled the men continuously, but not always with weapons. For half of each day they would exchange their rifles and bayonets for hammers and saws and shovels and axes. Then he divided them into teams of builders, earth-shifters and trench-diggers. If they grumbled that this wasn't what they'd joined the army for, he reminded them how a corps in adversity must be able to do its own engineering work—and for such straight talking they came to like and respect him, putting their backs into everything he asked them to do. Special attention was also paid to embarking and disembarking drills; and for these he would actually take them out to sea, keeping his land batteries on full alert against the possibilities of an English naval raid. 'One day soon,' he reminded everyone: 'We will call England to a severe account for three centuries of hostility.' But being so close to his troops he could not ignore their other recurring anxieties, in particular of what would happen to France if Napoleon died or was assassinated. Must they expect further upheavals, they asked him, perhaps ending with a restoration of the hated Bourbons? Ney did his best to reassure them; in such circumstances, he said, the army's commanders would not stand idly by. Yet the opinions he was assailed by from all ranks were unanimous: Napoleon should be allowed to form a dynasty . . .

Fortunately Georges Cadoudal's plot to kill Napoleon during the next parade on the Place du Carrousel never came to fruition. He had had hussar uniforms made; and dressed in these his picked men were to infiltrate the parade and 'as Napoleon passed down the ranks one of them was to present him with a petition, while the rest pulled out daggers and struck'.[40] It promised all the drama and elaboration of the death of Julius Caesar or Henry III's killing of the Duke of Guise at Blois. However, on the evening of 9 March Cadoudal, who had been lying low in the back of a Paris fruiterer's, decided to change his hide-out and Réal's police spotted him. In the ensuing chase and fight aboard a *cabriolet* Cadoudal shot one policeman dead and wounded another before being overpowered. Whereupon his network of agents collapsed, some forty-five of them falling into Réal's hands, including the two Polignacs. (General Pichegru had been arrested on 28 February.)

The capture of the plotters led to a spectacular trial beginning on 25 May; after which Moreau—because only partly implicated—was exiled to America and twenty conspirators were condemned to death. (Napoleon stepped in to reprieve ten of them, including Armand de Polignac.) Meanwhile it also led to the kidnapping from Germany, court martial and execution by firing-squad on 21 March of the Duke of Enghien. Under interrogation Cadoudal admitted: 'I was to attack the First Consul . . . when a Bourbon prince came to Paris', and at his summary torch-lit confrontation with Napoleon's officers at Vincennes, Enghien confessed to receiving '4,200 guineas a year from England in order to combat not France but a government to which his birth had made him hostile. I asked England if I might serve in her armies, but she replied that this was impossible: I must wait on the Rhine, where I would have a part to play immediately, and I was in fact waiting.'[41] The seven French colonels unanimously agreed that Enghien was guilty under Article 2 of the law of 6 October 1791: 'Any conspiracy and plot aimed at disturbing the State by civil war, and arming the citizens against one another, or against lawful authority will be punished by death'. Josephine pleaded for mercy, but on this occasion Napoleon sided with Talleyrand, who had pressed for the kidnapping in the first place:

> Did I do more than adopt the principle of your government, *he later announced to William Warden on St. Helena,* when it ordered the capture of the Danish Fleet, which was thought to threaten mischief to your country? It had been urged to me again and again, as a sound political opinion, that the new dynasty could not be secure, while the Bourbons remained. Talleyrand never deviated from this principle: it was a fixed, unchangeable article in his political creed. But I did not become a ready or willing convert. I examined the opinion with care and with caution: and the result was a perfect conviction of its necessity. The Duke of Enghien was accessary to the Confederacy; and although the resident of a neutral territory, the urgency of the case, in which my safety and the public tranquillity, to use no stronger an expression, were involved, justified the proceeding. I accordingly ordered him to be seized and tried: he was found guilty, and sentenced to be shot. The sentence was immediately executed; and the same fate would have followed had it been *Louis the Eighteenth.* For I again declare that I found it necessary to roll the thunder back on the Metropolis of England, as from thence, with the *Count of Artois* at their head, did the assassins assail me.[42]

By this time, of course, there was a loud clamouring throughout the nation for him to eliminate further assassination attempts and make France secure, by as Vincent Cronin puts it, 'enshrining his magistrature in an awe-inspiring title that could be handed on through his family'.[43] Why did he choose *Emperor* though? He already had all the worldly power he desired as First Consul. Why not just King Napoleon I? Well, and again to quote from Vincent Cronin's book:

> He looked at the matter from the point of view of a convinced Republican. The word *empire* was already in use to designate all French conquests outside France, and it did not conflict with the notion of a republic: indeed the famous song, *Let us guard the welfare of the Empire,* had been chanted by Republicans in the early years of the Revolution. As for the term the *imperium* on behalf of the people of the republic: hence coins displayed the emperor's head on one side, and on the other the word *respublica.* Napoleon, then, saw nothing objectionable to republican feeling in the word *emperor.* It was merely a change of title which would establish in the eyes of the world the legality and continuity of the Republic.[44]

As the top level debate continued so Ney at Montreuil was receiving the purely domestic news of his sister Marguerite's marriage to Claude Monnier and investigating the tiresome interruption to his training programme caused by the mysterious appearance of 'the bales'.

In the first days of March wild rumours spread throughout his corps that the English had instituted a new and dastardly form of attack. These had originated with Villatte, one of his brigadiers and commander of the guns at St.-Frieux, who at the same time expressed his alarm in a letter to Ney:

> The English, unable to conquer us by force, are turning to their last resort—the plague. Five bales of cotton have just been cast upon our coast; I hasten to inform you of it. From St.-Frieux to the mouth of the Canche all the troops are at their posts; patrols are moving along the beach accompanied by customs-house officers. In sight of this battery and almost within cannon-shot are a frigate and two sloops of war of the enemy; also several small fishing boats, which I presume contain other bales of cotton. As no one is allowed to take out any boat or vessel I have just received orders to fire at everything that may appear in the waters within range of our batteries.[45]

Once the news got around a degree of panic set in. Several veterans remembered the terrible effects of plague upon their comrades in Egypt, not pausing to ask themselves how the English could possibly have transported large quantities of the bacillus in those semi-scientific days, and Ney only quietened them by posting up orders. 'All are hereby forbidden to approach any boats or other objects that may be cast on the shore.' The alarm even reached the First Consul, who on 21 March expressed his concern in a note posted from Malmaison. 'I am informed, Citizen General, that the English have thrown bales of cotton upon our coast, which has led to the supposition that those bales are poisoned. Give me all the particulars you can collect on this matter. It would be very lamentable to think that every principle of humanity could be thus violated.'[46] So Ney rode over to St.-Frieux, poked about among the bales and then glared his displeasure at Villatte over what he clearly considered to be a false alarm. That same evening he wrote informing the Consul that 'the plague-infested bales' were nothing more than 'bundles of old hammocks' thrown overboard from some passing ship, not even necessarily an English one. It was to be his last communication with Napoleon before his momentous letter of 29 April when he spoke for the whole French army and added their wishes to those of the people.

Soldiers of all ranks at Boulogne and other centres were now massing in whole regiments, even brigades and divisions, to give voice to their enthusiastic approval of the plans for Napoleon's coronation. And Ney, since Moreau's disgrace widely regarded as the doyen of the old-style Republican generals, received a big batch of letters from Dupont, the former aristocrat, together with the following note:

> I enclose, my dear General, the personal addresses from the generals and colonels of the First Division to the First Consul. They all contain the same wish which we have already expressed at the head of the troops, that the hero of France should be invested with the Imperial dignity. May I ask you to lay these addresses before him as the most sacred pledge of the devoted attachment felt towards him by the division he has placed here under my command?[47]

Ney bowed before the inevitable. What else could he do? He was already under the personal spell of Napoleon, admiring the man, supporting him and now finding himself surrounded by a majority of his adopted fellow-countrymen who desired the French Republic to change its character. Accordingly he sat down and composed what is probably the most rhetorical, devoted and certainly one of the most important letters of his career:

The General-in-Chief, Generals, Officers and Soldiers of the Camp of Montreuil to the First Consul,
The French Monarchy has crumbled to pieces under the weight of fourteen centuries; the noise of its fall has alarmed the world and shaken all the thrones of Europe. France, abandoned to a total subversion, has during ten years of revolution undergone all the evils which could desolate a nation. You have appeared, Citizen General, radiant with glory and surpassing genius, and suddenly the storms have blown away. Victory has placed you at the helm of government; justice and peace are seated by your side.
The recollection of your misfortunes was already beginning to be effaced, and all the feelings of the French people were about to merge into that of gratitude alone, when a dreadful event has shown them the new dangers which they are about to encounter. Your life, mainly defended by thirty millions of men, has been threatened; and a single blow of a *poignard* would have thrown back the destinies of a great people, and revived among them the dreadful excesses of ambition and anarchy.
So appalling a prospect has dispelled every illusion, and the minds of all are divided between horror of the past and dread of the future. France with all its greatness and power, seeing that it might lose all in a single day, has been struck with consternation and dread. It is now like the Colossus with feet of clay. The time has come to put an end to such a state of anxiety, by making our powerful institutions secure for us a lasting prosperity. The same cry is heard from every part of France; be not, therefore, deaf to this expression of the national will.
Accept, General Consul, the Imperial crown offered to you by thirty million people. Charlemagne, the greatest of our ancient kings, obtained this from the hands of victory: do you, with still more glorious claims than this, receive yours from those of gratitude. Let it be transmitted to your descendants, and may your virtues be perpetuated upon earth with your name!
As for us, General Consul, full of love of our country and of attachment to your person, we devote our existence to the defence of both.[48]

Certain forms of Republicanism were still to be observed, however. A plebiscite was held and the vote was 3,572,329 against 2,569 'for Napoleon Bonaparte to assume the crown'. The Empire was officially proclaimed on 18 May 1804; and the following day Napoleon signed a decree reviving the ancient rank of *maréchal*, dating in France from 1047 but suppressed on 21 February 1793 by order of the National Convention. The Emperor decided on no less than eighteen marshals at this first creation: Berthier, Murat,

Moncey, Jourdan, Masséna, Augereau, Bernadotte, Soult, Brune, Lannes, Mortier, Michel Ney, Davout, Bessières, Kellermann, Lefebvre, Pérignon and Sérurier. The last four quite clearly were being honoured for their services under the Revolution (although Lefebvre had helped Napoleon during the *coup d'état* of 18 Brumaire). Slightly younger, and still on active service, Jourdan and Moncey also belonged with this group. But most of the others had served under Napoleon in Italy, even Bernadotte. Only Ney was promoted from the Army of the Rhine: Marshal of France at the age of thirty-five.

1   *Napoleon's Marshals.*
2   Vincent Cronin: *Napoleon,* London, 1971.
3   *A Dictionary of Napoleon.*
4   Ibid.
5   Ibid.
6   Ibid.
7   Ibid.
8   Carola Oman: *Napoleon's Viceroy, Eugène de Beauharnais,* London, 1966.
9   *A Dictionary of Napoleon.*
10   Margaret Laing: *Josephine and Napoleon,* London, 1973.
11   *A Dictionary of Napoleon.*
12   *Napoleon's Viceroy;* incidentally Caroline Bonaparte both loathed Mme Campan's and did marry Murat: see *Caroline Murat* by Joan Baer, London, 1972.
13   I. A. Taylor: *Queen Hortense and her Friends,* London, 1907; also *Mémoires de la Reine Hortense,* ed. Jean Hanoteau, Paris, 1927.
14   *The Memoirs of Marshal Ney.* This note, which was kept with the Ney family's papers, is typical of Josephine's warm-heartedness and generosity. No one could ever find anything really bad to say about her. Except Bonaparte's own relatives who deliberately invented things.
15   *Mémoires de la Reine Hortense;* also *The Memoirs of Marshal Ney.*
16   Ibid.
17   *A Dictionary of Napoleon.*
18   Sir John Marriott: *The Mechanism of the Modern State, A Treatise on the Science and Art of Government, Vol. 1,* Oxford, 1927.
19   *A Dictionary of Napoleon.*
20   *The Memoirs of Marshal Ney.*
21   *The Mechanism of the Modern State.*
22   *The Memoirs of Marshal Ney.* This letter is dated 18 October 1802; which means that after arriving in Geneva on 4 October Ney was left for a fortnight there with only the vaguest notions of what he was expected to do. The added mystery is why Talleyrand felt it necessary to divide his instructions into two letters and risked loss and confusion by posting the letters to separate cities. With Napoleon's thinking so clear, why be so devious? Unless, as some suggest, it had become second nature: see *Talleyrand* by Duff Cooper, London, 1932.
23   Ibid.
24   French Archives de Guerre; see Bonnal's *La Vie Militaire du Maréchal Ney.*
25   Ibid.
26   Ibid.
27   Ibid.
28   *The Mechanism of the Modern State.*
29   Ibid.

30   *The Memoirs of Marshal Ney.*
31   Ibid.
32   Vincent Cronin: *Napoleon;* also the Duke of Buckingham and Chandos: *Memoirs of the Courts and Cabinets of George III,* London, 1853-5.
33   *The Memoirs of Marshal Ney.*
34   This same snuff box is now said to belong to Les Invalides.
35   J. Lorédan: *La Machine Infernale de la Rue Nicaise,* Paris, 1924: also William Warden: *Letters on the Conduct and Conversations of Napoleon Bonaparte,* London, 1816.
36   Vincent Cronin: *Napoleon.*
37   Ibid.
38   Ibid.
39   *The Memoirs of Marshal Ney.*
40   Vincent Cronin: *Napoleon.*
41   A. J. Boulay de la Meurte: *Les Dernières Années du Duc D'Enghien,* Paris, 1886.
42   *Letters on the Conduct and Conversations of Napoleon Bonaparte.*
43   Vincent Cronin: *Napoleon.*
44   Ibid.
45   *The Memoirs of Marshal Ney.*
46   Ibid.
47   Ibid.
48   Ibid.

CHAPTER 4

The Awakening to Glory

From Elchingen (1805) to Friedland (1807)

> . . . in men's hearts, romanticism never dies. Pure reason may dethrone it; everyday life may leave it uncultivated; yet it remains—to trap men in their weaker moments, to spur them on in their stronger moments.
>
> ERIC AMBLER[1]

Ney's new rank meant that he received forty thousand francs a year over and above his salary as a general of division. And a few weeks later, when Napoleon awarded him the Grand Eagle of the Legion of Honour, he received a further twenty thousand. It also brought about changes in his life-style. From now on, apart from carrying the exclusive blue velvet covered baton studded with eagles, on any big occasion he had to get used to wearing a dress uniform with plumes and a lot of gold braid. It meant too that he had to spend a part of his time at the Tuileries on what can only be described as 'courtier' duties. On these occasions he was expected to bow to Napoleon and address him as 'Sire'; while in turn he was addressed by his equals and subordinates as *Monsieur le Maréchal.* (Later, when he was created a duke the Emperor and others at court would begin to use the formal *Monseigneur,* or 'My Lord'.)

For Aglaé as well there were changes, although not unwelcome ones. In the earliest part of their married life the Neys had been far from well off, and in the months at La Petite Malgrange Aglaé had had to share the most basic domestic chores with her sister-in-law Marguerite. She had never complained about this, but now suddenly she was returned to a world for which her family background and subsequent education at Madame Campan's had ideally prepared her.

It was no longer a question of poring over the household accounts and siding the dishes and tureens from the table. Once more she had servants; a visiting hairdresser; and an account with a leading *couturière.* At the Tuileries she came into her own: knowing when to curtsy, how to wear beautiful, low-necked gowns and again how to be both familiar and respectful towards Josephine. In addition she knew better than most of the marshals' wives the social advantages in holding her tongue if forced to sit stiffly but erectly on the edges of frail, uncomfortable chairs in the Tuileries anterooms. It was all part of a revived court ritual. The Emperor and Empress desired decorum above all. Bad behaviour was a thing of the past; Napoleon wanted elegance with a moral sense and he preferred those women at court who showed their attractions without any hint or offer of seduction.

At home Michel and Aglaé began to entertain more. 'Home' was still connected with his military posting: a rented house in the Rue St.-Pierre, within walking distance of his HQ inside the *citadelle* at Montreuil. But already the new marshal was negotiating for the purchase of a big house in Paris. It was in the Rue de Lille (formerly the Rue Bourbon), on the left bank and just along from the Hôtel de Villeroy which Eugène de Beauharnais, now a prince, had bought.[2] Berthier was installed in a former aristocrat's palace on the Rue St.-Honoré. Clearly it was the done thing for luminaries of the Empire to have a 'showcase' residence—and through their notary and agents the Neys would expend much money, care and attention upon the house's redecoration and furnishing. In the meantime though they did their entertaining at Montreuil: where they gained the reputation for keeping a good table and being very lively company, especially after dinner when Michel got out the card-table and poured his guests some special old *calvados*.[3]

Aglaé thoroughly enjoyed herself. Michel always treated her as his social equal, and when they had guests the ladies didn't automatically separate from their menfolk once the meal ended. To all her female friends and the wives of his officers the marshal displayed politeness and tact. On the other hand she was kept extremely busy. She had breast fed their two sons, Napoleon Joseph and Louis Felix, and afterwards remained a particularly fond mother. But now the boys had to grow accustomed to nurses and tutors while their mother played hostess to a constant stream of visitors: Ney's fellow-marshals, his subordinate generals, *préfets*, local dignitaries and their wives, emissaries from Paris. And soon she would have occasion to receive the most important visit of all.

Crosses of the Legion of Honour were pinned upon members of the Paris garrison by the Emperor himself during a parade to celebrate the fall of the Bastille. As a result, when a few days later Ney received two hundred and seven crosses for the men at Montreuil, he wrote back saying how disappointed his troops were not to be given their awards by the Emperor in person:

> It is my duty to acquaint Your Majesty with the feelings of regret expressed by the generals, officers and men at not receiving this glorious decoration from Your Majesty's own hands. They had flattered themselves that they would be as fortunate as those members of the Legion of Honour who were present at the ceremony on 14 July in Paris. This desire, in which I participate, arises from our great attachment to Your Majesty; and you will give an additional value to your favours if you confer them in person.[4]

His letter produced an immediate and satisfactory reaction. Napoleon replied that he would visit the Camp of Boulogne 'to preside at a military display and then distribute the crosses of the Legion of Honour to more than fifteen hundred officers and men of the Army of England'.

It turned out to be one of the most famous spit-and-polish parades in the history of the French army. By this time the *Grande Armée* had settled in on the Iron Coast, 'and there were broad highways, with signs set at intervals labelled *Avenue de Marengo, Rue de St.-Bernard*, etc. The veterans of Italy and Egypt had made themselves homes with rockeries, pyramids, vegetable gardens, aviaries and poultry yards. They were well-exercised, well-fed, bronzed and expectant.'[5] But they were first and foremost soldiers with pride, and the news of Napoleon's coming was sufficient to give the Camp of Boulogne the look of a disturbed anthill. Never had there been so much frenzied washing, pressing, polishing and hair-cutting; and never so many eager volunteers for extra drills until their movements were perfect.

Meanwhile the resident marshals were busy planning the actual parade in detail, together with the evening's entertainment. Soult had the task of co-ordinating their efforts, but they all agreed that two miles from Boulogne (on the coastal road to Calais) a platform should be erected: with upon it a throne facing the men. The throne was to be the apex of a huge semi-circle, and before it their 100,000 men would be drawn up in companies on the long, smooth beach—with as a backdrop the masts of the ships destined to transport them to England. Following the parade and distribution of crosses there would be races, wrestling-matches, boxing and a fencing competition. Then in the evening a tremendous banquet given by the marshals and their wives, while their men received double rations and free wine. Finally: a display of fireworks—at which, after the last rockets, fifteen thousand men would fire a *feu de joie*, their muskets lighting the sky with a rainbow of brightly-coloured stars.

16 August 1804 . . . and everything passed off superbly. '*Aux Champs* was beaten by 1,300 drums. After the men were decorated the massed bands of the Guard and Line paraded the troops to the strains of the *Song of Departure*, led by Méhul, and cries of *Vive l'Empereur!*[6] Nor did the Emperor himself disappoint anyone. Every time he pinned the ribbon on a man's uniform it was accompanied by a few words of praise: an allusion to where the recipient had fought before, whether in Italy, Egypt or along the Rhine. Obviously Napoleon's *aides* had done their homework. The roll was a long one. Michel Ney alone had to describe the achievements of seventy-seven officers in his corps and no less than a hundred and thirty other ranks. Nevertheless the Emperor charmed them all; and at the end, with no prompting from their officers, the men broke into a solemn oath to himself and to the nation: *Nous le jurons! Nous le jurons!*

Afterwards Napoleon rode out to Montreuil with Ney. (Where Aglaé would entertain him briefly, but the Empress Josephine and her old school chum Hortense at greater length.) Even the undecorated soldiers wished to pay tribute to their Emperor and Ney had posted detachments of them *en route*, where they greeted him with further shouts of acclaim. Napoleon was visibly delighted and complimented Ney on their appearance. 'Why, their uniforms are spotless!' he enthused. 'Their boots shine, their bayonets gleam. Without doubt they are as fine as any troops I have seen all day . . .' 'Sire,'

Ney replied, fully under his spell: 'France's military administration, like every other branch of our public service, has experienced and benefited from the happy effects of your generous government.'[7]

The memorable day was coming to an end. Having enjoyed an *apéritif* at the Neys' house, and noting how relaxed Josephine and her daughter were with Aglaé, the Emperor invited them all to travel back in his carriage to the banquet; and on the way to inspect the remarkable quarters which had been built for his short stay in Boulogne. Due to her researches into the lives of Hortense and her brother Prince Eugène, Carola Oman has given us a very detailed picture of what these were like. Napoleon had determined to respond to his army's parade with an equal show of pomp and *grandeur*; consequently, in addition to his HQ in a *château* at the village of Pont-de-Briques, about three miles from Boulogne, he had:

> on the cliffs a *pavillon*, built in 48 hours, on a site from which he could survey with a sweeping glance his four principal camps, the town and the quays. His Imperial Majesty's *baraque* did not look at all as if the timbers and glass had been brought to this salty grazing land ready cut and numbered. It was pearl-grey without. Inside there was a fine lofty Council Chamber, with a ceiling painted with an eagle speeding through gilded clouds in a cerulean sky. There was an enormous map of the Channel and coasts on the wall; and there was only one chair—of green morocco. Officers had to prop themselves on their sword-hilts if conferences lasted long.
>
> His bedroom presented the same combination of simplicity and luxury— an iron bedstead, a splendid dressing-case, a telescope adjacent which had cost twelve thousand francs, and through which could be seen quite clearly the walls of Dover Castle. Marines of the Imperial Guard and grenadiers of the old Consular Guard were on sentry duty night and day. Engineers had arranged in the foreground an ornamental sheet of water on which floated two black swans surrounded by flower beds and shrubs.[8]

Napoleon's good humour at Boulogne was somewhat spoiled by his knowledge that the British fleet was hovering just beyond his own; but he kept this fact to himself. What mattered now was French supremacy on land and it was something he undoubtedly possessed. Accordingly he left Boulogne 'well pleased with its every aspect' and rather sad at his going. In the weeks which followed Aglaé gave several dinner parties for Hortense: including one at which her friend's name was arranged in the garden with flowers and then illuminated by a series of flares. Also she kept Hortense with her long enough to help supervise their move from the Rue St.-Pierre to the Château de Recque, a few miles up-river. But the next big occasion in the Neys' social diary was the coronation; set for 2 December.

Their house in Paris was at last ready and the marshals a necessary pendant to the glittering scenes in Notre-Dame. Because security around the Cathedral was so tight, Aglaé and other first ladies of the Empire were obliged to leave their carriages in the vicinity of the Palais de Justice and walk the rest of the way. 'In low-cut gowns and thin slippers, through tortuous and dirty alleys, shivering in the icy wind. Judged not respectable by the wags of the Cité, they were showered with propositions *en route*.'[9] Once inside Notre-Dame though, such annoyances were quickly forgotten. Joining her husband in the gallery, when she looked down the sight took her breath away. There were diplomats, cavalry officers, administrators and other dignitaries, 'all crushed together, but strictly in order of precedence and appropriately costumed'. The only person who seemed to be moving about freely was the painter Louis David: charged with immortalising the spectacle on a canvas of exceptional size. For the moment he appeared to be studying the officers of the Imperial Guard, who occupied the tribunes near the throne. But Aglaé's attention had veered elsewhere. Amidst all the colour and riches on display one item caught her eye and filled her head with thoughts of the old monarchy. This was the sword and insignia of Charlemagne: brought to Paris from Aix-la-Chapelle specially for the occasion. Michel had noticed it too; although for him its particular significance was rather different. It appealed to his romantic sense—but it also represented, he felt, so much of the glamour which attached to the remote days of chivalry.

It was still only nine o'clock and as they waited there was much coughing, shuffling of feet and one or two faintings. But at ten the sound of the organ changed from a murmur to big crescendo chords, and the audience rose for the entry of the Pope: Pius VII, summoned from Rome to give the coronation an official Catholic blessing. He was carried towards the great altar and then everyone had to wait another hour before the arrival of Napoleon and Josephine.

Purposeful as ever, Napoleon strode in through the central Last Judgement portal; with his wife close behind and the scowling Bonaparte women forced to carry her train. For once—and wholly out of character—he had really splashed out on clothes: 1,123,000 francs-worth if one counts Josephine's jewels. Instead of his old faded green uniform he wore a maroon velvet coat cut in a style which recalled Henri IV; over this was a short cloak embroidered with golden bees: the symbol of his regime's activity. Around his neck was the collar of the Legion of Honour studded with diamonds and in the hilt of the sword he carried flashed the famous 'Pitt' diamond. Josephine in turn looked absolutely stunning. Her fine dark hair was surmounted by a diadem of pearls and diamonds, her shoulders were likewise decorated with jewels and she had a velvet train, lined with ermine over a dress of white satin, also embroidered with bees. The dress was Grecian in style, but gathered underneath the breasts to show off her slim and graceful figure to full advantage. Coming up beside her was Prince Eugène with the coronation ring: a twenty-carat emerald engraved with the arms of the Holy Roman Empire, plus a dove holding an olive-branch, 'a threefold symbol of divine revelation, Charlemagne and peace'.[10]

Pius, an aged figure in white, welcomed them at the altar. He anointed them with the sacred oil, gave them his pontifical blessing and offered up prayers on their behalf. Next he made as if to lift the two crowns; with intent to place them on the heads before him, but at this juncture 'a strange emotion thrilled all the onlookers—an emotion which had not been evoked in France for many centuries past'.[11] For now Napoleon got up from his knees, gently motioned His Holiness to stand aside and placed the crown of golden laurel-leaves proudly on his own head. Then he calmly crowned Josephine.

Aglaé, a scion of those high in royalty's service, didn't know what to think, although on the whole—deep down—she probably disapproved. But Michel was both stirred and mesmerised by this dramatic incident. He was a self-made man himself. And yet here he was watching another man of his own age, also self-made, and in the very citadel of Catholicism, deliberately setting aside the timeless power of Rome to insist upon being crowned *Emperor of the French* by his own hand. It was all the result of Napoleon's genius and indomitable energy, of course. But it aroused within Ney the first feelings of belonging to something like a magic circle. 'I am', he told himself, 'in the vanguard of what promises to be, after the Revolution, the greatest period of French history. And all due to this one man!'

However the self-crowning incident also revealed an interesting character development in the object of his allegiance. As First Consul Napoleon had been lively, informal, easy to get along with; always full of merriment and generous to a fault. His undoubted political ambitions were merely the other side of a very attractive coin. By contrast as the Emperor he would increasingly act out the part which he believed his responsibilities and position demanded. He continued to be a supreme administrator, a great general. But in manner he grew more serious, much more severe, even 'unamusable'. To those around him he displayed a firm, God-like authority even when he was most charming; and in the end this led those who were not one hundred per cent loyal to accuse him of tyranny. Perhaps they were right . . . although for the next decade what they thought or said scarcely mattered. For in addition Napoleon had discovered an undeniable truth about the gifted people he led, and he would exploit it to the full in maintaining his position; namely, their desire for greatness, success and a European reputation, or as he preferred to summarise it, *La Gloire*. It was an inspired divination by a human being used to ruling men's hearts; and if he had guessed correctly about the French nation then he was even more right in his judgement of Michel Ney.

Following the coronation the Neys stayed on in the capital for almost two months. It was another period of banquets and fêtes; of visiting the Tuileries, Malmaison, Versailles or Fontainebleau: wherever the Emperor decided to be; and was climaxed by a musical evening and dinner which the marshals gave jointly at the Paris Opera in honour of Josephine. The occasion proved a great success—even if Napoleon did pay rather too much attention to Madame du Dûchatel.[12]

Aglaé commented on the fact as they were driving home. 'Pay no thought to it,' Michel told her. 'Napoleon is Napoleon. He sometimes goes to another woman's bedchamber.' 'And you?' his wife asked him, colouring. 'Towards yours . . .' 'Tactfully put, at least.' 'It's true, *petite*. I'm not like the Emperor. He's an Oriental . . . he ought to live with a harem. He loves Josephine more than anyone else. Always will love her. But sometimes he has affairs with other women. The thing for Josephine to do now is have a child by him, to found his own dynasty.' 'That hardly seems likely,' Aglaé replied: 'And meanwhile we are expected to behave like saints while he does as he pleases!'

Ney sighed. He didn't know which was the worst about Paris—its social scandals or the intrigues of Talleyrand and his circle. It came as a relief in February when they returned to Montreuil and a soldier's life.

Over the next few months he brought VI Corps up to peak condition. Until, as he himself described it, they were 'ready for anything'. As a commander he was stern but just, a firm disciplinarian with human understanding. Moreover he never asked either officers or other ranks to do what he couldn't do himself. In overseeing their training sessions he would frequently get down off his horse and show a man how to adjust a bayonet; another how to reload more quickly. The generals of the corps remarked upon his skill in handling men, his great physical strength and his voice of command: which was confident, dynamic and of tremendous, almost tenor/bass-baritone range. As for the troops, quite simply 'they thought the world of him'. If it became necessary they were prepared to follow him to the death. For the time being though they drilled, went out on manoeuvres, continued to rehearse embarking and disembarking—and also built things. Not silly things, or 'bull' just to keep them occupied, but ones they could make use of and take a natural pride in. For the marshal was equally concerned about their welfare and the state of morale.

For instance, he had them construct new mess-halls, with bigger, better windows to allow in more light and fresh air; also new improved kitchens with particular attention given to hygiene. Then they built themselves recreation-rooms, with card-tables and other games provided for out of the marshal's own pocket. By the summer they had laundries, shower-rooms—even a tailor's shop. Only one building Ney demanded for himself, but it was a large one and housed his pet project: a special school for officers, where they were lectured on organisation and methods, tactics, strategy and finally where they could hold open discussions on the theories behind the fighting of war.

A key-figure in the latter was Ney's principal *aide-de-camp* at Montreuil, Colonel (later Baron and General) Antoine-Henri Jomini. A Swiss professional soldier, he had travelled to France hoping to make a career for himself in 1801; but it was only within the past year that he had entered the Neys' orbit. Jomini was a military historian and theorist, not really a fighting man. He was an expert on the campaigns of Frederick the Great and he seemed to know almost as much about Napoleon's tactics in Italy as their great originator. He was currently at work on a vast book, *Traité des Grandes Opérations Militaires*, which eventually stretched to eight volumes. However when he had shown the manuscript of the earlier sections to Murat, that prince (followed by several other marshals) gave him the brush-off. It was only after being introduced to Michel Ney that he found an interested, sympathetic reader. More than that. Ney advanced him money to pay for the book's publication, and took him on as a volunteer *aide* until a proper commission could be obtained for him. Once this came through he put the young Swiss in charge of his staff-room—and of the training school for officers.

What in particular intrigued Ney about Jomini's book was its revised evaluation of infantry units; and its suggestions as to how these might be used to greater advantage in future campaigns. To Ney, the ex-hussar, such suggestions came as a revelation—and would be the major influence upon his style of fighting during all the battles he took part in over the next three years: at Albeck (the testing-ground), at Elchingen/Ulm (their proving-point), then in turn through the Tyrol, at Jena, in the capture of Magdeburg, at Eylau, Güttstadt and finally at Friedland (his triumph).

So enthusiastic did he become over Jomini's theories that he wrote a paper himself for the school, partly based on his experiences in the Revolutionary wars, but also on the conception of infantry being used more for attack then defence:

> Rapid and skilful marches can usually determine the success or failure of a battle. Therefore the VI Corps' colonels of infantry must never neglect the progressive perfection of their troops in both ordinary and forced marches. Remember: one of the greatest difficulties in war is to accustom the men to regular marching and swift attacks. But the French have an advantage here, for they are more abstemious and generally tougher than say, the Austrians. It is in attacking that the French soldier excels; whether he is braving the fire of the enemy, which is often overrated, or whether he is manoeuvring as a result of bold and intelligent orders. However, when he adds the element of surprise he is likely to prove invincible.[13]

He went on to denigrate previous methods of warfare:

> The big European powers have usually concerned themselves exclusively with the drilling of infantry rather than teaching them to fight. In this they have overlooked one essential factor. Elaborate drills, invented in times of peace by over-systematic officers, do nothing to show the ordinary infantryman how to strike a blow at his enemy. And if the infantry fails in this then the whole war is lost. *For it must now be admitted by all military men that one's infantry is the great lever in battle.* Artillery and cavalry are simply indispensable accessories. Officers and NCOs must therefore strive to be as well-informed as possible. Our national genius offers enormous resources in this field, especially now when promotion is open to all. But two conditions remain absolutely imperative: the infantry must be good, swift marchers, accustomed to fatigue; also their fire-power must be effective.

A series of purely technical instructions followed, varying from the insistence that each division must have at least 100 good swimmers to the idea that even general officers should make the effort to march alongside their men. These indicated just how engrossed Michel now was in the skills of his chosen profession. However he concluded on a truly patriotic note; and one which indicated that his Republicanism had been absorbed into, rather than destroyed by the change to Empire. 'The French soldier,' he said, 'must never fight in any but a just cause. He must always fully understand why he is fighting. The object of going to war is to win, but this is no reason to be cruel. The real objective is to force the enemy to sue for peace.'[14]

For one last piece of information about the Marshal's training methods we are indebted to General Bonnal. Apparently, according to his *La Vie Militaire du Maréchal Ney*, VI Corps were ordered to do everything *at the double*, whether on the parade-ground or out in the fields: a fact which so impressed Joachim Murat that he swallowed his pride and reported to Napoleon that Ney's corps was the speediest and most efficient in the whole army. Strangely, General Bonnal adds, when the Empire fell in 1815 the idea of troops operating at the double disappeared from the French army until 1904. In the interim though it was adopted by the Prussians, and made a significant contribution to their victory in the Franco-Prussian war of 1870.

Not everything at Montreuil went smoothly of course. The military life would hardly exist without its share of clangers. But perhaps these are seldom so amusing as 'Marshal Michel's balloon incident'. He was known to be very enthusiastic about air-balloons and their possible uses in war, in a way anticipating their great effectiveness during the 1870/71 Siege of Paris. Consequently, when the officers of VI Corps wanted to give a fête for Madame Ney and asked her sister Adèle and Hortense de Beauharnais to help them arrange it, they planned for the main event to be a surprise balloon ascent. A large paper one, filled with coloured smoke. It was almost ready for launching when one of the staff-officers told them of their *faux pas*. 'My God, you can't do that,' he exclaimed: '*Monsieur le Maréchal* will blow up higher than the balloon!' It seems that several weeks before a young man had visited Montreuil and persuaded Ney he could build him an exceptionally good air-balloon for reconnaissance purposes. He was so convincing that the marshal parted with ten thousand francs for the job. Since when 'man, money and balloon had been neither seen nor heard of!'[15]

By summer the Emperor himself was back on the Iron Coast: again staying at Pont-de-Briques and frequently socialising with the Neys. At last it appeared the invasion of England was imminent. The Marshal was pleased to report that his corps had set a record at Boulogne, with over twenty thousand men embarking in ten and a half minutes. In turn Napoleon confided in him regarding his strategy. He intended, he said, to give the English fleet the slip. Admiral Villeneuve had left Toulon; other French vessels were sailing from Rochefort, Brest and Cap Ferrol; the Spanish fleet was in a state of readiness at Cadiz. They would all sail down past Gibraltar, drawing the English blockade-ships after them. Then they would turn about at night, beating the English back to the Channel and acting as a screen for the French invasion-barges.

One evening Aglaé was giving yet another party for Hortense. They were still dancing an hour after midnight, when suddenly someone dashed in with the news: 'Napoleon has embarked for England! The troops are sailing!' For several minutes there was utter chaos and flap—with at least one lady trotting down the road in her thin evening gown rather than miss seeing it. But then Napoleon himself marched in, smiling broadly. It was a deliberate false alarm; one of his ever-decreasing number of practical jokes.[16]

At first light he was back straining to catch a glimpse of the enemy coast through his telescope: deadly serious again. 'Thirty-six hours,' he murmured. 'Just thirty-six hours. That's all I need!'

However, as we know, 'The Enterprise of England' was destined never to take place. On 25 August 1805 Napoleon learned that Admiral Collingwood had bottled up the combined French and Spanish fleets in Cadiz harbour; also that Austria and the Russians had entered into a Third Coalition as England's allies. To be deprived of one's marine power and at the same time be confronted by a total war on land would have broken the spirit of a lesser leader. Not so Napoleon. Countering with what has been described as a *pirouette* of his forces along the Channel coast, he marched directly towards Central Europe for his first Imperial campaign: embarking on a series of masterly manoeuvres and battles which not only succeeded beyond his army's wildest dreams, but made him the dictator of the Continent and blazed the greatest glory-trail in France's history.

The *Grande Armée* as it came to be labelled marched faster than troops *en masse* had ever marched before. Upon receipt of his orders from Berthier Ney had VI Corps on the road even before he could leave himself. In under two weeks, thanks to his sound training, they had practically crossed France: via Arras, Reims and Châlons. The Marshal, after putting Aglaé and their sons in a coach for Paris, finally caught up with his men at St.-Dizier. Near Nancy he took time off to spend an hour with his father at La Petite Malgrange. Then their march was resumed. Even his thirty-six field guns were being moved at between thirteen and fifteen miles a day. At Seltz, facing the Rhine, his corps HQ had been promised a new bridge but when they arrived there it was hardly begun. In fifteen hours the men constructed a bridge of their own with boats and barges. In September they were already advancing through Bavaria and Napoleon made his remarkable prediction: 'If the enemy waits for me I shall catch him between Ulm and Augsburg'.[17]

Remember this was the first time Michel Ney had campaigned directly under Napoleon, and his frame of mind can be summed up in remarks he made to Jomini. The Swiss suggested a possible line of retreat if things went wrong. 'How can you imagine that French troops led by the Emperor would ever retreat?' the Marshal snapped. 'People who think of retreating before a battle has been fought ought to have stayed at home!'[18] Meanwhile his men were singing as they marched along, about what they were soon going to do to the Austrians.

They had their first taste of action near Albeck on 11 October where General Dupont's division came close to being beaten. The faults over this engagement (usually referred to as the Combat of Haslach) were largely Murat's, although he tried hard to put the blame on Ney. Briefly, Napoleon's tactics were to use Murat's cavalry as a false screen against the Austrian army commanded by General Mack von Lieberich, leading the enemy to believe the French had advanced through the Black Forest as in previous wars— whereas in reality the corps of Ney, Lannes and Soult were already beginning a great encircling movement, with Bernadotte, Marmont and Davout circling wider still to prevent the Russians joining in. From Stuttgart, and in very bad weather, VI Corps therefore moved steadily down towards Ulm and across the Swabian Jura: the pivot of a cart-wheel operation upon the Danube.

General Mack grew alarmed; but dithered. He found it a problem to believe his own intelligence which suggested there were French now approaching him from every direction! If so, then where were they weakest? Where could he most easily break through?

At this vital moment Ney and Lannes were instructed to take their further orders from Murat, who ordered VI Corps to cross from north to south of the Danube. Jomini pointed out how this would leave only Bourcier's dragoons to guard the north bank: hardly enough to block Mack's escape if he attacked in that direction. Ney agreed with him. 'It's ridiculous,' he said to Jean Lannes. 'And if the Austrians get through there it means they can cut our own supply-lines. Then we'll be the ones in trouble.'

'The trouble with Murat,' Lannes replied, recalling their previous arguments in Italy, 'is that he can't see anything on a grand scale. He's a marvellous cavalry leader but no tactician—and the Emperor ought to accept the fact.' 'Yes. It's become too much of a family affair,' Ney added: 'Unless the Emperor gives his brother-in-law an abundance of command then he gets nagged by his sister Caroline.'[19]

Together they rode off to Murat's headquarters, where Ney endeavoured to persuade him to change his mind. But his only concession was to let Dupont's division remain on the north side. 'That's not enough!' Ney told him angrily. 'Dupont leads six thousand men. The Austrians have over thirty thousand!' 'Don't talk to me about leading,' Murat barked back. 'I'm used to making my plans in the face of the enemy!' Ney turned pale; and his hand went to his sword-hilt, although fortunately he resisted the impulse to draw. 'Very well,' he said. 'I think you're mistaken, but today you are the commander. Your orders will be obeyed.' And he stalked out.[20]

Events were to prove him right. He brought his troops across to the south side, and the following day the Austrians attacked Dupont. They nearly broke through and Dupont's casualties mounted. But remembering their training the French fought back with amazing courage, convincing Mack they were a far larger number and forcing him to retreat into Ulm. Napoleon galloped to Murat's HQ, extremely vexed by what had taken place. Without the bravery of Dupont's division his entire campaign might have been ruined! At first he blamed Ney for leaving his subordinate in an isolated position; but then the facts came out and he rounded on his brother-in-law, standing silent by the window. 'In future stick with your horsemen,' he told him crushingly. 'From now on the individual corps commanders will be responsible to me alone!'

By 13 October the French were closing the gaps around Ulm; and on the morning of the 14th Ney led his celebrated attack across the bridge at Elchingen. The Austrian General Riesch had almost demolished the bridge, and strongly fortified a plateau which looked down on it with 9,000 men and a large number of cannon. On one flank he was protected by a dense wood, on the other by the village of Elchingen and a monastery, also full of troops. He hoped to keep the French pinned down long enough for Mack to attempt another break-out past Dupont. But the men of VI Corps and their commander were rather more than he'd bargained for! With Napoleon watching them, Ney's attack went in at daybreak. Ignoring heavy Austrian fire the Marshal was immediately in the thick of it, helping to throw makeshift planking across and then leading the first of his infantry over with fixed-bayonets. But the turning-point came when he returned to lead his cavalry across. 'It was', in the words of a Guard officer who stood near Napoleon, 'a truly magnificent charge!' With his infantry already up the slope and beginning to clear the plateau, he seized his horse, raced over the loose boards and threw his mounted-men around to take the village by storm. An hour later he was sitting in the monastery, calmly taking stock while an orderly polished his boots: 4,000 Austrians captured, several hundred horses and forty or so pieces of cannon. VI Corps had never faltered in their performance, fully justifying his efforts at Montreuil.

It also meant that General Mack's position was suddenly critical. 'There was no longer any chance of a general escape along the north bank; Marmont and the Guard were almost in the outskirts of Ulm to the south of the river, and Soult was steadily moving up the west bank of the Iller from Memmingen—blocking all chance of a break-out towards the Tyrol.'[21] The following day (15 October) Ney successfully stormed the Michelsberg defences, again with a fast, ferocious assault spearheaded by his vital infantry. All at once the city of Ulm lay exposed to French artillery-fire. It was the end for Mack. He hummed and ha'd, but eventually agreed to surrender on the 20th if nobody came to his assistance. (Some hope—once Napoleon had completed his manoeuvre!) The French had netted 25,000 enemy infantry and 2,000 cavalry, plus their commanding officer and the entire ordnance. The campaign was only twenty-six days old and already the Third Coalition was reeling: 'a triumph for Napoleon's system of *la manoeuvre sur les derrières*. The demoralisation consequent upon discovering a powerful enemy on his rear had played a decisive part in paralysing the victim . . .'[22] On 21 October VI Corps was given the honour of officially occupying Ulm: at just about the time when Nelson was defeating Villeneuve in appalling weather off Cape Trafalgar. But in the short-term England's finest naval victory could do little to halt the French glory-trail. As the remaining Austrian forces and the Russians were about to find out.

Ney took no part in the great battle of Austerlitz, when Napoleon brought Francis of Austria to his knees and so pulverised the Russians that the Czar Alexander left the field crying. The Marshal had been detached to invade the Austrian Tyrol, which went off without a hitch. Supported by a number of Bavarian units VI Corps drove back the Archduke John to take Innsbruck on 7 November. They made it their centre of operations, although Ney refused to occupy the Imperial Palace, finding it 'too grand'. Instead he preferred to enjoy his rapidly increasing reputation as a fearless commander with the most glamorous French army since Charlemagne's. Humble, and certainly not ostentatious in his life-style, nevertheless he had been born with a passion for glory—and Napoleon had succeeded in arousing it, binding him to the Empire for the remainder of its existence.

A week later the whole of the Tyrol had been conquered, wrested from forces four times as large as his own. Ney could afford to relax. The *Grande Armée* was driving on past Vienna. But what pleased the Marshal most was a remark attributed to the Emperor on the eve of Austerlitz (2 December, and the first anniversary of Napoleon's coronation): 'Oh, if only I had my Ney here now. He would soon give these ruffians a drubbing!'

The outcome of Austerlitz was yet another precarious peace (The Treaty of Pressburg) signed on Christmas Day. However already by this time Ney had been summoned to a meeting with the Emperor at Schonbrunn, where Napoleon embraced him and praised his talents 'most effusively'. He apologised for keeping him so long away from his wife and children, but requested him to stay on as governor of the Tyrol, 'at least until our policies in Central Europe bring about a new stability'. In lieu of leave he offered Ney a generous gratuity: part of the two million francs in gold he intended to distribute among the campaign's senior officers.

Ney had a new, young aide-de-camp by now, a Captain Raymond de Fézensac, who would serve him up to and including the Russian campaign. He became a great favourite of the Marshal's, but was far from being a yes-man; consequently his initial portrait of the VI Corps Commander is most revealing. He (Ney), he says:

> could sometimes be aloof and even brusque, but only in the cause of military discipline or to command obedience. Otherwise he was familiar, generous, warm-hearted and loved by everyone. He liked to dine separately; either with his close friend General Colbert, but more often than not talking war with Colonel Jomini, the Swiss who was allowed to wear his own uniform and appeared to us cold, very much the intellectual. Afterwards though the Marshal could be relied on to join in our mess-amusements. It was his natural instinct when relaxing to laugh, and he enjoyed our songs and jokes.[23]

Fézensac concludes by stating that he only found Ney difficult about two things. 'He was extremely sensitive of the dignity due to his rank. Also he was jealous of his reputation as a brave, successful fighter.' And this ties in with a later portrait of him in society by Junot's wife, the Duchess of Abrantès.

'He could be the liveliest of companions in a small group, where he felt able to behave naturally. But he was less at ease in larger gatherings, often imagining that he might be snubbed by some high official or made to look silly.' She cites as one example a reception (by this time a great rarity) given by Napoleon at the Trianon. Michel and Aglaé were invited to dine first at the house of an acquaintance in Versailles, together with the Junots, Marmont's wife (the Duchess of Ragusa) and the Count Lavalette. It proved an excellent meal and Michel, after several glasses of wine, appeared in the best of humours. Until it was time to change for the reception . . .

Aglaé broached the subject as delicately as possible. 'We haven't a great deal of time,' she said. 'Now if your costume requires any attention—'

'Attention?' her husband replied, laughing. 'But why should it? I only wore it yesterday!'

'No, no, dearest, not your uniform. You know the Emperor likes all of you to wear Court-dress . . .'

Ney's expression changed. 'What! Not that masquerade again! Well, *I* won't get myself up in fancy-dress. Making myself a laughing-stock like so many of those other fools. No, no, NO!'

'But, my dear, you really must. The Emperor expects it.'

The Marshal was suddenly in a dangerous mood. 'And why does he expect it? Because he's been lobbied by the manufacturers of Lyon and the people who make embroidery. Aren't their profits big enough already. Very well, I'll buy ten costumes if it will please him. But that doesn't mean to say I have to wear them.'

On this occasion Aglaé persisted however. Knowing what her husband was like, she had secretly bought him a costume and now ordered her maid to bring it in for all to see. To Ney's horror the coat was embroidered with flowers, but everyone else praised it and said he must wear it. With an expression of anguish he seized the coat, thrust the maid's arms into the sleeves and then stood back. There, he cried, did they seriously believe he would deck himself out in these *mardi gras* clothes?

His good humour only returned with the arrival of Junot in full Court-dress. At last Ney saw the funny side of things. 'Not you too, Junot!' he roared. 'How can you bear to be seen in those trappings?' And he joined his hands in an attitude of prayer.

Everyone burst out laughing. Nevertheless when they left for the Trianon Ney was wearing his marshal's uniform.[24]

He returned to Paris on 13 September 1806—but for a mere ten days before being recalled to the field. Austria was beaten and cowed. Prussia had taken her place in the Coalition and Napoleon was determined to move with even greater speed than in the campaign of 1805.

The Prussians had over a hundred and fifty thousand men under arms, and were still discussing plans for an offensive when they discovered in October that the French had arrived on their doorstep, sweeping up from southern Germany with six corps, the Imperial Guard and Murat's cavalry all advancing in a giant square. King Frederick William III of Prussia was another ditherer, but motivated by greed; his wife wore the trousers and was extremely pretty, but with no experience of war. Between them they could be relied on to issue contradictory orders. Nor was their Commander-in-Chief, the Duke of Brunswick, any more decisive. He thought Napoleon's thrust might be against Leipzig and drew up the Prussian army in two big blocks between Jena and Auerstädt, hoping to fall on the French flank. But Napoleon was heading directly towards the Prussians 'and bent on destroying them'. Although Brunswick didn't yet know it, there was no French flank; it had turned into a broad, menacing front.

However in the important opening double-battle of the campaign—on 14 October 1806—even though he helped to crush the Prussians and hastened their detachment from Russia, Ney made his first serious mistake. The intrepid victor of Elchingen, seeking a further instalment of glory, grew impetuous and attacked without waiting for final orders. It proved a chastening experience and earned a deserved Imperial reprimand.

The French had crossed the mountains in three columns, then turned in a wide arc and divided so that two battles were fought on the same day. At Auerstädt Davout defeated the larger Prussian block, while at Jena, Jean Lannes, Ney and Augereau overcame the other. But both battles were fought following an early morning mist, through which Ney struck prematurely and at one stage was nearly cut off. It forced Napoleon to order the general attack before he was really ready—although fortunately this turned out well and by early afternoon the victory was secure. 'The Emperor was very much displeased at Marshal Ney's obstinacy,' Savary records. 'He said a few words to him on the subject—but with delicacy.'[25] Perhaps too delicately—because it would require a second mistake and a full upsurge of wrath before Ney accepted that Napoleon demanded obedience above all. But after Auerstädt/ Jena the Emperor's wrath was directed towards Bernadotte, whose men had not fired a shot all day, leaving Davout with 26,000 men to face 63,000 Prussians. The fact that Davout had pulled off a most spectacular victory, in six hours bringing the Prussians to a standstill and then routing them, was in Napoleon's view no excuse for the behaviour of I Corps' commander. In consequence he awarded Davout and III Corps the honour of entering Berlin at the head of the army; whereas he nearly had Bernadotte court martialled.

David Chandler suggests Bernadotte ignored Davout's appeals for support out of 'sheer professional jealousy'.[26] And Marbot remarked at the time how 'the whole army expected to see the Marshal severely punished'.[27] But in the end Napoleon relented; if only because, as he wrote, 'This business is so hateful that if I send him before a court martial it will be the equivalent to ordering him to be shot'. Instead he made a permanent enemy of Bernadotte by giving him the most humiliating verbal dressing down of his life.

Again to quote David Chandler:

> If the completeness of Napoleon's and Davout's joint victory was seriously affected by Bernadotte's intransigence and jealousy, this does not detract from the scale of the joint victory. In the course of one day no less than three field armies had been almost irremediably shattered, and over 25,000 prisoners, 200 guns and 60 colours and standards had fallen into French hands. The remnants of the once-proud forces of Hohenlohe, Brunswick and Ruchel converged along their various routes to the west of Apolda—where the ensuing chaos can be imagined.

It was a situation Napoleon now exploited to the full.

Ney, anxious to redeem himself, played a somewhat unusual role in the pursuit. But at least he was included. To begin with he passed through Weimar, where he decided to sleep at an inn again while Murat occupied the Grand Duke's palace. Then from 17 October at Erfurt, he was given the task of sorting out the 14,000 Prussians who had nominally surrendered to Murat's cavalry the previous day. One of them was Marshal Mollendorf, eighty-two years of age and an officer whose career stretched back to Frederick the Great. Upon hearing that he was hospitalised due to a bullet-wound received at Auerstädt, Ney visited him and delighted the old man by treating him with the utmost *politesse*. 'I come as a young marshal to pay my respects to a very senior one,' he is reported to have said.[28]

From Erfurt VI Corps pushed on northwards. They were free now from accompanying Murat and proceeding towards their ultimate prize of the Prussian campaign: the capture of Magdeburg. It was the most strongly fortified city in Germany—so that Napoleon had preferred to by-pass it on his march to Berlin. However Ney upon his arrival made such threatening gestures, and his corps displayed so much activity in preparing what was intended to be a ferocious assault, that on 6 November General von Kleist opened negotiations. Four days later Magdeburg was surrendered, together with 22,000 troops, 600 guns and a vast quantity of supplies and ammunition. Napoleon on hearing the news sent Ney his warmest congratulations and invited him to Berlin, 'where VI Corps were duly decorated'.

Prussia was all-but finished. The legend of her military invincibility so painstakingly built up by Frederick the Great was hardly worth a candle. Only fifteen thousand Prussian troops under Lestocq had escaped to link up with the Russians, while Frederick William and his family had taken refuge in the distant city of Königsberg. In the meantime Napoleon began his advance through eastern Prussia towards Poland, which he hoped to make independent, although he recognised that 'there is no simple solution to the problem'. For centuries the Poles had been the victims of expeditions and expansion either by the Teutonic Knights or Russia. The French could give them their freedom, but for how long would Napoleon be able to help them maintain it?

Ney and his VI Corps marched through Posen, to Thorn and after three weeks they were at Soldau. Then there was the bad weather. Every day it rained. Men and horses sank to their knees in the mud. It promised to be a miserable Christmas—and for Ney a worse New Year, because without his even realising it he was about to make his second serious mistake of the campaign. Not a mistake in battle this time, but one guaranteed to bring a Jupiter-like Imperial thunderbolt down about his ears.

After struggling along the terrible rutted roads beyond Soldau, 'across a mournful landscape of swamp and dripping pines', they eventually bivouacked around Niedenburg: 'in a wretched collection of hovels' and very short of food. Napoleon (from the palace of Stanislas Augustus in Warsaw) issued strict instructions that there was to be no further forward movement until the spring. Instructions which Ney, in the interests of his corps, now chose to ignore. The French were strung out in a long line of camps from Warsaw to Danzig, but on 2 January Ney pushed eastwards, scouring the Polish lakeland areas around Allenstein and advancing as far as Heilsberg in search of food. In his defence it has been claimed that he was more sinned against than sinning. For his supply-wagons had been raided by other French commanders and his men were starving. But in all probability it was his sudden action which decided the Russian General Bennigsen to open up a winter offensive against the French: exactly the sort of campaign Napoleon had hoped to avoid. The Emperor was enjoying the pleasures of his relationship with Marie Walewska. Now he would have to leave Warsaw and fight. Therefore when Soult acting in his courtier's capacity 'sneaked' on the full extent of Ney's foraging Napoleon didn't spare his marshal's feelings. 'Your duty is to obey me, not to act independently,' he raged. 'Furthermore you have no right to anticipate what may or may not be passing through my mind at any given time!' Berthier sat beside him, composing an official reprimand.

Unlike Bernadotte, Ney didn't try to talk his way out of it. He just stood there and took it on the chin. Moreover not being a man to bear grudges his loyalty remained the same: based on a genuine admiration for Napoleon as

France's leader and Europe's greatest general. As the harangue ended he offered his sincere apologies. 'If my previous record as a soldier is not good enough to atone for my recent errors, then I beg Your Majesty to dismiss me. I only reproach myself. Perhaps of late I have been too ardent in pursuing what I felt to be my duty. But I accept that my first duty is to Your Majesty.' He would, he concluded, hope to make it up to the Emperor in the final battles of the campaign. And he did.

His contribution to the bloody battle of Eylau, fought on 8 February 1807, came late in the day but proved decisive. The fighting mostly took place in a snowstorm and represented the first decided check to the *Grande Armée*. For once the French attack, although spirited, lacked co-ordination. As the blizzard intensified the men could hardly see twelve paces ahead and by late morning Augereau's VII Corps was shattered. 'Through the snow and smoke their broken squares could just be made out retreating towards the cemetery, with the Russians in hot pursuit.'[29] At this particular crisis-point the French were saved by their cavalry leaders: Dahlmann (who received a mortal wound), Murat, Bessières, Grouchy and the brilliant Lasalle. By their prompt action and the bravery of their men, what looked like becoming a rout was finally averted. 'Heads up, for God's sake!' General Lepic roared as his horse-grenadiers awaited their turn to charge and he noticed several of them ducking to avoid the bursting shells. 'Those are bullets—not turds!' Twice the squadrons 'overcame the Russian masses, knocked out their artillery, then ran over them in the opposite direction. Returning to the attack they forced them back and broke their resistance.'[30]

But the cavalry had also lost heavily and in the afternoon another crisis developed. Soult's corps on the French left were badly mauled, while 4,000 Russians stormed the cemetery which the Imperial Guard only retained after a fierce hand-to-hand combat. On the other side Davout had been endeavouring to turn the Russian east flank; but instead at four o'clock he found himself being turned by the arrival of Lestocq's Prussians. All along the front the situation now looked very grim indeed.

Ney's vital intervention came at 7.00 p.m. He had been far away to the north-west, ordered to shadow Lestocq and report if he made any move. Through the hours of the morning he had no idea that a major battle was in progress at Eylau. This was largely due to the deadening effect of the snowstorm upon the distant gunfire; but also because Napoleon's summons to join him, although sent at 8.00 a.m., did not reach VI Corps HQ until 2.00 p.m. Its arrival coincided with the Marshal's receiving intelligence that Lestocq had been slipping units away south-east for the best part of three hours.

He marched for Eylau immediately, through the swirling blizzard and harassed by Lestocq's rearguard, but making good progress nevertheless and arriving in time to restore France's fortunes. It wasn't that VI Corps did all that much fighting, but the appearance of his troops—fresh and eager for action—put new heart into the French. Their resistance stiffened all along the line, and when another Russian attack failed Bennigsen lost his nerve. Half an hour earlier he had been claiming victory, but now he ordered a retreat, leaving the *Grande Armée* in possession of the field.

On this occasion there was no question of a pursuit. The French had lost as heavily as their enemies: a casualty rate of one man in three, but in Augereau's VII Corps even higher. The Emperor gave orders 'for a bivouac in position without breaking ranks', while medics moved about picking up the wounded. Ney looked in on Chief-Surgeon Larrey, who would be amputating and performing operations for the next forty-eight hours without a break. And when he rode around the battlefield men lifted up their arms to him and pleaded to be killed. 'What a massacre!' he remarked to Fézensac, surveying the red-stained snow, the riddled bodies. 'And not a damned thing to show for it!'[31]

The French fell back towards the towns and cities along the River Passarge to reorganise, with Ney in command of the rearguard. At Güttstadt on 1 March he halted and settled down with his 14,000 men to face Russian troops estimated at 60,000. For the rest of March, through April and May, he held the line: frequently at bayonet-point, but never showing fear. And when he finally retreated upon Deppen, defending every step of the way, the *Grande Armée* was ready again. 'Lend me a pair of trousers,' he said to Soult when he reached Imperial HQ. Then he went to see Napoleon.

'*So*: I see you've let yourself be beaten by the Russians!' the Emperor said to rib him. But he embraced him affectionately, and in similar vein went on to explain the importance of taking the offensive. 'There has to be one more decisive battle,' he stressed. 'Then, I believe, we may have peace.'

The French blockbuster began at the start of June, driving north-eastwards. It was nearly stopped at Heilsberg by the Russians on June 10, but afterwards continued until the decisive battle around Friedland on 14 June. The Russians retreated upon Friedland to avoid being outflanked and began to cross the Alle there—although keeping some of their troops west of the river to attack the French forward units under Marshal Lannes before he could be reinforced. However Lannes held on until Napoleon arrived with the main army; and this largely determined the fact that the Russians either had to stay and fight or chance losing their rearguard.

Friedland today occupies both banks of the Alle but in 1807 it existed entirely on the western side. It was ringed by isolated copses and approached from the west through gentle, rolling pastures. Lannes' operations on the morning of the 14th, after fighting off the Russian rearguard, had been to capture a series of hamlets within two to three miles of the town. But by midday he had intelligence that Bennigsen had returned to occupy positions before Friedland in force and therefore hesitated to proceed further.

Soon after midday Napoleon arrived in person: alert to the possibility that this could settle the war for him and anxious not to waste a moment. Calling for his telescope, he surveyed both the enemy and the terrain. Although the Russians were drawn up in formidable array, their forces were divided by a strategic millstream and its lake: making it very difficult for one wing to support the other. Also, if the French could once throw them back upon the town then they would be fighting with the river at their backs. Berthier counselled waiting until the next day when the French could assemble more men, but the Emperor was adamant. By four in the afternoon he had 80,000 troops and considered that number sufficient. The battle began shortly after five with Ney's troops as the key-factor.

Napoleon's initial instructions are worth quoting at length here, because not only do they explain the French line-up, they also show how Ney successfully carried them out to the letter: thus demonstrating both his obedience and his courage:

> Marshall Ney will form up on the right between Posthenen and Sortlach, in support of General Oudinot's present position. Marshal Lannes will hold the centre, in a position extending from Heinrichsdorf to near Posthenen. Oudinot's grenadiers, now forming the right of Marshal Lannes, will turn slightly to their left to attract the enemy's attention. Marshal Lannes will also close up his divisions to form two lines by this movement. The army's left will be formed by Marshal Mortier . . . but he will never advance. *All movement must be on our right, using the left as a pivot.*
>
> General d'Espagne's cavalry and General Grouchy's dragoons, together with the horsemen of the left wing, must manoeuvre so as to inflict the greatest possible harm on the enemy once he tries to retreat, forced back by the attack from our right. General Victor with the Imperial Guard—horse and foot—will form the reserve . . . La Houssaye's dragoon division will be placed under him; Latour-Maubourg's dragoons will obey Marshal Ney; General Nansouty's heavy cavalry will be at Marshal Lannes' disposal.
>
> I shall be found with the Reserve . . .
>
> *The advance must always be from the right, and the initiation of the movement must be left to Marshal Ney, who will await my order. As soon as the right advances against the enemy, all the guns of the entire line will redouble their fire in the direction which will be most useful to protect the attack by the right wing.*[32]

Considering he had only made up his mind that afternoon the disposition was very cleverly thought out and extremely practical. In fact, Napoleon at his best. But his final orders to Ney were brief and blunt. 'Go in at the double! Take the town and seize the bridges behind!'

When Ney's men first poured out of Sortlach Wood they also had the advantage of surprise. Bennigsen, in common with most of the Russian generals, was getting on in years. He simply couldn't conceive that the French might mount an attack so late in the day; and he ignored the increased artillery fire between 5.00 p.m. and 5.15 p.m.

The next thing he knew the French were racing towards him, heading directly for the central clock-tower of Friedland, with bayonets fixed and driving in the Russian outposts before them. This was Marchand's division, which so terrified some of the Russians that they didn't stop running until they fell into the Alle.

To Bennigsen's credit he recovered sufficiently to order a logical counter-attack. But his Cossacks were charged down and dispersed by Latour-Maubourg's cavalry, and when his gunners opened up on Ney's leading columns they were devastated themselves by Dupont's artillery firing from across the Eylau road. Even the longest-serving Russian officers couldn't recall an attack against them of such ferocity. Bennigsen launched a second counterattack with his infantry units, only to have these too pushed into the Alle by 'the mad, red-faced, sword-swinging marshal at the head of his fearless corps'. With the Russian artillery effectively silenced, Dupont was now free to switch his fire across the millstream upon the Russians to the north. In his place General Victor rushed up guns to within sixty paces of the remaining unprotected Russian infantry on Ney's side. Meanwhile Ney himself was within a short stone's throw of Friedland's main gateway.

At this juncture, and amid scenes of growing disorder, Bennigsen attempted his third and final counterattack: with the Russian Imperial Guard, every man a bearlike giant over six feet in height. VI Corps overran them as they might have done a race of pygmies, although as Lieutenant Norvins who took part noticed when he surveyed the corpses: 'Most of them had bayonet wounds in the chest, the highest point our soldiers could reach'.[33] Another of Ney's aides (Bechet) recalled how the Marshal had appeared 'like the God of War incarnate'.

Soon after 6.00 p.m. Ney reported the capture of Friedland, the unblocking of its streets filled with dead and his seizure of all four bridges. Any Russians left now were trapped. It was the signal for Napoleon to order an almost leisurely 'rolling up' of those to the north of the lake by the troops under Lannes and Mortier helped by Grouchy's cavalry.

When the last bullet was expended around ten o'clock the French had suffered less than seven thousand casualties. Russian losses stood at 25,000, plus 80 guns and virtually every piece of their field equipment. Friedland forced the Russians to open negotiations, led to the disarming and partial dismemberment of Prussia, gave most of the Poles their freedom and at last offered hopes for a European peace. *There must soon be peace*, Napoleon cried, visibly relieved. And at this moment he undoubtedly believed it. I have a facsimile of the medallion struck after the battle. On its reverse side there is an olive-tree growing up from the rows of dead while Mars puts away his sword. *However Napoleon was not the first European leader to dream of peace through victory.*

There can be no argument about who was the hero of the occasion though. In taking Friedland Michel Ney had led the greatest, most successful attack of his career. And Napoleon was quick to single him out for praise. 'That man is a lion!' he observed; while Berthier wrote to Cambacérès: 'You can form no idea of Ney's brilliant courage—equalled only in the Age of Chivalry. It is to him that we owe so much on this memorable day.' The Marshal in his turn was appropriately modest. He wrote to his father-in-law: 'I feel sure peace will come now. For myself, I just want to be with Aglaé and my sons. I feel worn out.' But he added an interesting postscript. 'How glorious it is to be a Frenchman!'

If he was now content with his share in *La Gloire* then rewards were also coming his way. On 30 June the Emperor gave him an annual increase of 18,000 francs, payable by the exchequer from the Grand Duchy of Warsaw. Three months later he was awarded the huge gratuity of 300,000 francs in cash and a further 300,000 in bonds on the Bank of France. Finally, at the beginning of 1808 he figured in the Emperor's list of ducal creations. Jean

Lannes became the Duke of Montebello, Soult the Duke of Dalmatia, Augereau Duke of Castiglione and Davout Duke of Auerstädt. Michel Ney was named after his first success under the Emperor: Duke of Elchingen. He celebrated the event by ordering for Aglaé a magnificent double row of pearls.

So, he was rich. He was a peer of the Empire. He was regarded by everyone in Paris as a national hero. And it wasn't long before he was adding something else to his life-style. Also in 1808 Aglaé heavily pregnant with their third child, accompanied him by coach to view a mansion and country-estate, much larger than La Petite Malgrange. It was called Le Château des Coudreaux and situated at Marboué, a farming community near Châteaudun in the department of Eure-et-Loir. The Neys liked the place immediately, decided to buy it and Michel at least hoped to spend a good part of each year there. 'If the peace lasts it will be our place to escape to from Paris and its politics,' he told his wife.[34]

But the peace would not last, and before very long Marshal Ney was called back to the field of action.

[1] *The Dark Frontier*, London, 1936.
[2] The Rue de Lille was (and still is) an important feature of the Faubourg St.-Germain, or 'noble suburb'. It runs parallel with the Quai Anatole France and consists chiefly of houses built in the seventeenth and eighteenth centuries by royalty and leading aristocrats. Such houses present to the street a mere high wall and high wooden gate, but behind are usually courtyards, gardens and what has been described as 'a noble wastefulness of design'.
[3] Ney enjoyed card-games but seldom played for money. He also liked wine, especially Burgundy and the *rosé* of eastern France; plus certain fine brandies. But he was not a heavy drinker.
[4] *The Memoirs of Marshal Ney.*
[5] *Napoleon's Viceroy.*
[6] *The Anatomy of Glory.*
[7] *The Memoirs of Marshal Ney.*
[8] *Napoleon's Viceroy.*
[9] *The Anatomy of Glory.*
[10] Ibid.
[11] *A Dictionary of Napoleon.*
[12] *Mémoires de la Reine Hortense.*
[13] *The Memoirs of Marshal Ney.*
[14] Ibid.
[15] A. Hilliard Atteridge: *The Bravest of the Brave*, London, 1912.
[16] *Mémoires de la Reine Hortense.*
[17] A. J. M. R. Savary, Duke of Rovigo: *Mémoires sur l'Empereur Napoleon*, Paris, 1828.
[18] General Baron A. H. Jomini: *La Vie Politique et Militaire de Napoleon*, Brussels, 1841.
[19] *The Memoirs of Marshal Ney.*
[20] Ibid.
[21] *The Campaigns of Napoleon.*
[22] Ibid.
[23] General R. A. P. J. de Fézensac: *Souvenirs Militaires*, Paris, 1863.
[24] *Mémoires de Madame la Duchesse d'Abrantès*, Paris, 1831-4. The other dress-uniform can be seen at Les Invalides. It consists of white silk stockings, white linen trousers tied just below the knee and white rosettes on the black, heeled shoes. The coat is of

dark blue velvet, tailed and with much gold braid and a high braided collar. He wore a cream-coloured cummerbund and a scarlet sash over the right shoulder to the waist, with the Grand Eagle of the Legion of Honour over his heart. White, lace-edged *jabot* and lace cuffs. Also he held a stick which resembled an elongated baton.

25 Savary: *Mémoires sur l'Empereur Napoleon.*

26 *The Campaigns of Napoleon.*

27 Baron J. B. A. M. de Marbot: *Mémoires*, ed. General Koch, Paris, 1891.

28 *The Bravest of the Brave.*

29 *The Anatomy of Glory.*

30 Ibid.

31 Fézensac: *Souvenirs Militaires.*

32 Napoleon: *Correspondance.*

33 Lt. J. de Norvins: *Souvenirs d'une Histoire de Napoleon*, Paris, 1897.

34 Following the Marshal's death his family suffered considerable economic hardship. Not only were they excluded from public life, but they were harassed, many of their assets seized and Ney's pension rights as a general officer in the service of France cancelled. In fact at this moment in time only Bernadotte came to their rescue financially. Despite differences with Napoleon leading to his joining the Allies, the former Prince of Pontecorvo, now the Crown Prince of Sweden, still regarded his former colleague Michel Ney with feelings of affection. Upon learning of Aglaé's difficulties he offered her two eldest sons commissions with the Swedish army. It wasn't until after the 'July' Revolution of 1830 that the family's problems were finally eased. The lawyer Dupin (a member of the defence-team at Ney's trial) stood close to the new 'Citizen King', Louis Philippe. He obtained a state pension for Aglaé of 25,000 francs per annum; also she was allowed to place a bust of her husband in the Panthéon and his portrait was hung in the Musée de Versailles. In the interim though Coudreaux had had to be sold. After being seriously smashed up by the Prussians—to the point of being unliveable in—it then became a serious liability. It passed into the family of General Count Honoré Reille, Ney's subordinate at Waterloo and a Bonapartist who survived the proscriptions of the 'White Terror' to become a marshal himself in 1847. The *château* and estate still belong to his descendants; and the present count is also the Mayor of Marboué.

CHAPTER 5

Disgrace in the South

Spain and Portugal, 1808-1811

'The Hero', and also by this time the idol of his much-decorated, battle-hardened VI Corps, was directed towards the Iberian Peninsula in Napoleon's orders dated 2 August 1808.

On the face of it there appeared no reason why during the coming campaign he should not add to his military reputation. Spain's regular armies were suspect under fire and in any case badly-led; the bands of *guerrilleros* had become a nuisance certainly, but he would teach them an unprecedented lesson. As for the British expeditionary force, now thought to be concentrated upon Lisbon: well, its troops had never met the *Grande Armée* at full strength and the most recent intelligence said that its commanders were at logger-heads. Obviously the possible connotations of *La Gloire* stayed uppermost in his mind as he travelled to a roll-call and inspection of his men beside the Pyrenees. Perhaps even to the extent of receiving a kingdom once the fighting was over. (Why not? After all, the various brothers of Napoleon had them—and just a month ago that prize show-off Murat had been created King of Naples!)

Alas: the future realities of the situation would prove an exact opposite to all his hopes. It was France's marshals who quarrelled, not the British; and Michel Ney's years in Spain and Portugal were destined to be the worst, the most abysmal, of his entire career. The most widely-known fact about them is that on 22 March 1811, after acts of gross (if provoked) insubordination he suffered the ultimate humiliation for any general officer: he was relieved of his command. But two other events of these three years are of equal significance in any understanding of Ney's character. The first was an act of deliberate inhumanity. It's the only one I can find in his lifetime as a soldier; nevertheless it is there as a permanent blot upon what was otherwise an exemplary service record. The second event involves his clash with Wellington at Bussaco. This was a short, sharp battle which the British won. At the time it had been fought merely to delay the French, and did little to reduce their strength across the Peninsula. But the methods Wellington employed at Bussaco were to exercise a powerful influence over Ney; and very much later they would affect his judgement at Quatre Bras, with grave consequences for the outcome of Waterloo.

However, the results in war so frequently defy prediction. And while being coached smoothly down through France, aided and abetted by a season of fine weather, the Marshal displayed much evidence of being keen to reach the new theatre of operations. Everything, he felt sure, would go well.

Now though we must look back briefly to a number of things which had helped write his present set of orders.

Napoleon, it emerges, was clearly interested in the political affairs of Iberia even during his First Consulship. Yet he could never give the area all the attention it warranted until he controlled Central Europe: which meant after the Austrian and Prussian armies had been knocked out of the firing-line and Czar Alexander I of Russia, also defeated, had signed a peace treaty on the raft at Tilsit in 1807. Then at last the Emperor and a large portion of his *Grande Armée* were free to intervene in Spain, Portugal or anywhere else they chose.

Tilsit was really the linchpin of France's overall strategy against her persistent and traditional enemy across *La Manche*. Napoleon described this strategy as his 'Continental System'. Quite simply, its aim was the slow strangulation of Great Britain's economy by closing Europe's ports to her manufactured goods. Only thus, with French supremacy on land cancelling out Britain's at sea, did it appear possible to revive the Peace of Amiens— which the British, jealous of France's growing prosperity, had broken. When the Czar also agreed to join the system, aside from the odd loophole of convenience (British boots for French infantry?) and the inevitable petty smuggling, there was but one European capital and port left open for the British to use: Lisbon. War of some description in the Peninsula therefore became a certainty.

'Decadent' Spain and 'treacherous' Portugal! This was how they stuck out to Napoleon in his triumphant appraisal of Europe after the signings of Tilsit. That elegant raft on which an overawed Czar had vowed to be his friend represented far more stability than either of them. Spain was France's ally, in spite of losing her battle-fleet under the command of a French admiral at Trafalgar; Portugal was by a mediaeval treaty England's oldest ally. Consequently Napoleon's initial task seemed straightforward. With Spain's acquiescence he would march French troops across her and crush the Portuguese at source. Afterwards he might give their country to Spain—or better still leave a Frenchman to rule it. ('If Portugal does not do what I wish the House of Braganza will not be reigning in two months!' 27 September 1807.) On 18 October, not prepared to wait even this long, the Emperor despatched General Junot at the head of 24,000 troops to make for Lisbon. Prince John, the Regent of Portugal, first of all dithered and then turned coward: fleeing with his court to join the British fleet. As a result, although Junot mismanaged their march, France's soldiers met no opposition and were occupying Lisbon by December. A reserve French force took up stations just inside Spain at Vitoria.

It might have been the end to all Britain's efforts and aspirations regarding the Peninsula. It should have been. But now Napoleon himself was guilty of the most colossal blunder—due to the fact that he simply lost confidence in his Spanish allies.

The Bourbon King of Spain, Charles IV, was totally incompetent. For years the country had been manipulated by Queen Maria Luisa's lover, Manuel de Godoy. The heir to the throne, Prince Ferdinand, was not much brighter than his father; although he had the advantage of being preferred by the people, whereas Godoy was detested by everyone.

(What a family! They stare out at you from Goya's paintings like a collection of dressed-up gargoyles. . .) In foreign affairs Charles IV didn't really know what he wanted. Godoy favoured the alliance with France for as long as it buttressed his own position. Prince Ferdinand hoped the French would guarantee his succession if Godoy tried to seize the throne. Meanwhile each one intrigued against the other, leaving a ramshackle apparatus of government to muddle on as best it could. Which at first upset the administrator in Napoleon and then started to worry him militarily. It required a strong Spain if the British were to be kept out of the Peninsula. And so, from what he considered was becoming urgent necessity, a remarkable new plan evolved. He would sweep aside all of these dodos, place his brother Joseph at the head of government in Madrid and give to Spain's people the same civil liberties that Frenchmen enjoyed!

Quietly, in February 1808, Napoleon secured the frontier fortresses at either end of the Pyrenees. Also he despatched Murat towards the Spanish capital with a 'friendly' force of cavalry. Then in April he invited Charles, his family and Godoy to a conference at Bayonne. Once there they found themselves under house-arrest. The French offered them lavish retirements; meanwhile Joseph Bonaparte was packing his bags to become King of Spain. Everything passed off as smooth as silk; just like someone closing a drawer. Yet Napoleon's blunder lay in his very pragmatism. At the Tuileries, among smart, sophisticated and—on the whole—logical people, he believed the Spaniards would prefer prosperity and increased freedom under a Bonaparte to grinding poverty and near-starvation under the Bourbons. But he was wrong. Nationalism holds more appeal than reform when the latter is offered by an interfering neighbour. The Spanish royals, although previously despised, quite suddenly were regarded in their own country as martyrs.

On 2 May 1808, the infamous *Dos de Mayo*, Madrid rebelled against the French garrison. Murat crushed these rebels effectively but bloodily. Over four hundred Spaniards died, either in the streets or in front of hurriedly-organised firing squads.

Events now moved swiftly. By the end of May nearly every province of Spain was in revolt—and the British had been invited back as benefactors. Spain's regular troops fought the French in somewhat haphazard fashion. At the outset they were usually beaten, but where the terrain suited them, in the high sierras and especially in the south, they began to make headway. Particularly when they had assistance from local resistance units, the *guerrilleros*. Any small or isolated group of Frenchmen was likely to be fallen upon and annihilated; every captured despatch-rider tortured for information before being shot. 'It was . . . a fight with rules that none of them (*the French*) knew.'[1] Meanwhile the British had returned under Sir Arthur Wellesley and Junot found himself bounced out of Portugal. By August a Spanish army under General Castaños was preparing to re-enter Madrid. 'King' Joseph Bonaparte fled.

Napoleon, unfortunately for the Spaniards, was moulded out of stronger stuff. 'Ungrateful, Bourbon-loving' Spain would now be made to feel the gale force of the *Grande Armée*! And under the direction of the Emperor himself. 'I have sent the Spaniards sheep whom they have devoured. I shall send them wolves who will devour them in their turn . . .' Michel Ney and his VI Corps were merely the steel cutting edge. By the October of 1808 seven more marshals had received Napoleon's orders to hasten with their troops into the Peninsula. They were Victor, Soult, Moncey, Lefebvre, Mortier, Lannes and Bessières. In other words, the stage was filling up for those episodes in Spanish history which are well-known today, due to an artist's eyes and genius (Goya's), as *The Disasters of War.*

By 30 August Ney had reached Irun on the Spanish frontier. From here he moved to Vitoria. Having thus secured passage for the remainder of the *Grande Armée*, his next task was to reconnoitre, probe and—if need be—skirmish in an ever-widening crescent: gathering all the intelligence he could pending the Emperor's arrival with his main force.

The first reports seemed averagely encouraging. At the summer's end Navarre and the southern Pyrenees were still very green. There was ample provisioning for the incoming soldiers and some good forage for their horses. Meanwhile, although Spanish opposition in the area was sizeable, its regulars were incompetently grouped and led by a man singularly lacking in inspiration: General Joaquin Blake—the same commander that Bessières, helped by Lasalle's cavalry, had hammered in front of Madrid when otherwise all was going wonderfully well for the Spaniards.

The only disquieting intelligence concerned the *guerrilleros*. Some of these were no more than bandits intent on plunder; men, often escaped criminals, who lived in the hills and who, if the French hadn't been there, would have attacked and robbed their fellow-countrymen. Others were bands of genuine patriots, usually well-officered by ex-regulars. *All* of them, however, fought along similar lines: with cunning, audacity and an ever-increasing viciousness. It is estimated that in the year's earlier campaigns they were costing the French a loss of three hundred men killed or wounded *per day.*[2]

Clearly it was their methods which worried Ney. In fact, during those first few days at Vitoria they filled him with a gloomy foreboding—which in turn switches a powerful light upon the twinned facets of his personality.

As a fervent French national (if only by adoption) the Marshal could appreciate the need of certain Spaniards to resist any alien invasion; and especially one by men who had kidnapped their royal family. This love of *patrie*, appealing to the romantic part of him, did nothing but earn respect for the genuine *guerrilleros* in his eyes. Also the realist in Ney could understand their earnest desire to win. As a soldier with France's beleaguered Republican armies, and later under the severe tutelage of Napoleon, he had come to accept the basic idea that wars were fought to be won. They were no longer an aristocratic tournament, an affair for best uniforms and a limited campaign while the good weather lasted (with an amicable treaty at the end of it). No, they had become a total involvement; continuing until the issue was properly decided and the field swept clear of one's enemies.

On the other hand, it must be pointed out that the *guerrilleros'* 'means' alternately disgusted and distressed the Marshal. As a romantic warrior, as a collector of *La Gloire*, he had always followed a soldier's code that intermingled behaviour and protocol with honour. Like Jean Lannes, he could be a ferocious front-line fighter—but only against bodies of men who were either striking at him or defending themselves with weaponry. He did not kill prisoners. He did not authorise torture as a way of obtaining information. To a vanquished enemy he was magnanimous: which meant he showed every courtesy and consideration to the officers, gave food to their men and medicine to those who were wounded. Finally, the discipline in his own corps was invariably strict—with heavy penalties, including the death-sentence, for those convicted of looting, brutality and rape.

Now though the Spanish resistance groups were fighting without the handbrake of any code whatsoever. Ney himself had killed men and witnessed dreadful carnage on the battlefield; he had survived almost a dozen wounds. But when he saw the first French corpses in Spain: castrated, sometimes with their ears, noses, hands and feet gone too—or with blackened heads after being hung upside down over slow fires; in effect, it turned even his strong stomach over. It was an affront to his finer feelings, but also it distressed the realist in him. For he envisaged the future and that terrible moment when an exasperated *Grande Armée*, hitherto famed throughout Europe for its soldier's code, would suddenly go SNAP!: and afterwards begin to take reprisals along similar lines. Accordingly, he impressed upon his officers 'the need to keep VI Corps concentrated', never dispersed to unit points which the *guerrilleros* might feel strong enough to attack. Supply lines and communications were to be doubly protected. (It's an undeniable fact that a majority of the corps commanders who followed him into Spain paid far less attention to such details. The atrocities against them continued to mount, their men's patience did give way and the reprisals were at times horrific: compounding the error of unrestrained barbarism and thus providing *The Disasters of War* with a grisly *dénouement*.)

A secondary problem for the Marshal at Vitoria was the depressing huddle there of King Joseph and what remained of his entourage after the precipitate flight from Madrid. Ney, never much good at playing the consummate courtier, put in a number of duty appearances. He even offered military advice—based upon his reconnaissance in the area. But the reception given him was chilly. (Even by Marshal Jourdan, his former commander, who held the unenviable post of Chief-of-Staff to the disestablished royals.) Joseph, unfortunately no soldier himself, evidenced a jealous dislike of all who were; but more especially Ney, because he was the friend and part-*protégé* by his marriage of Josephine, still a hated interloper to every Bonaparte bar Napoleon.[3]

Happily, within a few days Marshal François-Joseph Lefebvre arrived with the IV Corps. Stalwart old Lefebvre, now also Duke of Danzig, was on excellent terms with Ney and his supporting troops provided the latter with a valid reason to quit Vitoria and launch his own corps upon a minor offensive. On 26 September, after moving north-west at what was lightning speed for an army corps, Ney captured the city and port of Bilbao. Which in turn both surprised Blake's troops and drove them due east: straight into the path of Lefebvre and the IV Corps, who thoroughly trounced them at Durango on 31

October. However this successful foray was terminated on 5 November by news that the Emperor had reached Vitoria. Marshals Victor, Soult, Moncey, Lannes and Mortier were with him, together with Generals Junot, Gouvion-Saint-Cyr and a *Grande Armée* numbering well over 150,000 men. Without due ceremony King Joseph was pushed into the background and Napoleon now summoned all of the French corps commanders in order to outline his strategy for the lasting conquest of Spain.

Victor was to take over the right-wing (and he inflicted a final defeat on Blake at Epinosa, 10/11 November). Soult had the central command, and on 10 December he all-but obliterated Belvedere's army at Gamonal—which led to the fall of Burgos and should have cleared the way for Napoleon himself to strike directly at Madrid. Except that things were not going according to plan on the French left, and Marshal Ney was the person being blamed.

Verdicts vary here: from Richard Humble ('. . . it was Ney who spoiled the plan to destroy Castaños. . .'),[4] to Peter Young ('the Emperor blamed him, unjustly, for the escape of Castaños. . .')[5] The established facts though are as follows. The Emperor's orders were for Lannes, supported by Moncey and also Junot, to attack General Castaños head-on, then continue south-east along the Ebro, breaking the Spanish reserves under Count Palafox and halt-ing finally with the capture of Zaragoza. In the meantime Ney was to detach from the right, swing across to an even deeper south-eastern position via Calatayud and so take what remained of Castaños' forces from behind. All in all a most imaginative plan; which promised to emulate Napoleon's masterly manoeuvres at Ulm and Austerlitz.

On 23 November Lannes routed Castaños at Tudela, chasing the Spaniards southwards in headlong flight through the passes of Sierra Moncaya. But where was Ney to close the trap and finish them off? Instead of being at Calatayud he was still miles behind his ordered schedule at Soria. As a result, Castaños escaped more or less intact, Palafox shut himself up in Zaragoza and both Lannes and Moncey were to spend many months in a costly seige of that city. Napoleon, needless to say, was furious.

In Ney's defence several points ought to be made here. The French were new to the area; they lacked maps, any bank of intelligence, and there was no time to do a proper reconnaissance. The terrain itself was appalling, with some of the worst roads in Spain and a local population as hostile as it was poor. All of the barns, flour-mills and silos had been emptied by *guerrilleros*, who infested the surrounding hills and harassed the VI Corps at every twist along their way. Worse still, when at last they marched into Soria, Ney received a false, but plausible message that Castaños with 80,000 men had defeated Lannes and pushed him right back to the Ebro. Discovering the truth kept him a further two days in Soria; two vital days during which Castaños made his escape. Even a marshal of France can run out of luck! *And yet*, the pedal-point remains: that Ney failed in his (and his Emperor's) objectives. And a Napoleon grown accustomed to uninterrupted military success could add that his 'favoured' strike-commander had overcome much more difficult situations in Central and Eastern Europe.

The only good to come out of an otherwise spoiled opportunity was that Ney lost very few men to the *guerrilleros*. His belief in concentration, staging his march as one tight and well-protected formation, had minimised the possibility of ambush or for groups of stragglers to be pounced upon. Which obviously impressed Napoleon, because his bad temper now abated to such an extent that Ney got away with a mere scolding. By 4 December the Emperor had broken through the Somosierra Pass, Spain's last natural advantage, and occupied Madrid—immediately reinstating 'big' brother Joseph there as his puppet ruler. On 14 December VI Corps (14,000 men) was ordered into the capital, reviewed by Napoleon, praised on its appearance and several of the veterans received decorations. Ney, growled at one moment for his hesitancy, in the next sentence was being complimented on his prudence! The Emperor's normal optimism and high spirits, his *aides* agreed with relief, had been fully restored.

But the time for parades and some re-equipping was necessarily brief. On 19 December '. . . a despatch-rider brought the Emperor a note from Marshal Soult'. Apparently, '. . . the British cavalry, with 5,000 infantry of Moore's army and a crowd of delirious peasants, had overrun Soult's posts at Rueda and Tordesillas. Valladolid was in upheaval and Franceschi's cavalry had been forced to evacuate it.'[6] The British! But of course! Hadn't they been quiescent and out of sight and mind for far too long? In truth, a revitalised British force under Sir John Moore had sped out of Portugal to Valladolid and so linked up with another 17,000 men embarked at La Coruña under the command of Sir David Baird. They thus threatened the whole of Marshal Soult's hold on north-western Spain.

Napoleon's reaction was instantaneous. 'Ney was ordered to advance to the Guadarramas, followed by Lefebvre-Desnoëttes and the cavalry of the Guard. The foot Guard was to march with the Emperor. The departure was scheduled for 21 and 22 December. The weather was fine, the skies clear.'[7] But not for long.

> The moment the troops reached the Escorial, snow began falling and the wind blew a gale force. The higher they climbed into the mountains the harder the ground froze underfoot. Ney's troops got over the pass with the greatest difficulty. Lefebvre-Desnoëttes followed, but the weather became so bad that it was impossible to advance. The squadrons lost their way and a whole troop of dragoons disappeared into a gorge. Worse, in the gusts of wind and sand that penetrated to their very bones they were hurled back upon the infantry who were climbing on foot, and upon the artillery, throwing it into a welter of confusion.[8]

It was terrible, but cross the Guadarramas they did—and suddenly, on 23 December, Moore heard that the French were coming at him. 'We shall have to run for it,' he said: words which signalled that one of the most remarkable chases in any army's history had now begun.

Michel Ney did not take part in the whole of this. But it was he who got the vanguard rolling. Crossing the Guadarramas Marshal and Emperor slogged it out more or less side by side, the former forcing his troops up through the higher passes (and despite some awful weather) as if his very life depended upon it. Napoleon was equally eager to push on. 'His Majesty marched . . . in the midst of his soldiers, his hat pulled down over his eyes, his boots full of snow, leaning first on the arm of Duroc, then of Savary.'[9] At one point the Emperor even took over the lead. He might have been captured; and Ney, now back in favour, was bold enough to chide him; 'Sire, I thank Your Majesty for acting as my advance guard!'[10] By Christmas Day they were well beyond the mountains. Well beyond Valladolid too, and racing for the Esla bridges. 'It was the feast of peace and joy—also a reminder of man's duty to God. Nevertheless, at 5.30 p.m. the horse and foot Guard got under way behind Ney for Medina-del-Rio-Seco.'[11]

Along the Esla, on 29 December, the British rearguard decided to make a covering stand. It was a brief, brutal affair around the little town of Benavente. Lord Paget's cavalry regiments and a reserve infantry division, 'outnumbered by well over two to one, fought like madmen to hold the French'.[12] They faced frontal assaults by the Chasseurs of the Imperial Guard: killing, wounding or capturing a hundred and sixty-five of them. The French also lost a standard-bearer and a lieutenant of the Mamelukes killed, while Lefebvre-Desnoëttes was captured. But Ney still got across and his hot pursuit of Moore continued.

By New Year's Day, 1809 the Marshal was another 60 kilometres on in Astorga, where the British had set fire to the Duke's palace and the magazines, pillaged the town and fled in disorder. 'I would never have believed that an English army could fall apart so quickly,' Moore wrote to Lord Castlereagh. 'Its conduct during the last marches has been infamous beyond description.' 'We will string up the French with the guts of the British,' local Spaniards grumbled.[13]

However, in Astorga Napoleon was met by a courier with disturbing news from Paris. There were rumours of plots being hatched by Fouché and Talleyrand, while the Austrians had again started to make threatening noises in Central Europe. 'Le Tondu'[14] realised he must go back. The following day, and until departing on 17 January, he worked ceaselessly to reorganise the army in Spain and leave it with an effective high command. Ney and the VI Corps had earned a rest. From Astorga Marshal Soult with fresh troops took up the pursuit of Moore. Ney's new job would be to police Galicia and the Asturias. But not before he had been drawn (as support to Soult) into one final clash with the British. This was at Calcabellos on 3 January—where his cavalry commander and personal friend, General Colbert, took a sniper's bullet through the forehead.

Soult chased Moore all the way to La Coruña, only to suffer the ultimate frustration there. After a series of brilliant defensive actions—comparable with those of Ney later on in Portugal and Russia—the British commander succeeded in embarking nearly the whole of his army. It cost him his life. 'I hope the people of England will be satisfied,' he groaned, split open by a French cannon-ball. 'I hope my country will do me justice!'[15] On the other hand, escaped though they might be, his British were well and truly kicked out of Spain—and looked like staying so in the conceivable future.

In the months that followed Napoleon was writing himself back into the victory columns. Moving troops of the line as well as the Guard with typical skill, at Essling in mid-May the Emperor stalemated an Austrian army under the Archduke Charles which outnumbered his own by four to one. (Jean Lannes was mortally wounded here, the first marshal created under the Empire to die.) Still east of Vienna, in July at Wagram and Znaim the Austrians were so heavily defeated they felt obliged to call for an armistice. The intrepid Lasalle died in the last charge at Wagram; and Marshal Bessières was wounded. But in one short season all of Central Europe was French-dominated again.

Echoes of these battles reverberated throughout the royal courts of France's disguised and undisguised enemies, putting fear into everyone who heard them. At the same time though, and most regrettably, such sounds tended to drown out the fresh alarms coming from an Iberian Peninsula where the struggle was by no means over. The Portuguese and Spaniards simply refused to be pacified—a fact which now encouraged the British to hazard a further expedition to Lisbon under Sir Arthur Wellesley. Moreover the French themselves were weaker than they believed in the Peninsula. Many *Grande Armée* veterans had been withdrawn for the Austrian campaign; while Napoleon's command structure, so excellent on paper, had turned those marshals he left behind into potential rivals, united by only one thing: a complete lack of respect for their new commander, King Joseph.

Wellesley (not yet the titled Wellington) arrived in Lisbon on 22 April. His immediate opponents, he knew, would be the forces of Marshal Soult, now based on Oporto and along the River Douro. Marshal Ney, he wrote in a despatch dated 24 April, would be 'left to the war of the peasantry which has been so successful.' This illustrates, I think, how Ney, further north in Spanish Galicia and the Asturias, was finding his 'policing' job there well nigh impossible. The terrain, although rough, he had learned to cope with; and by spring (in fact as Wellesley landed) he was preparing to deliver *le coup de grâce* to those remaining regular forces of Spain still in the Asturias. But the *guerrilleros*, aided by a hardy local peasantry, were revealing themselves to be some of the most fanatical fighters he had ever encountered. They were like the mythical, many-headed Hydra. Immediately he put them down in one district, so they would rise up again in two more!

To a commander of Ney's abilities it must have seemed an empty, needling kind of warfare. He was almost permanently bad-tempered and in a highly volatile state of mind. Yet he remained capable of gallantry; even of the occasional romantic gesture. When Major Napier (the elder brother of England's famous soldier-author) fell wounded near La Coruña, he had been passed back through the French lines and into the custody of Ney:

> The latter, treating him rather with the kindness of a friend than the civility of an enemy, lodged him with the French consul, supplied him with money, gave him a general invitation to his house, and not only refrained from sending him to France, but when by a flag of truce he knew that Major Napier's mother was mourning for him as dead, he permitted him, and with him the few soldiers taken in the action, to go at once to England, merely exacting a promise that none should serve until exchanged.

The younger Napier concludes: 'I would not have touched at all upon these private adventures, were it not that gratitude demands a public acknowledgement of such generosity, and that demand is rendered more imperative by the after misfortunes of Marshal Ney.'[16]

In early May, supported by Generals Kellermann from León and Bonnet from Santander, Ney finished off the Spanish regulars. The Marquis de la Romana's soldiers broke and fled—which delivered Oviedo, the main city of the Asturias, into French hands. Ney now felt able to devote himself entirely to the insurrectional problems of Galicia. Imagine his surprise, therefore, when the area was suddenly 'invaded' by the tatterdemalion remains of Soult's II Corps, beaten and then hounded out of Portugal by a newly-created Viscount Wellington! Consider too how the Marshal's surprise turned to anger after Soult began to regard himself as the senior commander, demanding arms and food for II Corps that Ney's own men badly needed in their war against the *guerrilleros*. The result was a violent altercation during which Ney half-drew his sword and threatened Soult with a duel; whereupon their staff-officers hastily intervened. Calming down following this, and smothering his pride, Ney next proposed that II and VI Corps should join forces in a maximum effort to stamp out the resistance across Galicia. An idea Soult, also calm by now, fully endorsed.

What happened to trigger off the subsequent events persists as a mystery. Largely because Soult, the root cause, studiously avoided giving any explanation for his behaviour: either to an impatient Emperor, to the worried, vacillating King Joseph or in any form of historical *mémoire*. Having agreed upon the plan, which was for a two-pronged march right over Galicia towards the sea, winkling out all resistance bands along the way, both corps commanders reviewed their men and set off. But then Soult, for no apparent reason, about-turned the II Corps, marched them right out of the province and—again for no apparent reason—headed towards León. Ney unsuspecting, was still force-marching his troops in the opposite direction when a courier caught up with him and passed on the news.

News? It was the last straw! The Marshal flew into a rage the like of which his *aides* had never seen before. I keep reminding myself that, while Ney possessed a quick temper, it was normally brought on only by someone else's incompetence; also that his anger tended to diminish just as swiftly as it had reached boiling-point. Nor did he ever bear a grudge. (A fact illustrated by his sincere willingness to co-operate with Soult again when the latter was Napoleon's overworked Chief-of-Staff at Waterloo.) But this particular raging in Galicia knew no bounds. It threatened to consume the man! One could try to defend him, of course, arguing that it was simply the 'letting-go' after months of fighting a kind of war to which he was unsuited and perhaps incapable of adapting. Nevertheless what he did strikes me as being unfair, inhumane and totally unworthy of the gentlemanly character I associate with the remainder of his life.

Soult, the logical target for his outburst, was no longer around. So the Marshal hit out at those other authors of his six months' discontent: the Galicians. But now he was indiscriminate. Whereas previously he had taken the utmost care to separate *guerrilleros* from peasants and townsfolk, this time he ignored any such distinction. Ordering a general movement out of Galicia in the direction of Valladolid and Madrid, he allowed his men to devastate their line of retreat. 'His VI Corps left a trail of twenty-six blazing towns and villages as it marched out of the province: a permanent disgrace on Ney, and on the French Army as a whole. No military purpose whatsoever was served by their brutality. It was an act of pure, vindictive spite.'[17]

At least there was no homicide or recorded rape. The VI Corps officers ordered the populace outside before any firings took place. Moreover it occurred in a century when generals were used to shielding their departures with a grand conflagration. But it made nonsense of Ney's earlier severe discipline within the VI Corps. And its connection with a person usually so magnanimous, so well-behaved whether in victory or defeat, comes down to us through history as an unpleasant shock. We know he could never be a retaliatory killer. We know too that as an ex-hussar it upset him to order the shooting of horses and mules to prevent them falling into enemy hands. Again, eye-witnesses report how the loss of a friend (i.e. General Colbert) invariably distressed him for longer than it would most men. All of which gives the colour of his rage in Galicia an increasingly awesome, even a frightening hue.

Probably he realised this himself—and accordingly set out to master his own emotions. Because from here on, although still subject to bursts of rage, we have no firm evidence that he allowed their outward expression to deteriorate from the verbal to the physical . . .

In reality the opening phase of Ney's involvement in the Peninsular War had almost touched upon its conclusion. Did he but know it the more momentous part was yet to come. For the present, however, and not being a crystal-ball gazer, it seemed to the Marshal as if an impasse had been reached. Everything about the war had become dreary, inconclusive. Whatever Wellington decided on, there were no equivalent moves from Paris and certainly no inspirations out of Madrid. As the summer of 1809 ground on King Joseph, supported by Marshals Victor, Mortier and Soult, blundered through a whole series of engagements north and south near the Portuguese-Spanish frontier. But he called upon Ney only at Baños in August, when the Marshal charged and won the day. Otherwise it was 'support, support, support' and no action.

By the autumn of 1809 though Joseph had indulged in a ludicrous political manoeuvre that now rebounded upon him. Having quarrelled with Victor, the King of Spain suggested his former crony ought to take 'a spot of leave in France', meanwhile asking Ney to command I Corps in addition to his own. Victor, his suspicions aroused, declined to go on leave. Ney, maybe at last waking up to the Bonaparte family's penchant for intrigue, refused to take over I Corps without a signed order from the Emperor. Anyway the story leaked its way back to Paris, and Napoleon—who for months had shown only desultory interest in the Peninsula—reacted in style! How dare Joseph interfere with his command structure! On 4 October Ney received a concise message: *HE was to go on leave, with General Count Marchand taking over the VI Corps!*

In the absence of reliable sources or any letters one can but guess at the Marshal's mood as he journeyed back into France to take his enforced spell of leave. He would not have disobeyed an order from the Emperor in any case. But, if he had doubts about the wisdom or relevance of a particular order, then he was neither afraid nor hesitant to speak his mind. *In this instance he maintained an absolute silence.* My own considered opinion, therefore, is that he quit Spain with a great sigh of relief. Whatever regrets he must have experienced at being separated from his prized possession, namely the VI Corps, they were amply compensated for by his getting away from the snake-pit of King Joseph's military council. Victor and Soult now intrigued against each other to become the dominant influence upon Joseph—and in Ney's view the one who succeeded was welcome to it.

He spent most of his leave with Aglaé and the boys at Coudreaux. Their visits to Paris were notably brief; and this time they paid no social calls on the Tuileries, held no glittering receptions at their house in the Rue de Lille. For Napoleon—desperate to produce an heir—was in the throes of divorcing Josephine and marrying Marie Louise, daughter of the Austrian Emperor. The Neys had no say in the matter, of course. But they were clearly unhappy to see their good friend and matchmaker being packed off to the rose gardens of Malmaison. They sensed too that as known familiars of Josephine and Hortense de Beauharnais they would be unlikely to receive an invitation from 'the new set'.

However if Ney kept away from Napoleon, it was not very long before the Emperor came back to him. The Peninsular War continued to be 'a running sore'. Marchand, after failing dismally with the VI Corps and losing Salamanca in the process, had been replaced by Kellermann, who recaptured the city. But Wellington still held Portugal, out of which he now made increasingly bold forays into Spain itself. And so Napoleon decided upon a fresh strategy. He would leave King Joseph, Victor, Soult, etc., to garrison Spain and form an additional 'Army of Portugal' to attack Wellington directly: the latter consisting of Ney, back as commander of the VI Corps, General Reynier with the II Corps and Junot with the VIII Corps. A reserve force, nominally dubbed the IX Corps, was assembled under the command of General Drouet—better known to history as the Count d'Erlon and at Waterloo the last senior officer Ney ever addressed on the field of battle.

Again the Emperor of the French was displaying a truly remarkable initiative. This additional army *ought* to have won the protracted struggle for Iberia. Except that Napoleon could make mistakes just as easily as he prepared for victories. On 17 April he appointed to be his overall commander in Portugal Marshal André Masséna.

In 1810 Masséna was fifty-two years of age and had no experience of the war in the Peninsula. (To be fair, he tried to wriggle out of going there at all.) He was a small, dark man of exceptional cunning, born at Nice, and with only one eye: the other having been shot out by Napoleon during a hunting-party, for which Berthier obligingly took the blame. His military record was dazzling. It included storming the bridge at Lodi and therefore being the first Frenchman into Milan; then a great success over the Russians and Austrians at Zurich (September 1799)—where Ney served under him. In the following year, 1800, he became a national hero after his extended defence of Genoa, when his men survived on bread partly made with sawdust, starch and hair-powder—thus giving Napoleon time to cross the Alps and crush the Austrians at Marengo. More recently though he had fought with equal distinction at Essling and Wagram; which again brought him into the Emperor's line of vision. The latter created him 'Prince of Essling' and now ordered him to command in the Peninsula. 'Your reputation alone will suffice to finish the business,' he insisted.[18]

Unfortunately this was not the case. Iberia had already shown itself to be a grave-digger for reputations and Masséna's was hardly the last to suffer there. Nevertheless his conduct of the war proved a disappointment because it was so lazy and unimaginative. Given the benefit of hindsight, and in spite of a later flattering remark by Wellington, 'The ablest after Napoleon was, I think, Masséna',[19] one can only look back upon him as a major factor in the Allies' ultimate triumph over the French. At the outset, of course, Ney had no inkling of this. He esteemed Masséna the soldier on account of Zurich and was perfectly happy to serve under him. On 10 May, learning that the new commander had reached Valladolid, he dashed off a letter of greeting.

But even if Masséna had continued to be a good soldier there were two traits in his make-up guaranteed to bring him into conflict with Ney, and the first of these revolved about money.

Michel Ney tended to be rather high-handed where money was concerned. He liked having plenty, was generous with it and considered that a 'Marshal of France' ought to be suitably rewarded, allowing him to maintain a fine house in Paris and a place in the country. However he believed it should all be done officially. In other words, with a big salary and various gratuities to be determined at the end of a successful campaign by the Emperor himself. Every franc must be clearly shown in the nation's account-books. Otherwise a man's public name and his *gloire* were at risk! It followed that he opposed every form of plunder or racketeering, and could be relied upon to deal severely with any instances arising from within his own corps.

In contrast Masséna was one of the stingiest, most avaricious men in the history of the French marshalate. His meanness became a legend in his own lifetime. At Wagram, after a nasty fall from his horse, he was driven about the battlefield in a light carriage. Once the fighting was done, to everyone's surprise the Marshal announced his intention of bestowing upon his coachman and postilion—who had shared all the dangers—200 francs apiece. 'You mean in the form of an annuity?' one of his younger *aides-de-camp* ventured. Masséna, already a millionaire, nearly threw a fit! 'I would sooner see you all shot and get a bullet through my arm,' he screamed. 'If I listened to you I should be ruined. Ruined!'[20] However, the story reached Napoleon's ears, who ordered that the two civilians be given a decent reward.

Far more serious from the army's point of view though was Masséna's unashamed acquisitiveness. He looted, extracted perquisites, misappropriated funds and indulged in all kinds of racketeering with stores and equipment. Napoleon, a meticulous accountant, reprimanded him frequently; and on more than one occasion only a new victory saved the Marshal from dismissal.

The Emperor was also much concerned that he set a bad example. In a letter to Joseph Bonaparte dated 15 November 1805, he refers to 'the disorders . . . growing up' in Masséna's Army of Italy. 'I hear that they have imposed a contribution of £1,600 on the Austrian sector of Verona. I intend to reward so richly the Generals and other officers who have served me well that they do not need to dishonour by their cupidity the noblest of the professions and arouse the disgust of their soldiers.'[21]

In the interests of the coming campaign Ney at first showed a preparedness to distinguish between Masséna the soldier and the man of greed. But an early incident reveals how there were clear limits to what he would and would not accept. After reviewing his VI Corps near Avila, the Marshal discovered that some of the men's pay had been withheld by Headquarters: which, in effect, meant Masséna. He promptly rode into Avila and extracted the amount in question from local tax-officials. This money, he informed them with studied politeness, 'ought to be returned very soon either by Marshal Masséna or His Majesty, King Joseph'. The latter dashed off a heated protest to Napoleon—but the Emperor rightly assumed that Ney was teaching HQ a lesson and in consequence left him alone.

The second reason for conflict between Masséna and Ney involved the army commander's sex-life. The diminutive, one-eyed Masséna was a notorious (and apparently highly-successful) womaniser. Under normal circumstances it would not have been important. Michel, although happily-married himself, was hardly the sort of person to sit in moral judgement upon the peccadilloes of a fellow-officer; especially his own commander. After all, he had seen the world. But to take a mistress on campaign, and dressed up as a soldier: well, it appeared to him unthinkable! Or at least it did until Masséna's arrival at Valladolid . . .

The new overlord:

> who intended to deprive himself of no personal amenities during his conquest of Portugal, had secured the appointment to his staff of no less than fourteen *aides-de-camp* and four orderly officers. One junior ADC was a Captain Renique, whose attractive sister Henriette (incidentally married to an officer named Leberton) was Masséna's mistress, disguised as an extra ADC.[22]

At the sixteenth century Palacio Real in Valladolid the Duchess of Abrantès, Junot's wife, had prepared a special dinner to welcome the Marshal. When he drove up accompanied by 'Madame X' (as the army soon nicknamed her) the Duchess was first of all horrified, then outraged; and in the end she turned her back on them both. A few evenings after this Ney was invited to a dinner at Masséna's HQ. As the second most senior officer present, he found himself seated next to Henriette and looked to for some entertaining small talk. He preferred to concentrate upon his food.

Such snubs should have put a brake on the affair. But Masséna 'insisted on taking his mistress everywhere, even to the front-line, dressed as a cornet of dragoons and wearing the Cross of the Legion of Honour.'[23] Over one hundred and fifty years later these scenes perhaps appear amusing. At the time though their impact on relations between the headstrong marshals was deadly. Ney could have put up with Masséna's greed. He could have ignored the presence of 'Madame X'. But when Masséna began to make military mistakes as well, Ney's distrust of his immediate superior grew in proportion —until eventually the whole command structure of the Army of Portugal was threatened.

At the beginning of June, Masséna moved his HQ forward to Salamanca and called the three corps commanders to a meeting.

The ways into Portugal, as each of them knew, were comparatively few and very difficult. Most roads ran high up around the sides of mountains, because the more obvious routes, the river valleys of the Tagus and Douro, passed through such steep gorges that a thunderstorm could raise the water-level by fifty feet in as many minutes. Naturally these mountain-roads were easier to defend than to force. While the best road of all, and certainly the only one suitable for moving an army and its equipment along at any reasonable pace, was straddled by the Spanish fortress of Ciudad Rodrigo and the Portuguese frontier-town of Almeida, both at present in enemy hands. Capture these and a march down towards Lisbon became feasible.

Far away in Paris the Emperor had assessed the situation nearly as well as his men on the spot; and the directive he sent off to the Army of Portugal was quite specific. 'The Prince of Essling will have 40,000 infantry and 9,000 to 10,000 cavalry . . . with which he will besiege first Ciudad Rodrigo and then Almeida, and will then prepare to march methodically into Portugal, which I do not wish to invade until September, after the hot weather and in particular after the harvest.' (Dictated to Berthier at Le Havre, 29 May 1810.) By early June Masséna was repeating this order to his corps commanders and inviting their views on how best to implement it.

Now Ney, while waiting for Masséna to arrive at Salamanca, had taken it upon himself to reconnoitre all the border areas to the west and south-west. Ciudad Rodrigo, he discovered, was garrisoned by five and a half thousand Spanish regulars under the veteran General Andrés Herrasti; moreover it had plenty of durable food and ammunition. Nearby, strung out in a 'thin red line' on the left bank of the Agueda river, was the British Light Division under General 'Black Bob' Craufurd. But the main Anglo-Portuguese force (with Wellington) had stayed deep inside Portugal at Viseu. Based upon this information Ney therefore wrote to Masséna with the suggestion that they 'mask' Ciudad Rodrigo and Almeida with one division and some field guns, punch through Craufurd's line and then launch a full strike directly at Wellington's HQ. An idea he again argued strongly for at Masséna's conference, only to find that his army commander was just as strongly against it. 'Masséna decided that the fortress (Ciudad Rodrigo) must be reduced by regular siege operations, as Napoleon had ordered.' He called up the siege-train, 'previously delayed by heavy rains, and instructed Ney to invest the fortress closely'.[24]

Following this decision Ney got on with the job. He placed his three infantry divisions in a tight circle around the fortress: which topped a cone-like hill and faced the besiegers with strong, medieval ramparts. At the same time he deployed the VI Corps' cavalry as a screen against any surprise attacks by Craufurd. Then Major Conche, a most reliable 'sapper' commander, was ordered to start his men digging. By 15 June their first parallel had been completed. Five days later the batteries were ready and the guns opened up, a bombardment that was duly rewarded on 25 June when the town's central powder-magazine took a direct hit and exploded. 'Wellington, over 20 miles away at Almeida, had heard the blast!'[25] Meanwhile Major Conche promised the second parallel would be ready by July and Ney began to consider his final assault plans.

However, these extremely satisfying prospects were unpleasantly jolted one morning later by the appearance at his command-post of a Lieutenant-Colonel Valazé: an officer on Junot's staff. Valazé brought a letter from Masséna stating that he was to take over the engineering work from Conche. Quite why the army commander wished to replace this successful officer we have never discovered. It might be '. . . that Junot had sent his man to obtain credit for the victory which would be due to the VI Corps.'[26] Anyway, Marshal Ney bridled. And he packed Valazé off with a stinging rejoinder. 'Tell Marshal Masséna that I don't want the Duke of Abrantès to trouble me with his *protégés*. If they are so good, let him keep them for himself.'[27] Masséna naturally was furious. 'I am going to remove Ney and send him back to France!' he shrilled. At which point Junot intervened to calm him. Ney was just 'being thoughtless,' he said. 'Very well,' Masséna relented, 'I shall allow him to remain, but you will see. This proud fellow will upset all our plans with his stubborn self-will and foolish vanity.'[28] Poor Valazé was despatched to VI Corps yet again.

At the mere sight of him this time Ney thumped the table in anger. Then he sat down and wrote Masséna what is now a famous letter. It was an unusually long letter for him; but its length came about due to sheer aggravation rather than any attempt at eloquence or a love for words. Its tone was insolent, even insulting. Its message contained insubordination. The Marshal had been unusually provoked—and by an act which militarily made no sense whatsoever. 'Monsieur le Maréchal,' he began:

> I am a Duke and a Marshal of the Empire like you; as for your title of Prince of Essling, it is of no importance outside the Tuileries. You tell me that you are Commander-in-Chief of the Army of Portugal. I know it only too well. So when you tell Michel Ney to lead his troops against the enemy you will see how he will obey you.
> But when it pleases you to disarrange the staff of the army appointed by the Prince of Neuchâtel, you must understand that I will no more listen to your orders than I fear your threats. See here, just ask the Duke of Abrantès what he did and I did a few weeks ago when we received from that other—*Soult*—(who is major-general to the King and who did such pretty things there where we are going) orders differing from those that came to us from Paris and were therefore the orders of the Emperor. Do you know what we did? We obeyed the orders from Paris, and we did right, for they praised us for it, and praised us warmly. I had letters from Madrid in which I was called—I believe—a rebel. As that is about as absurd as calling me a *poltron* (poltroon), I took no notice of it, and General Junot would certainly have acted in the same way. Adieu,

Monsieur le Maréchal. I esteem you and you know it. You esteem me, and I know it. But why the devil sow discord between us over a mere caprice? For after all, how on earth are you to know that your little man can throw a bomb better than my old veteran, who is, I assure you, a reliable fellow. They say your man dances prettily; all the better for him, but this does not prove that he can make those Spaniards dance, and that is what we want. I remain, Monsieur le Maréchal, etc. NEY[29]

Masséna was beside himself. 'You see,' he shouted at Junot. 'It is impossible to do anything with that man! Am I then but a sham Commander-in-Chief? Am I to disregard such a letter as this, am I to be spit upon by this damned Ney? No! I mean that this young man shall conduct the siege, and by the devil in hell Monsieur Ney shall bend his knee before my will, or my name is not Masséna!'[30] He hurried off to Ciudad Rodrigo, taking Valazé with him, and offered Ney a choice: either accept the new engineer or resign his command. Faced with this, the Marshal backed down. But he did so with a cold, hostile courtesy; at the same time informing Masséna that Major Conche had finished his work on the second parallel.

The fortress was now near to giving up. Repeated requests to Wellington had failed to produce any help. The defenders were grumbling quite openly about their British allies. On 9 July Ney succeeded in breaching the ramparts, and at six o'clock in the evening sent a message to General Herrasti that he intended to take the town by storm. Nobody inside Ciudad Rodrigo thought that he wouldn't. 'The white-haired Governor was compelled to surrender on 10 July. Ney gallantly gave him back his sword, congratulated him and shook hands. At that moment there was no doubt which leader commanded the Spaniard's greater respect, Ney or Wellington.'[31] By 21 July the French were on the move towards Almeida.

Craufurd, against Wellington's orders, had deployed his Light Division on open ground to the south of this fortress. On 24 July Ney (with 24,000 men, in fact the whole of his VI Corps) caught him and almost smashed him there. Craufurd got away across the Coa river, but it cost him 400 officers and other ranks to do so and it left Almeida completely isolated.

The town 'had an important role in Wellington's plans. He had put 4,500 Portuguese troops into the place—regulars and militiamen—under a British general, Brigadier Cox.' It was an extremely solid fortress, sited on a granite plateau which made the digging of siege trenches an unenviable task; it was well-stocked with provisions and Wellington hoped that it could be held at least until the bad weather of autumn set in to embarrass the French further.[32] In addition Ney had a great problem in bringing up his guns and siege-equipment. There was a chronic shortage of draught animals for the vehicles: yet another black mark against army HQ. Nevertheless, by dawn on 26 August Almeida was thoroughly invested and the bombardment opened up.

On this occasion though it was only for fourteen hours. Again one of Ney's gunners found the defenders' powder, stored inside the Cathedral; but it was a real stroke of luck to happen so early in the siege. A stray French shell had ignited a trail of loose powder leading directly to the main magazine and '70 tons of gunpowder went up with a terrifying explosion—*like a volcano erupting*, recalled one of Ney's awed staff officers in the French lines.[33] Cathedral, castle and half the town-centre just vanished! Cox surrendered what was left the next day. By 15 September, 'Reynier's II Corps had joined up with the corps of Junot, Ney, and General Montbrun's Reserve Cavalry; and the Army of Portugal, 65,000 strong and fully concentrated at last, was ready for the march on Lisbon.'[34]

The unexpectedly early fall of Almeida confronted Wellington with an appalling dilemma. For nearly a year his teams of engineers had been at work between Torres Vedras and Mafra, just north of Lisbon, preparing the 'Lines' which their commander hoped would become an impregnable defence against the French. They were damming or diverting the streams in front of certain chosen positions, fortifying the various heights and setting up a chain of 'intervisible' signal-stations. This 'master plan . . . within the next twelve months was to form the turning point of the Peninsular War.'[35] By mid-September the Lines were almost, *but not quite, ready*—hence Wellington's problem. Whether to retreat inside and chance the remaining weak spots; or whether to fight a delaying battle further north? With a fine blend of Ney-like defiance, his own matured prudence and what is best described as happy foresight he opted for the battle. But where to stage it?

As events turned out Marshal Masséna helped to decide this for him. In Almeida the French had discovered the raw materials for nearly half a million pounds of bread. Yet Masséna, instead of using this to feed his army during a straight thrust towards Coimbra, Portugal's last important city before Lisbon, elected to take the more circuitous (and difficult) route via Viseu. 'There are certainly many bad roads in Portugal,' Wellington commented, 'but the enemy has decidedly taken the worst in the whole kingdom.'[36] Furthermore the French Commander then remained in Viseu for six days: a town with no extra bread supplies. 'To Napoleon Masséna sent the excuse that he needed to wait in Viseu for the artillery to catch up.'[37] But Marbot, his increasingly critical senior ADC, recalls that he wasted most of the time finding suitable lodgings for and dallying with 'Madame X'. At one point 'Ney had to shout the results of a very inadequate reconnaissance through Masséna's bedroom door.'[38] The net result was that France's soldiers ate up all their rations and Wellington had ample time to prepare what had become an obvious defensive point along the Coimbra road: the ridge of the Serra do Bussaco. By 26 September, when Masséna called his leading officers together in Mortagoa, a village just to the east, their opponents were waiting and confident. It was the French now who would have to make the running.

Sir Charles Oman has called the Bussaco ridge 'one of the best-marked positions in the whole Iberian Peninsula'.[39] And the Countess of Longford has given us the following marvellous description of how the Anglo-Portuguese forces were occupying it. 'All along Bussaco's towering hog's back, running north and south for almost ten miles, Wellington stationed his sixty cannon and army of 51,000 men, half British and half Portuguese— a somewhat thin barrier of men perched upon a mighty barrier of mountain. A staff officer, on being asked by an artillery officer for a map of Bussaco to send to England, replied: *You only have to draw a damned long hill, and that will be sufficiently explanatory.*

'Past Bussaco's southern end wound the beautiful Mondego, under a strange, perpendicular cliff of grey granite, its surface fractured into countless shelves and known as 'The Library'. Deep heather covered its sides and summit, broken by spiky aloes, boulders of black basalt and pink and grey limestone, or pine trees springing from precipitous ravines. Solid stone windmills stood here and there on high plateaux. At the loftiest point of all, two miles from its northern end, Wellington established his headquarters in the wooded and walled Convent of Bussaco. Every room was occupied by his staff, except for the Prior's and that of one monk who managed to fill his up with lumber. From a narrow, cork-lined cell with whitewashed walls and brick floor, Wellington issued his final orders of 26 September 1810. His soldiers, most of them concealed behind the *massif's* crest, were to eat a cold evening meal and lie down without fires in total darkness and eerie silence. He knew that Masséna would force a head-on collision next day. No flanking movement to north or south. Head-on, against the mountain. It had been almost too much to hope for, but it was going to happen.'[40]

How could he know that Masséna would opt for a frontal attack? The French Commander didn't even reach this decision until late afternoon on the 26th! Never mind. Either by inspired guesswork, or a shrewd assessment of his enemy's tactics, Wellington was one hundred per cent correct about what 'was going to happen'.

On the morning of 26 September, Reynier informed Ney that 'his *tirailleurs* (sharpshooters) were in contact with the enemy on the lower slopes'. The Marshal, using his telescope, thought 'they were withdrawing north-west, perhaps to Oporto, but holding on to the monastery with a semi-circle of guns.' And he sent this message back to Reynier: 'If I were in command, I should attack without a moment's hesitation.'[41] After further scrutiny, however, he concluded that Wellington had 'dug in'—and this remained his opinion when he attended the afternoon's conference.

'It is worth pointing out,' Richard Humble writes:

> that on the eve of this battle Masséna was the first French Commander-in-Chief to have the opportunity of seeking the advice of generals who had had the bitter experience of attacking British troops in position. These were Junot (at Vimiero, in 1808) and Reynier (at Alexandria during the Egyptian expedition, and at Maida in 1806). Even Ney had had a brush with Moore's rearguard during the retreat to Coruña.[42]

The big question though was how Masséna would react to their advice. Ney, despite being second-in-command, was hardly on speaking terms with his chief. Still, he put forward his views. 'Ney has often been dismissed as an unthinking advocate of the heads-down attack. But at this conference he argued against an attack on the ridge. He wanted Masséna to accept the fact that a surprise attack was now out of the question, pull back, and make sure of Oporto before resuming the advance on Coimbra and Lisbon.'[43]

Junot agreed with him, likewise Fririon (the Chief-of-Staff); and so too did

Lazowski, the artillery commander, because his guns would not have the angle of elevation to sweep the ridge at its highest points. Only Reynier and finally Eblé, the chief engineer, favoured a frontal attack. But immediately Masséna seized upon the idea. His one brown eye glared at Ney, and he seemed to be ignoring the latter's advice almost as a matter of principle. They would attack, he ordered, in two places. Reynier with the II Corps must go for the British right (which proved to be their middle and so meant the French could be turned). Ney and the VI Corps were to attack the strongest part of the line, around the monastery to the north. 'It was an astonishing decision,' Richard Humble concludes. No ideas for out-flanking Wellington had been discussed; nor any thought given to the French cavalry superiority, although this might have succeeded in a manoeuvre around the northern slopes. 'And the Prince of Essling, having put his signature to this thoroughly unimaginative sheaf of orders, rode back the eight miles to his Headquarters and 'Madame X' for the night. The attack was scheduled for dawn on the morrow.'[44]

27 September came up in a swirl of mist and fog patches. Reynier launched the first of his three attacks just after 6 a.m.—and his men were grateful for the morning's elemental cover as they toiled up the boulder-strewn slopes. By 7 a.m. however, when Ney's own attack was scheduled to begin, the mists had dissolved to almost nothing. The Marshal could just make out Reynier's skirmishers up near the crest (on their way to defeat as it happened, either blasted by grape-shot and canister or driven in by bayonet charges).

Setting off along the Coimbra road from the village of Moura, Ney had divided the VI Corps into separate strike forces: twelve battalions under General Loison to the north, eleven more to the south under General Marchand. It still meant going upwards, of course; and against hellish fire—including shrapnel—from Craufurd's Light Division and General Pack's Portuguese. At least though the French believed in their corps commander and to begin with they made real progress. Loison drove an Allied force of 1,500 from the village of Sula. Then he attacked the strong-points higher up: Ross's battery of twelve guns and 'Black Bob' Craufurd's command post at Sula Mill. In spite of devastating fire they reached the crest and overran the guns—only to be confronted by Wellington's reserve: seasoned troops with fixed bayonets. Craufurd, a little man on a big horse, and with a stentorian voice, ordered the counter-attack. 'Now, Fifty-Second! (The Light Division) Avenge the death of Sir John Moore! Charge! Charge! Huzza!'[45] Seconds later 'eighteen hundred British bayonets went sparkling over the brow of the hill. Yet so brave, so hardy were the leading French, that each man of the first section raised his musket and two officers and ten soldiers fell before them. Not a Frenchman had missed his mark! They could do no more.'[46]

The British fired at ten paces. 'Three terrible discharges.'[47] Many Frenchmen of the first wave had their heads blown off. The next line felt cold steel and went tumbling down the hill. 'Down, down into the valley, sprawling and slithering over boulders streaming with blood.'[48] Ney 'threw forward his reserve division, and opening his guns from the opposite heights killed some of the pursuers'.[49] But when he called off the attack Loison had suffered over a thousand casualties.

Meanwhile Marchand's division, south of the Coimbra road, had fought to within sight of the monastery. They 'gained a pine wood half-way up the mountain . . . and sent a cloud of skirmishers against the highest part.'[50] But eventually here too they were stopped: by four battalions of 'the despised' Portuguese. 'On that steep ascent . . . Pack's men sufficed to hold them in check, and half a mile higher up Spencer showed a line of the footguards which forbad any hope of success. Craufurd's artillery also smote Marchand's people in the pine wood.'[51] Ney, who had rushed over to join the division, 'after sustaining (*Craufurd's*) murderous cannonade for an hour, relinquished that attack also'.[52] By midday Marchand, Loison and their corps commander were back in Moura. Whatever Masséna might do to save face (such as not officially abandoning the fight until 4 o'clock in the afternoon) Bussaco Ridge had been the worst French defeat in the Peninsula so far. (4,500 casualties as against 1,200 on the Anglo-Portuguese side.)

Ney didn't bother to report in person to army HQ. He feared losing his temper again; and really what was the use? Instead, after seeing that the wounded were being properly cared for, he shut himself away to write next-of-kin letters to the families of his veterans. Some of these had fought for him at Ulm and Friedland, only to die now in an idiotic assault on easily-defended positions. Well, he vowed—the iron suddenly reappearing through his upset—it would never happen again! At least not on Masséna's orders!

He must also have given considerable thought to the opposition on Bussaco's hog's back, and in particular to the man who organised it. Admittedly the geography had favoured its defenders. But they were deployed with obvious skill; and by a commander who then put heart into them to stand firm against some of the toughest infantry in the *Grande Armée*. Clearly 'this Wellington' was a person to be reckoned with. He had a cool head, stubbornness, an instinctive feeling for terrain and position—*plus* the ability to inspire a body of soldiers long regarded as the war's underdogs. Ney realised (or at least he decided) that Wellington was the most gifted commanding officer he had fought so far; and from this grew the mingling of respect with caution destined to affect his judgement on future occasions. He would never be known to shrink from an engagement with the British general. But his famous zeal for attack would be tempered by a degree of carefulness whenever Napoleon's hitherto 'emaciated leopard', now proven a lion, was rumoured to be in the vicinity . . .

I will merely summarise events of the next few months—since Michel Ney took little part in them.

After Bussaco the Anglo-Portuguese slipped away to their now fully-prepared Lines of Torres Vedras. Masséna and Junot started to give chase, but paused on the way to plunder whatever the British had left in Coimbra. Ney, back with the rearguard, tried to put a stop to this; but when other marshals were helping themselves his orders to the rank and file went largely unheeded. (At one point he sent the gift of a valuable telescope back to Masséna with a caustic note saying that 'he was not a receiver of stolen goods'.)

From Coimbra the French then continued south until their advanced units probed the outlying strength of Torres Vedras: which both Junot and Ney refused point-blank to attempt by direct assault. For once Masséna did not press them. Rather he decided to camp in front of the Lines and wait. Altogether for five months; through an awful winter, with dwindling provisions and a mounting sick-list. (Wellington continued to be supplied through Lisbon by the Royal Navy.) On 5 February, when a new directive arrived from Napoleon, the besiegers were weaker than the besieged. The Emperor rebuked Masséna 'for having attacked the Bussaco ridge frontally without previous reconnaissance.'[53] Clearly now he must by-pass Torres Vedras by bridging the Tagus, then approach Lisbon from the eastern direction. Mortier's V Corps would be sent to help. If this proved impossible however, 'then he should retire to the Mondego . . .'[54] Masséna knew what *he* wanted to do—and it didn't involve waiting around for Mortier. But he was wily enough to discuss it first with his officers at an informal luncheon: thus partly sharing out the responsibility:

> When lunch was over Masséna pointed out the impossibility of the army remaining longer in its present position, owing to the exhaustion of all local supplies. Only fifty rounds of ammunition per man remained. He then asked Foy (*General Maximilien Foy*) to recount his recent conversations with the Emperor. When Foy had outlined Napoleon's observations Masséna asked each in turn their views on the following three alternatives:
> (a) Hold on to the present position in the hope that V Corps would advance to support them.
> (b) Force the passage of the Tagus at Punhete and occupy the Alentejo to gain touch with V Corps.
> (c) Withdraw to the Montego valley, where fresh supplies might be found, and regain touch with Almeida.
> Ney and Reynier were in favour of crossing the Tagus; Junot thought that to do so would be to abandon north Portugal to the British. The army commander decided to retreat to the Montego.[55]

The retreat got under way during the night of 5/6 March 1811 and now Ney comes back strongly into focus. For he was to command the rearguard: consisting of the VI Corps, part of IX Corps (Conroux's division) and Montbrun's Reserve Cavalry. He also had to ensure that Eblé destroyed all the bridging equipment. Though he couldn't yet know it his retreat was to become the Marshal's interim revenge over Wellington. For while the British chased him hard, he eluded or frustrated them at every turn—thus saving the Army of Portugal and indirectly causing Wellington another year's fighting across Spain. Ney too had his moments of anxiety. But he handled everything with dexterity, even panache; and in certain respects this first fighting retreat prepared him for the increased difficulties he would have to cope with on the way home from Moscow.

At the time the French set off Wellington was in fact prepared for a surprise strike against them. But thanks to another thick fog—and some sentries made out of straw and topped with shakos—they were thirty miles along the Coimbra road before the British set off in pursuit. Ney took his first stand at Pombal on 10 March, where his gunners 'took immense delight' in bringing Craufurd's Light Division to a bloody halt. Then, under cover of darkness, they slipped away to Redinha—where on 11 March they inflicted another two hundred casualties on the British.

Now it was Wellington's turn to be cautious. For once, he decided to stalk rather than give chase. When Ney moved the nine miles to Condeixa, by-passing Coimbra, Wellington endeavoured to send Picton's division by a mountain track around the French left. But again Ney got away; this time five miles to Casal Novo, and so fast that it nearly resulted in Masséna's own capture!:

> The Marshal and his staff were dining in the open air on the bank of the stream at Fonte Coberta and had just received Ney's message that he was retiring from Condeixa, when a patrol, (of Arentschildt's Hussars) rode up and would have captured him had he not mounted quickly and galloped back to the protection of Loison's division, which was camped on the other side of the village.[56]

'Madame X' was thrown from her horse during this excitement, and Masséna became convinced that Ney had withdrawn so quickly in order to expose him to danger. 'Such conduct is inexcusable!' he insisted to his Chief-of-Staff, who was now despatched 'to find some decent coffee and sugar for Madame'.[57]

Not that HQ's opinions of him really troubled Ney any more. Madame Junot has a story—which certainly rings true—of how news (untrue) reached Ney via an ADC that Masséna was taken by the enemy. The Marshal was snatching a few hours' sleep. But he jumped up, fully awake and shouted: 'Taken prisoner is he? By God! So much the better, for now the army is saved!'[58]

At Casal Novo Ney took another stand, and with all three divisions. 'On 14 March Erskine's advanced guard blundered into Marchand's outposts; as the morning mist lifted, five battalions found themselves confronted by eleven French battalions and suffered nearly a hundred casualties.'[59] Then the French retreated in successive lines; seven miles eastwards to Miranda do Corvo and during the night to the gorge of the River Ceira at Foz de Arouce. It was here that Wellington came nearest to a victory:

> Ney's corps held the heights on the left bank, with Junot's corps on the right bank, separated by a single narrow stone bridge. The river was in flood, and a thick fog hid the landscape until well on in the day. Wellington came up late in the afternoon, observed the awkward position which the French were in and decided to attack at once. The French were completely surprised and a panic ensued at the narrow bridge, where they suffered nearly 300 casualties and lost a regimental eagle. The British loss was 71.[60]

However, when darkness fell again, and by an effort that seemed superhuman, Ney whipped the remainder of his corps across the bridge and Wellington had missed his chance.

It was also to be the last chance. For his own supply lines were now dangerously extended, and from this point on Ney marched his troops at the rate of twenty miles a day all the way to the Portuguese-Spanish frontier. They were ragged, their boots were falling apart and usually they marched hungry. But they made it nevertheless. Soon they expected to be in Ciudad Rodrigo for a much needed rest and refit.

Then, on the afternoon of 22 March, a courier rode up to VI Corps' headquarters with new orders from Masséna. Ney could hardly have registered more surprise if they'd been fired at him out of a British cannon! The whole army, Masséna directed, would proceed south through Plasencia to Spanish Extremadura. There they were 'to link up with Mortier's V Corps and make a joint advance upon Lisbon from the south-east'.[61] It may well be that this bold, but wholly unrealistic plan was intended to placate Napoleon after the failures at Bussaco and in front of Torres Vedras; we shall never know. But the effect of these orders upon Ney—who, like Napoleon, had wanted an easterly attack upon Lisbon much earlier and from much nearer—was instantaneous.

That same afternoon he dashed off three letters to Masséna by successive couriers. In the first he stressed the difficulties of such a move, for in Extremadura (one of the poorest provinces of Spain) the army 'would starve'. In the second he pointed out that, if they abandoned Almeida and Ciudad Rodrigo, Wellington could invade Spain 'by that same route'. And in the third—sent at 6 p.m.—he specifically refused to obey the orders 'unless the Emperor himself had sent fresh instructions to the Army of Portugal'.[62]

At 8 p.m. Masséna sent an ADC back with the following message: 'Please inform me whether you persist in refusing to obey my orders and in ignoring the authority conferred upon me by the Emperor; in that case I shall have to take steps to enforce it. I await your reply by the hand of my ADC.'[63] By nine Ney had replied: 'I persist in refusing to allow VI Corps to march . . . on Plasencia, as ordered today by Your Excellency, unless you can show me the Emperor's orders authorising that move. I have already explained to Your Excellency why I am determined to march tomorrow towards Almeida.'[64] The Marshal had 'rightly called it madness and wrongly defied his superior'.[65] At 10 p.m. Masséna relieved him of his command, gave the VI Corps over to Loison and ordered Ney back into Spain, there to await instructions from the Emperor. Colonel Pelet, a senior ADC, rode off at midnight for Paris. He carried copies of Ney's 'insubordinate' letters and had orders to repeat Masséna's version of the quarrel, 'directly to Berthier'.[66]

I can think of only two suitable postcripts to these deplorable events. The first is yet another quote from General Sir James Marshall-Cornwall's book on Masséna. 'Reynier and Junot shared Ney's objections to Masséna's new plan but, apart from a warning from Reynier that he knew the country through which they had to march and that it was destitute of supplies, they obeyed the order.'[67] The campaign towards Extremadura petered out after just one week.

The other comes from the diary of a contemporary French soldier, Denis-Charles Parquin (later commissioned Captain in the Imperial Guard). It was written after Masséna's trouncing by Wellington at Fuentes de Oñoro on 5

May. 'Unfortunately for the army Marshal Ney was no longer there. A difference which the Prince of Essling had deprived us of his talents and of his sword.'[68] I find this an interesting commentary on the state of morale. For Fuentes de Oñoro saw the VI Corps, which Ney had commanded since 1804, cut to pieces in yet another series of futile frontal attacks. And five days later Masséna himself was dismissed. 'Prince, the Emperor charges me to inform you that he expected more from your energy, and from the opinion of you which he formed as the result of those glorious episodes in which you have so frequently taken part.' Signed: ALEXANDRE (*Berthier*), Prince of Wagram and Neuchâtel, Chief of the General Staff.

[1]   Richard Humble: *Napoleon's Peninsular Marshals*, London, 1973.
[2]   Ibid.
[3]   For more about the intrigues and disillusion at Joseph's Spanish court see Michael Glover's *Legacy of Glory: The Bonaparte Kingdom of Spain*, London, 1971.
[4]   *Napoleon's Peninsular Marshals.*
[5]   *Napoleon's Marshals.*
[6]   *The Anatomy of Glory.*
[7]   Ibid.
[8]   Ibid.
[9]   Ibid.
[10]   *Napoleon's Peninsular Marshals.*
[11]   *The Anatomy of Glory.*
[12]   *Napoleon's Peninsular Marshals.*
[13]   *The Anatomy of Glory.*
[14]   Literally 'The Shorn One': Napoleon's nickname to the men of his Imperial Guard.
[15]   Major-General Sir William Napier: *History of the War in the Peninsula*, London, 1826-40.
[16]   Ibid.
[17]   *Napoleon's Peninsular Marshals.*
[18]   *Napoleon's Marshals.*
[19]   Philip Henry Stanhope: *Notes of Conversations with the Duke of Wellington, 1831-51*, London, 1888.
[20]   General Baron Marcellin de Marbot: *Mémoires*, Paris, 1891.
[21]   *Lettres Inédites de Napoleon Ier, 1799-1815*, Paris, 1897.
[22]   General Sir James Marshall-Cornwall: *Marshal Masséna*, London, 1965.
[23]   Ibid.
[24]   Ibid.
[25]   Ibid.
[26]   Legette Blythe: *Marshal Ney, A Dual Life*, London, 1937.
[27]   Ibid.
[28]   Ibid.
[29]   Ibid.
[30]   Ibid.
[31]   Elizabeth Longford: *Wellington, The Years of the Sword*, London, 1969.
[32]   *Napoleon's Peninsular Marshals.*
[33]   Ibid.
[34]   Ibid.
[35]   *Marshal Masséna.*
[36]   Wellington: *Despatches*, ed. Lt.-Col. Gurwood, London, 1834-8.

37 *Marshal Masséna.*
38 *Wellington, The Years of the Sword.*
39 *History of the Peninsular War,* Oxford, 1902-30.
40 *Wellington, The Years of the Sword.*
41 Marbot: *Mémoires.*
42 *Napoleon's Peninsular Marshals.*
43 Ibid.
44 Ibid.
45 *Wellington, The Years of the Sword.*
46 *History of the War in the Peninsula.*
47 Ibid.
48 *Wellington, The Years of the Sword.*
49 *History of the War in the Peninsula.*
50 Ibid.
51 Ibid.
52 Ibid.
53 *Marshal Masséna.*
54 Ibid.
55 Ibid.
56 Ibid.
57 General Jean-Baptiste-Frédéric Koch: *Mémoires de Masséna,* Paris, 1848-50.
58 *Mémoires de Madame la Duchesse d'Abrantès.*
59 *Marshal Masséna.*
60 Ibid.
61 Koch: *Mémoires de Masséna.*
62 Ibid.
63 Ibid.
64 Ibid.
65 *Wellington, The Years of the Sword.*
66 Koch: *Mémoires de Masséna.*
67 *Marshal Masséna.*
68 Parquin: *Mémoires,* Paris, 1842.

CHAPTER 6

Rehabilitation

Into Russia with the III Corps, 1812

Who is there but knows, that from the depth of his obscurity the looks of the fallen man are involuntarily directed towards the splendour of his past existence—even when its light illuminates the shoal on which the bark of his fortune struck, and when it displays the fragments of the greatest of all shipwrecks?

PHILIPPE DE SÉGUR
History of the Expedition to Russia

Napoleon normally ignored the wrangling and petty jealousies of his marshals. On the whole he took the view that they were a talented lot and therefore just as likely to be temperamental and thin-skinned. Provided they obeyed his *own* orders, and in their private lives did nothing to bring disrepute upon the Empire, he preferred to let their individual rivalries work themselves out over a period of time. Also, of course, he paid them extremely well: which combined with their disliking one another made them think twice about standing against him—at least until Fontainebleau, 1814. As always with Napoleon, the skilful politician was hard at work inside the great soldier; and certainly he manipulated the operational marshals as no one else would have dared: in turn flattering, bullying and deceiving them—but making them sink their differences whenever there was a battle and then expecting each to give of himself to the utmost.

However, for a marshal to be relieved of his command was a serious matter indeed—and one which threatened to have its repercussions in other parts of the army. The Emperor was justifiably furious; far more than he had been at Jena or over the Tudela affair. It even crossed his mind to retire Ney on half-pay.

But in the end it was Masséna he retired. As more evidence accumulated about the Commander-in-Chief's sloth and ineptitude (all dutifully presented by Berthier), so the Imperial wrath was diverted away from his 'unruly' subordinate. Ney's disobedience, otherwise inexcusable, Napoleon now equated with undue provocation and the gradual deterioration of orders coming out of Spanish HQ. The Marshal got off with a severe reprimand (by letter) and was then ordered to his country estate 'to cool down'. From there, at the end of the summer, Berthier instructed him 'to proceed at once towards the Camp of Boulogne, where it is His Majesty's intention for you to assume command'.

The time spent at Coudreaux provided him with a much needed rest, and led to Aglaé's conceiving the fourth and last of their sons.[1] It was an opportunity too to meet the people on the estate, 'good men and women, who do their work honestly and happily'; also to put in a surprise appearance at the harvest, his first farming since La Petite Malgrange. But the Camp of Boulogne struck him as the Napoleonic equivalent of being sent to Siberia. He went there late in September and immediately found it 'a dreary place, half empty and forgotten . . . decidedly bad for the spirits . . .' Most of the barrack-huts and corps HQs had been left abandoned after the invasion of England was called off. Many had weathered badly and were falling to pieces, thus adding to the bleak, forlorn aspect of the landscape. Only the coastal defences remained, together with some smaller camps for training purposes. For Ney it was a time and place to brood: on the valid reasons for his hot-temper in the Peninsula, and yet the shame and virtual banishment to which his outbursts had now reduced him.

Consequently when he was summoned to take part in the Russian campaign its effect was like a tonic. It would make almost impossible demands upon him as a soldier. On his patience, technical knowledge, determination, physical fitness and instinctive courage. Nevertheless he joined the expedition without a second's hesitation. It promised to be a war across open country of manoeuvres and decisive battles rather than the sporadic, hit-and-run tactics of Spain and Portugal. As such it must, he believed, favour the *Grande Armée*—as Austerlitz and Friedland had done—and he went off full of enthusiasm that France would win. Again, as he explained to Aglaé: 'I intend to restore my good name and my honour as a general with France's army in this, the coming clash between the two Emperors who at present divide Europe . . .'

For at least two years Russia had been breaking the Continental System while openly proclaiming 'eternal friendship' with France. Her merchants found a hundred and one different ways of trading with Great Britain, but the Czar Alexander still sought Napoleon's agreement to his expansion in the Balkans, towards Constantinople and—which the Emperor could never accept —further into Poland. Meanwhile behind this two-faced attitude great efforts were being made to reforge the Russian army (devastated at Austerlitz and Friedland) with the idea of striking at the French once their hold upon Central Europe showed its first signs of weakening. At Tilsit, Alexander, completely defeated, had been overawed by Napoleon; since when he had harboured a growing resentment against his 'Imperial' brother. He affected to believe in liberal values, and gave this as his reason for wishing to see Napoleon's supremacy terminated. But he regarded submission to the Continental System as an affront to his honour, while on the darker side of his mind he persistently dreamed of conquest.

As for the Emperor, by the 'System', Philippe de Ségur tells us, 'he had declared eternal war against the English; to that system he attached his honour, his political existence, and that of the nation under his sway.' On the other hand, he could not successfully maintain it without the unanimity of the Continental nations; in other words by their keeping to the terms of the treaties he had either made or forced upon them. Now, therefore, when France received dozens of reports of British vessels entering Russian ports, together with further intelligence concerning Alexander's immense new army, her leader's reactions were suitably predictable. 'Influenced by his position and urged on by his enterprising character, he filled his imagination with the vast project of becoming the sole master of Europe, overwhelming Russia, and wresting Poland from her dominion.'[2]

By 1812 the die was cast. Everyone in Europe knew the two giants were going to measure their strength against one another. On 27 January Napoleon sent a list of his grievances against the Czar to his partners in the Confederation of the Rhine. Then he set about assembling the army of over half a million men he considered necessary for an expedition towards Moscow. Many of the regiments were already complete, and the Imperial Guard had been expanded and put into a state of readiness. Even so, the preparations took him a further six months. For he was determined to create a truly 'European' army, involving all of his partners and even certain unwilling allies. By May he had about 450,000 men. Of these, 'about half the infantry and one third of the cavalry were foreign, and they included Spaniards, Poles, Swiss, Croats, Portuguese, Württembergers, Italians, Dalmatians, Bavarians, Saxons, Illyrians, Neapolitans, Prussians and Westphalians'.[3] Macdonald's mainly-Prussian X Corps would be guarding the left flank (by the Baltic and into Lithuania). While the right flank was covered by a separate Austrian army of 40,000 men under Prince von Schwarzenberg. This left the King of Denmark to garrison the rear, and France herself to be defended by a thinly-spread umbrella of the Young Guard, a few line regiments, the National Guard and even customs-officials who were holding rifles in their hands for the first time. Only the armies fighting in the Peninsula were not drawn upon.

Ney received his orders during the last week of January. Eventually he, Davout and Oudinot would be called upon to produce the main, central thrust of the *Grande Armée* against the Russians commanded by Barclay de Tolly, Bagration and Tormassov. First of all though he was ordered to Mainz, with the immediate task of whipping III Corps into shape: thirty-five thousand infantry with two thousand four hundred cavalry. Again his bilingual ability came in useful, because many of these troops were German. By the end of May he confidently expressed the opinion that III Corps would prove the equal of VI Corps when called upon to give battle.

Already they were beginning to move across Germany and into Poland. On 29 May they were at Thorn on the eastern bank of the Vistula. From there they proceeded towards the River Niemen which since Tilsit had become the Russian border, and on 22 June (at a farm called Nogurisky) they listened to Napoleon's official declaration of hostilities:

Soldiers, the second Polish war has commenced. The first was concluded at Friedland and at Tilsit. At Tilsit, Russia swore eternal alliance with

France, and war with England. She now violates her oaths. She will give no explanation of her capricious conduct until the French eagles have repassed the Rhine; by that means leaving our allies to her mercy. Russia is hurried away by fatality; her destiny must be accomplished. Does she believe us to be degenerate? Are we not still the soldiers of Austerlitz? She places us between war and dishonour; the choice cannot be in doubt. Let us advance then; let us pass the Niemen and carry the war into her territory! The second Polish war will be as glorious for French arms as the first; but the peace we shall conclude this time will carry its own guarantee. It will put an end to the fatal influence which Russia has exercised over the affairs of Europe for the last fifty years.[4]

At ten o'clock the next evening the first crossings were made. General Morand moved three companies of infantry over the river in boats, and during the night Eblé's engineers constructed two wide pontoon bridges. By mid-morning on 24 June the immense central phalanx of the *Grande Armée* was streaming across, 325,000 men altogether, with Murat's cavalry in the lead and Poniatowski's squadrons of Polish lancers serving as reconnaissance patrols. Lieutenant Heinrich Vossler, a Württemberger under the command of Brigadier-General Ornano, tells us 'it was a glorious morning, but in the afternoon a violent thunderstorm began to gather and soon the rain came pelting down in buckets'.[5] Napoleon promptly took shelter in a convent. Then he galloped on to join his vanguard at Kovno where the bridge was down and several Poles were drowned in the swollen Vilia before a crossing could be made. The town was occupied without opposition, but found stripped of everything. Meanwhile no sign of Barclay's army. Just a group of Don Cossacks who broke and fled into the nearby forest. It was a first, ominous sign of Russian tactics. Of drawing the French on, extending their lines of communication, leaving the towns and villages empty or on fire and giving battle only in the last resort. This way, Barclay reasoned, disease, exhaustion and a general lowering of morale would do their fighting for them.

Ney crossed the Niemen on 25 June, with Prince Eugène's Army of Italy behind him and Prince Jerome's laggardly Westphalians still at Grodno. III Corps were being held in a support role, but this did not spare them the difficulties of the march itself. In fact, according to Ségur, it was usually the centre of the column (Ney and Eugène) which suffered most. 'It followed the road which the Russians had ruined, and of which the French advanced party had just completed the spoilage.' Often they went hungry because their own bulky supply-convoys fell so far behind. But even worse was the climate. Heinrich Vossler leaves us with a detailed description of this and its results as the army's centre passed through Kovno, Vilna and marched on towards Smorgoni:

Before crossing the Niemen . . . we had been thoroughly parched by the persistent, oppressive heat. Thereafter we endured three days of continuous torrential rain followed by alternating periods of unbearable heat and downpours the like of which I had never experienced. To sum up, our situation was this: we were embarked on a strenuous campaign entailing frequent forced-marches along abominable roads, either smothered in sand or knee-deep in mud and frequently pitted by precipitous ravines, under skies alternately unbearably hot or pouring forth freezing rain.[6]

On the days of excessive heat and choking dust there was also the problem of finding suitable drinking water. 'Our drink,' Vossler goes on to say:

Consisted—not even of inferior *spirits* or at least wholesome water, but of a brackish liquid scooped from stinking wells and putrid ponds. Under these circumstances it was not surprising that within two or three days of crossing the Niemen the army, and in particular the infantry, was being ravaged by a variety of diseases, chief among them dysentery, *ague* (malarial fever) and typhus.[7]

The horses too suffered from the impure water; and from the effects of being given green fodder. Already by Vilna hundreds of them had died.

Despite their growing sick-list Ney and Eugène succeeded in keeping the French centre on the move. In the event it was Jerome who let the Emperor down. Out ahead the Russians were still not offering battle, but Davout had managed to place his corps as a wedge separating the forces of Barclay in the north from Bagration's troops to the south. Moreover the latter were caught in a very tricky position, with difficult river-crossings dividing their units and marshland along their left-hand flank. If at this juncture Napoleon had taken the decision to switch Ney from the centre to strike south-eastwards, then he might well have caught and finished off Bagration. Or by detaching Oudinot from Davout he could have done the same thing. That he did neither was entirely due to his expecting Jerome to come up and accomplish the task. He had sent daily orders to the King of Westphalia 'to hurry along his advance on the southern sector'. Also, of course, by being free of the centre's baggage-trains, it meant that any army commander marching on the southern side could advance with far greater speed. However Jerome botched it. His messy crossings of the Niemen took four days. Then he tried to make up for lost time by forced-marches, which in the heat merely exhausted those troops who still remained fit.

Davout straddled the road at Minsk, waiting for Jerome to push Bagration towards him. At this moment, '40,000 Russians were cut off from the army of Alexander, and enveloped by two armies. Napoleon exclaimed, *they are mine.* And in fact only three marches were necessary to enclose Bagratian within an impassable ring.'[8] Jerome, alas for the French, was not even in touch. Bagration, having found the Minsk route blocked began to make his escape over a long and narrow causeway 'across the marshes of Shlutz, Glusck and Bolruisk'. Even now it was not too late to fall upon his rearguard. Napoleon, very angry with Jerome, placed him under Davout at this point and the Marshal requested the King of Westphalia 'to move up and push the Russians briskly into the defile, the issues of which at Glusck he was about to occupy'. Bagration was in a desperate situation still. 'But the King, already irritated by the reproaches which his uncertainty and dilatoriness had brought upon him, could not suffer a subject to be his commander; he left his army, without anyone to replace him, or without even communicating, if one is to believe Davout, to any of his generals the order which he had just received.'[9] As a result, over the next two days, and by fighting two or three successful skirmishes while his main force walked two abreast across the causeway, the lucky Bagration escaped towards Smolensk and the opportunity to become part of another Russian front.

Ney nearly saw some action in the week which followed. There were still Barclay's forces to the north at Drissa, and Napoleon—incensed at Jerome's incompetence and defection—ordered III Corps to support Murat's cavalry in a push over to that direction. Again though the Russians withdrew, this time south-east to Vitebsk. Ney (supported by Oudinot) thought he had caught them halfway along the road to Vitebsk; but it proved to be just the rearguard, and his movements were confused by a storm 'which flung at them marbles of ice capable of denting breastplates'. He would not, as it turned out, have a real *go* at the Russians until the affair near Krasnoe, preparatory to the capture of the ancient and holy city of Smolensk.

By this time Napoleon had ordered III Corps south again, with instructions 'to proceed towards Smolensk' where the forces of Barclay and Bagration 'had at last become united'. Ney's troops were intended to become part of a grand manoeuvre, not only swinging away from Vitebsk and seizing Smolensk but also encircling the Russian army while it was still in one place.

> On the night of (August) 13th-14th, General Eblé completed the throwing of four pontoon bridges over the Dnieper near Rosasna and . . . by dawn, no less than 175,000 troops were safely across. The weather was dry, the roads were good, and by three in the afternoon the leading elements had reached the town of Krasnoe, some 30 miles west of Smolensk, and there encountered the first signs of Russian opposition.[10]

Ségur, now Napoleon's *aide-de-camp*, takes up the story:

> On the 15th of August, at three o'clock, we came in sight of Krasnoe, a town constructed of wood, which a Russian regiment made a show of defending; but it detained Marshal Ney no longer than the time necessary to come at and overthrow it. The town being taken, there were seen beyond it 6,000 Russian infantry in two columns, while several squadrons covered their retreat. This was the corps of Neveroski.
> The ground was unequal, but bare, and suitable for cavalry. Murat took possession of it; but the bridges of Krasnoe were broken down, and the French cavalry were obliged to move off on the left, and to defile to a great distance across bad fords, in order to come up with the enemy. When our troops were in sight of the latter, the difficulty of the passage they had just left behind, and the bold countenance of the Russians made them hesitate; they lost time in waiting for one another and deploying, but still the first effort dispersed the enemy's cavalry.[11]

This seems to have been the moment when Murat completely lost his head.

Instead of allowing III Corps to move up through the cavalry and attack the Russian infantry, now formed into a huge square, he deliberately blocked their movement and set about attacking the square himself. According to David Chandler, he launched no less than forty piecemeal cavalry charges—all without success—while keeping Ney's men bottled up in the narrow Krasnoe defile. Eventually, after much urgent pleading, a way was cleared for the Marshal to bring up his artillery and this did the trick. His first salvoes cut a great gap through the Russian square and immediately Neveroski ordered a retreat. Ney's aides counted 1,200 Russians dead, they had 1,000 prisoners and had captured eight cannon. But it could hardly be called a French success; and Ségur adds that although fought on the Emperor's birthday there were few celebrations. 'In our situation, there was no other festival than the day of a complete victory.' As an afterthought Ney fired a 100-gun salute, and then hurried off to explain to Napoleon 'that he was merely using up captured Russian powder'.

The occupation of Smolensk on 18 August, after two days of endeavouring to take it by storm, was seen as much more of a victory. But again the main Russian forces escaped: and the psychological value of holding the city did nothing to lift the Emperor's mood of frustration at their thus eluding him. Ney, Davout and Poniatowski had led bayonet charges on the 16th to clear the outer suburbs. Following which they set up artillery to blast their way through the mediaeval ramparts into the citadel. Barclay moved out under cover of darkness on the 17th, covered by a very thin rearguard and at the last moment setting fire to his stores. It was these fires, and a burning bridge over the Dnieper which at 2.00 a.m. on the 18th alerted a sleepless Ney to what was happening. With his neck bandaged up where a bullet had grazed him at Krasnoe, the Marshal strode off to wake Napoleon and deliver the bad news.

At a conference later that morning Ney, Davout, Mortier, Grand Marshal of the Palace Duroc and General Lobau all advised the Emperor not to proceed beyond Smolensk. Their supply-lines, they pointed out, were now dangerously extended—while the Russians continued to burn or otherwise destroy everything they had to abandon. But Napoleon would not listen. 'The Russians are women,' he snapped, 'they acknowledge themselves vanquished!' And he outlined the next moves. They would begin their pursuit the following morning, this time with Ney in the lead and Murat's cavalry divided to cover him. If the Russians still refused to fight, well, in less than a month they could be in Moscow; and in six weeks the Czar would be suing for peace. Having made these intentions clear, and without waiting to hear their individual views, he then mounted his horse. Nevertheless as he rode off Duroc expressed what each person there had been thinking: 'If Barclay has committed so very great a blunder in refusing battle, the Emperor would not be so extremely anxious to convince us of it.'[12]

Ney almost gave the French their longed-for field victory on the first day of his pursuit. Leaving a Smolensk still in flames on the morning of 19 August, he proceeded with caution because the terrain was divided up by the Dnieper's many tributaries. There was a brief skirmish to take the defile of Stubna, but then quite unexpectedly he came upon Barclay's rearguard and centre trying to extricate their baggage and equipment from the defile of Valoutina on to the plateau beyond. Ney had 12,000 men with him. He promptly attacked with these and dashed off a request for reinforcements.

That these did not arrive in sufficient quantities in the end lost the French their chance. Ney's men fought with the utmost determination, engaging more and more Russians and gradually, gaining a yard at a time, they debouched upon the plateau. This allowed Murat to bring up some of his horsemen to operate on the two wings and during the afternoon Davout arrived with a further 8,000 men. But the larger designated reinforcements, the Westphalians (now commanded by Junot) failed to materialise. General Gudin had just received a mortal wound, and Ney was holding on with his last ounce of courage. In a fury, Murat left his cavalry 'and crossing the woods and marshes almost alone . . . hastened to Junot . . . upbraiding him for his inaction'.

Junot put forward his excuses: 'I am waiting for new orders . . . my cavalry are not ready . . .' '*Nonsense!*' Murat shouted. 'Go in and finish this business: your glory and your marshal's baton are still before you!' Junot did nothing though. 'Too long about Napoleon, whose active genius directed everything, both the plan and the details, he had learned only to obey: he lacked experience in command; besides, fatigue and wounds had made him an old man before his time.'[13]

By nightfall Ney and Murat were forced to admit that their units had reached the point of exhaustion. They had certainly hurt Barclay, but in the course of the night the Russians were able to proceed with their retreat. The Emperor rightly blamed Junot for this lost opportunity—'he has spoiled our whole campaign'—but he himself was not entirely blameless. At 5.00 p.m. feeling tired, he went back to Smolensk to rest: just at the moment when his presence was required to urge III Corps and Murat's cavalry on to their final effort.

Admittedly the following morning he did his best to make up for it. He galloped off towards Valoutina with a case of decorations:

> Ney's troops, and those of Gudin's division, deprived of their general, had drawn up there by the corpses of their companions and the Russians, among the stumps of broken trees, on ground trampled by the feet of the combatants, furrowed with balls, strewed with the fragments of weapons, tattered garments, military utensils, carriages overthrown, and scattered limbs; for such are the trophies of war, such the beauties of a field of victory!'[14]

Gudin's battalions had been cut down to mere platoons. Napoleon 'could not pass along their front without having to step over, or to tread upon carcasses and bayonets twisted by the violence of the shock'. He realised it was high time to forget his own chagrin and bestow words of praise and rewards.

'Never, therefore, were his looks more kind; and as to his language, he proclaimed: *This battle was the most glorious achievement in our military history.*' On the 12th, 21st and 127th regiments of the line and the valiant 17th (Gudin's men) he conferred no less than eighty-seven decorations and promotions, plus gratuities for all who had taken part in the conflict. Also, until this engagement the 127th had marched without an eagle, for by an Imperial decree not only was it necessary for a regiment to earn its colours in battle, but it had to prove in a second battle that it knew how to preserve them. The Emperor demanded an eagle from his aides and delivered it to the last surviving senior officer of the 127th with his own hands.

Ney's corps was next singled out for favour. Napoleon gave out the awards as if in Republican days:

> He was successively surrounded by each regiment as by a family. There he appealed in a loud voice to the officers, subalterns, and privates, enquiring who were the bravest of all these brave men, or the most successful, and recompensing them on the spot. The officers named, the soldiers confirmed, the Emperor approved: thus, as he himself observed, the elections were made instantaneously, in a circle, in his presence, and confirmed with acclamations by the troops. These paternal manners, which made the private soldier the military comrade of the ruler of Europe . . . delighted them. He was a monarch, but the monarch of the Revolution. In him there was everything to love and he did nothing to reproach them.[15]

Just as he was distributing the last promotions so a staff officer rode up to say that Marshal Ney was fit and well; that he had in fact been out on reconnaissance that morning and now awaited the Emperor's presence at lunch on the far side of the battlefield. When Napoleon arrived there he found the table had been laid in a little shelter made of branches. But the food was superb—'the best since Paris, and from where I don't know!'—after which the Marshal suddenly shouted 'to horse' and dashed off on another reconnaissance. Napoleon watched him go with something like wonder. 'These have been the finest feats of arms in French history,' he murmured, '. . . and in this heat! With such soldiers a man could conquer the whole world.'[16]

But when at last the French obtained the great battle which their leader so desperately wanted it was at a time and in a place of the Russians' choosing. Technically Borodino was a victory for France and once more Michel Ney covered himself with glory. But it took the form of an assault—or rather a series of assaults—upon the most formidable defensive position the *Grande Armée* had ever encountered; resulting in heavy casualties. Also, and due to a serious error on Napoleon's part, the victory was then rendered inconclusive. It could, in fact *it ought* to have put an end to the campaign: crushing Russian resistance and forcing the Czar Alexander to seek yet another humiliating peace-treaty. Instead of which, by a single wrong decision Borodino became the prelude to an overwhelming defeat for French arms and the decline, leading to the fall, of Napoleon's Empire.

By the second half of August the expeditionary forces were not the only ones anxious for a decisive battle. The Czar and his entourage had been horrified by Barclay's tactics of retreat, and abandoning sacred Smolensk proved to be the last straw. Barclay was retained, but the overall command and strategy now passed to Mikhail Golenischev Kutuzov: seventy years old, very fat from overeating, frequently prostrate after his bouts of drinking and the hero of Russia's wars against Turkey. The oddity of this particular appointment is that Kutuzov was by inclination just as much a 'retreating' general as Barclay; at least whenever Napoleon and the French were facing him. In fact he had never before allowed himself to come to grips with the *Grande Armée*. However, he was more obviously *Russian* than Barclay, a professional soldier of Scots and Lithuanian extraction. Also in taking up the command he did promise Alexander a battle to save Moscow.

His choice of area near the little village of Borodino was highly predictable. Both photographs and contemporary engravings show it to be a beautiful, rolling countryside of open spaces interspersed with woods of birch and pine, small streams and—to the south—a certain amount of marshland. Of real importance to Kutuzov though: it lay *between* where the old and new Smolensk roads converged at Mojaisk; while down its eastern side, behind a tributary of the River Moscowa called the Kalotcha, ran a broad ridge featuring several dramatic outcrops. These, 'the old fox of the north' judged, once fortified would be impregnable. And any force attacking them from the west would become the perfect target for good artillery-fire.

Immediately therefore he began to convert the site. Towards the northern end of the ridge, on the steepest outcrop, he placed a large quantity of cannon. Their arc of fire was designed to interlock with that from batteries near the village of Gorki and with the fire from Bagration's positions in the centre and to the south. Around the guns were two sideways *épaulements* (or breast-works), and two long, forward-slanting parapets which met in a 'V' of one hundred and sixty degrees. The position became known as the Raevsky Redoubt after the general who mainly defended it. It was open at the rear, but in front further protected by a 22-foot wide ditch and as the ground fell away it offered the defending gunners every possible advantage.[17]

Bagration's sector had fewer natural advantages, and in the end its defence would come to rely very heavily on the stubbornness of the Russian troops. The outcrops were smaller here; but three of them had been crowned with *flèches*, or arrow-shaped fortifications, and further south still the village of Utitsa was garrisoned. Finally, in front of the whole position, Kutuzov had constructed an artificial, earthen redoubt on a low mound near the village of Schivardino. In its outlying position, this could only be regarded as a buffer, although if held it meant the attackers would have to operate on an extremely narrow front, bottle-necked between the Kalotcha and a wood that was almost impenetrable.

After inspecting his men's handiwork on 3 September, the Commander-in-Chief felt able to write to the Czar that 'the position I have taken up at Borodino four miles from Mojaisk is one of the best you could find in this weak terrain. I have resorted to artifice in order to remedy the one weak sector of the line which lies to the left. I only hope the enemy will attack us in our position: if they do, I am confident we shall win.'[18] And with this prediction he returned to his hampers and his vodka . . .

The French, meanwhile, were now approaching their enemy at an increasingly rapid pace. The terrible heat was beginning to turn into rain (and on the day of the battle itself a howling gale blew). But at least they had got their order of marching sorted out, and they met no opposition other than the hit-and-run Cossacks. Michel Ney was back in the 'rest' position, with Davout marching ahead of him, Prince Eugène to his left and Poniatowski on the right. On the whole too the men marched more easily because they had reduced their necessary amounts of equipment.

Ahead of everyone though, even his own cavalry, galloped the *Grande Armée's* principal extrovert, Joachim Murat. Undeterred by a series of tactical quarrels with Davout, the King of Naples was determined to corner the applause. 'He looked as if he had just raided a theatrical wardrobe, with his great golden spurs, huge riding boots *à la* Gustavus Adolphus, his sumptuously-embroidered jacket of light blue, and his cascading black ringlets surmounted by a white-plumed Jacobean hat.'[19] An entire cart was needed to transport his spare uniforms, pomades and toiletries. Little wonder that the taciturn Davout looked upon him with distaste. On the other hand, given the right ground for manoeuvring or leading a charge he remained one of the best cavalry officers of his day.

'Murat's Cossack enemies loved him for his panache, and lived in hopes of taking him alive.' Not that there was much chance of this. 'For day after day they hovered just in front of the advance guard, clearing and burning the villages as they went in a ruthless application of scorched-earth tactics. The first sign of the work of the incendiarists was usually a column of thick black smoke which climbed perpendicularly towards the sky, then flowed away in the direction of the wind; thereupon other columns of the same kind welled up in succession and soon the flames would spread through the entire village, consuming it in less than a quarter of an hour.'[20]

The French army arrived at Borodino on the afternoon of 5 September, at which point the Cossacks suddenly disappeared; away behind the 18,000 Russians who surrounded and were putting the finishing touches to the Schivardino Redoubt. At last they were going to stand and fight then! Napoleon ordered a general halt while the rest of his troops came up (in all making 124,000 infantry, 32,000 cavalry and nearly six hundred artillery crews). In the meantime, and ordering Murat to regroup the cavalry, he rode forward himself to survey the position. Particularly what was happening around the forward redoubt.

It wasn't much of a barrier, he suspected. But with his experienced eye he noted how it could restrict his front. Also it prevented normal reconnaissance and the opportunity to survey what Kutuzov had laid out behind. Accordingly, he decided it must be taken before any assault was conducted upon the main enemy positions. And the sooner the better. Ney and the other commanders had come up to join him. Immediately they noticed how the Russians had followed the old Prussian tactics at Schivardino of putting 'all their troops in the shop-window'. Whatever boost to morale they gained from this courageous stance, nevertheless they presented excellent targets for the French gunners. Napoleon agreed, and this determined the start of the action.

Even as the *Grande Armée* was concentrating on the plain south of Borodino, so Poniatowski's V Corps of Poles had begun a flanking move south of Schivardino, while the French artillery opened up prior to General Compans' division of I Corps making a frontal assault. The redoubt was defended with great bravery by the Russian 27th Division under Prince Gorchakov, and its subsequently being taken was certainly no walkover. But the first salvos cut wide gaps through the standing defenders.

'Compans skilfully availed himself of the undulations of the ground,' Ségur informs us:

> Its elevations served as platforms to his guns for battering the redoubt and screened his infantry while drawing up into columns of attack. The 61st marched foremost; the redoubt was taken by a single effort, and with the bayonet; but Bagratian sent reinforcements by which it was retaken. Three times did the 61st recover it from the Russians, and three times were driven out again.

Whereupon Poniatowski came storming up the thinly-held flank and Gorchakov gave in.

Ney had no part in this fight. He was busy drawing up his men and supervising their bivouac to the south. III Corps had now recovered from its fierce fight at Valoutina and we have specific French army records of its composition at Borodino. 10th Division (commanded by General Ledru) consisted of sixteen battalions including two Portuguese and two German. 11th Division had eighteen battalions including two more Portuguese and four of Illyrians under General Razout. 25th Division was mostly the Royal Württembergers commanded by General Marchand. Attached were two brigades of cavalry, one French, one German under General Wöllworth and the corps' artillery under General Foucher. Ney was perfectly satisfied with everything he saw as he reviewed their lines, and after Valoutina he had no reason to doubt their fighting ability.

The next day (6 September) was one of monumental, but deceptive calm. Hardly a shot was fired; and yet Napoleon had risen at dawn, examined the Russian positions several times through his telescope and then spent the rest of an exhausting day deploying his troops for a massive assault on the Russians' left-wing the following morning. At the same time the Russians were making prodigious efforts to add to the Raevsky Redoubt, over on their already stronger right-wing.

To deal with the French first. Napoleon was far from well. He had a septic throat, an upset tummy and was finding it difficult to pass water: a recurrence of his old bladder complaint which always left him feeling torpid and depressed. Yet he didn't want to let the Russians give him the slip again; and so he drank some hot punch, rode out to scrutinise the terrain and the enemy positions, then drew up his plan of attack.

He promptly rejected the idea of a frontal assault upon the Russian right, Ségur states, because of the way the Kalotcha curled round to protect it. Instead he ordered Prince Eugène (supported by Grouchy's cavalry) to bridge the river and capture the village of Borodino. From there they could move on to shell the Russian flank; and if all went well even recross the Kalotcha and begin to fight their way towards the massive redoubt (the Raevsky) where Kutuzov's right-wing stretched towards Bagration's positions. For it was over against the latter the Emperor was resolved to mount his main attack. The

Raevsky, as he rightly decided, would be the toughest nut to crack—but much easier once the French could work their way around it. 'Eugène shall be the pivot!' he exclaimed: 'It is his side that must commence.' Then the big punch would be delivered against the Russian left. A colossal artillery barrage first 'to open their ranks and redoubts' before Davout and Ney rushed upon them, 'supported by Junot and his Westphalians, by Murat and his cavalry, and lastly by the Emperor himself, with 20,000 Guards'.[21] At the same moment Poniatowski with 5,000 men would advance down the old Smolensk road to take Utitsa, 'turning the wood on which the French right wing and the Russian left were supported. He would flank the one and annoy the other. . . .'

By late afternoon Russian observers must have noticed the first signs of this development. It couldn't all be left until the cover of darkness. But if they informed Kutuzov then he did nothing. Instead of Bagration and Utitsa being reinforced the only activity was by Barclay in response to a message from Raevsky on the big redoubt. Raevsky had decided the redoubt was vulnerable to cavalry, whereupon Barclay despatched all his pioneers to lend a hand:

> Raevsky had already dug a chain of bone-breaking wolf pits a hundred paces in front of the battery, but he told Lt. Bogdanov: *there is still one very important thing that will have to be seen to; while we are holding out in the battery the enemy could easily sweep around our flank and take the fortification from the rear. We must put more formidable obstacles in their path.*

Bogdanov got down to work immediately.

> He extended the parapet laterally by two twenty-five-yard-long breast-works, one on either side. Using timber and spikes from dismantled houses, he planted a double palisade around the rear of the battery; the inner palisade rose vertically to a height of eight feet, while the outer one sloped menacingly outwards for six and a half feet. Gaps were left at the two ends of the palisade to permit troops to pass between the palisade and the breastworks.[22]

As evening came on Raevsky carried out his final inspection. 'Now, gentlemen,' he announced to his officers, 'we may rest secure. When daylight comes Napoleon will espy what seems to be a simple open battery, but his army will come up against a virtual fortress. The approaches are swept by more than two hundred cannon, the ditch is deep and wide and the glacis (bank) solid.'

7 September . . . and already a high wind was blowing. The breaking of a dawn which would soon see 260,000 men locked in what Kutuzov later called 'the most bloody battle of modern times'. Before the world saw another dawn 94,000 men and horses lay dead; while the survivors—after fighting for almost fifteen hours—would remember how 'the ground trembled for a dozen miles around'. It was, in one soldier's words: a fair idea of what hell must be like. . . .

Ney shared a pre-dawn breakfast with Davout; then they wished each other *good luck!* and by five o'clock were in position with their troops. Napoleon's proclamation had to be read out by torchlight: a difficult thing in a strong wind:

> Soldiers! Here is the battle which you have so ardently desired. The victory will now depend upon yourselves; it is needful for us; it will give us abundance, good winter-quarters, and a speedy return home! Behave as you did at Austerlitz, at Friedland, at Smolensk and afford to remotest posterity occasion to cite your conduct on this day: let it be said of you, *He was in that great battle* under the walls of Moscow![23]

The Emperor himself had been up since three, after a troubled, almost Shakespearean night. First he was anxious lest the Russians should escape him again. Next he insisted on sending for Marshal Bessières, to find out whether his precious Guards were comfortable. Shortly after this he sent for Ségur, who found him supporting his head with both hands. *What is war?* he muttered. *A barbarous profession whose art consists in being stronger than the enemy at any given moment!* 'He then complained *of the fickleness of fortune,*' his *aide* recalls, '. . . which he said, he began to experience.'[24]

At three he finally dressed, called for more hot punch, and endeavoured to shake himself out of this gloomy mood. He recognised the needs of the day; and when General Rapp arrived at four a note of optimism had returned to his speech. 'Today we shall be at grips with the notorious Kutuzov. You must remember him during the Austerlitz campaign. He stayed three months in the same room in the same fortress. Didn't even get on his horse to tour the fortifications!'

At last, 'at five o'clock, one of Ney's officers came to inform him that the Marshal . . . requested permission to begin the attack. This news seemed to restore his strength. He arose, called his officers and left the tent announcing: *We have them at last! Forward! Let us go and open the gates of Moscow!*'

On the strokes of six the 'softening up' volleys thundered out. First, from the Guard batteries on the extreme right, then all along a line of 100 guns in front of I and III Corps. It was the signal for Davout to put in the initial hammer blow against Bagration. Ney for the moment remained in a position more or less facing the end of the Raevsky Redoubt. Meanwhile the sound of gunfire to the south indicated that Poniatowski was on the move.

However, the day's first success went to Prince Eugène. Over on the extreme left his General Delzons took the village of Borodino with only one volley from the horse-artillery and a bayonet charge, whereupon Eugène placed twenty guns near the river and soon could even reach the northern side of the Raevsky Redoubt with his fire. Which was just as well, because in a bloody clash on the other side of Borodino the pugnacious Davout had suffered one of his rare repulses. To be fair, he had severely mauled the Russian left and is not always given credit for starting the job which Ney eventually finished on that side of the field. But Bagration's men fought like demons. After taking a terrible pounding from Davout's forward cannon they then recovered sufficiently to catch I Corps with a storm of canister fire. The

French infantry marched without firing: it was hurrying on 'to get within reach of and extinguish that of the enemy' when the canisters burst. Compans, the leading general, fell; so too did Rapp, although he reached the nearest fortifications and even occupied them for several minutes. Worst of all, however, Davout's horse was killed under him and in the resulting fall the Marshal was severely concussed. His being carried from the field—thought to be dead—represented a crisis in the battle, and cancelled out the good news from the south where Poniatowski, after a hard fight, had taken Utitsa.

Ney now entered the fray for the first time—overcoming the crisis, seizing the initiative and occupying a position of prominence which he did not relinquish until the battle was over. (A prominence, incidentally, that was also visual: Napoleon having ordered him to fight in his dress-uniform and to wear his decorations. Incredibly, although an obvious target, he managed to get through the day without attracting a sniper's bullet.) Upon the Emperor's urgent request he dashed across the plain with III Corps to help:

> whereupon the 57th (regiment) of Compans' Division, finding itself supported, took fresh courage; by a last effort it succeeded in reaching the enemy's entrenchments, scaled them, mingled with the Russians, put them to the bayonet, overthrew them and killed the most obstinate of them. The rest fled, and the 57th maintained itself in conquest. At the same time Ney made so furious an attack on the other two redoubts (Bagration's *flèches*) that he wrested them from the enemy.[25]

Such was the power of *Le Rougeaud* to put fresh heart into the troops!

It was now late morning. But this wasn't the end of the Russian left. Murat moved forward with his horsemen to clinch the affair: and ran straight into Bagration's counterstroke . . . which caught the French 'in the disorder of victory'. The King of Naples was forced to throw himself into one of the redoubts, 'where he found only some unsteady soldiers whose courage had forsaken them . . . running round the parapet in a state of panic'. Bagration had thrown the whole of his reserves forward: eleven fresh regiments, including General Duka's *cuirassiers*; while the French also came under murderous fire from the Raevsky Redoubt. The *flèches* changed hands several times, mostly in fierce hand-to-hand combat, 'with the cavalry, infantry and gunners all thrown together . . . and striking out with bayonets, musket butts, swords and rammers, trampling the fallen underfoot'.[26] But eventually Ney's infantry got, and this time *kept* the upper hand—even those French and Germans who had hesitated and seemed likely to run, turning and fighting on with a new spirit. Murat operated with great skill using small formations in and around them. Also Poniatowski's contingent fought its way around the wood from Utitsa to join in.

The struggle for the *flèches* raged altogether for five hours; and on several occasions, in recognition of French bravery the chivalrous Bagration clapped his hands and shouted *Bravo!* However, just before 11.00 a.m. the Russian commander was wounded (fatally) in the leg and hip. Following which his troops' will to resist 'simply petered out'. At long last Ney and Murat could switch their attention and resources to the awesome problem of the Raevsky

Redoubt—trying to make up for Prince Eugène's shock defeat there.

Eugène had crossed the river again, silenced the batteries near Gorki and advanced upon the redoubt in good order. Despite a hail of canister the French pushed upwards on the northern side and climbed in through the embrasures. Then, after a hand-to-hand combat, for ten glorious minutes they actually found themselves in possession. Ten minutes! In fact, for as long as it took Barclay's reserves to fall upon them via the two back entrances and hurl them in red ruin out and down the hill. Barclay estimated that the attack had caused France over three thousand dead and wounded.

So the responsibility for taking the redoubt now passed to Ney and the King of Naples—plus whatever help Napoleon could send. Their own troops in the Bagration *flèches* were elated and clearly prepared to fight again—but the next time they would tire more quickly. Urgently, therefore, Ney requested the Emperor to give him support; particularly by moving up the Imperial Guard. Meanwhile he began to make use of every piece of artillery he could lay his hands on.

Except during Eugène's 'ten minutes' the Raevsky gunners had had something of a field day, especially against the massed cavalry. (Their latest victim was General Montbrun. One of their shell splinters hit him full in the stomach. 'A good shot!' he cried, then sank dying from the saddle.) But Ney transformed the situation. In spite of the difficult elevation, and with some assistance from Prince Eugène on the other side, the French gunners now opened up with at least 150 pieces. The Russians were forced to sit inside the redoubt, suffer terrible casualties—and wait.

Ney's assault force went in shortly after three o'clock, using up the whole of I and III Corps' infantry and again Murat's cavalry. Still no arrival of the Guard though. Napoleon, when General Belliard galloped up with the request, had hesitated; and then consulted Bessières, a personally brave, but overcautious marshal with limited experience of command. Belliard, on Ney's behalf, pleaded with the Emperor. Even if the Guard did not join the final assault on the redoubt, at least let it swing around and catch the Russians from behind. From Ney's present position, he declared:

> . . . the eye could penetrate without impediment as far as the road to Mojaisk; at the Russian rear they could see there a confused crowd of flying and wounded soldiers, and carriages retreating; that it was true there was still a ravine and a copse between them, but that the Russian generals were so confused, that they had no thought of turning these to their advantage; that in short, only a single effort was required to arrive in the middle of that disorder, to seal the enemy's discomfiture, and terminate the war!'[27]

The Emperor, however, continued to prevaricate; and in the end—because he was ill and no longer his original decisive self—he took Bessières' advice to be prudent. Belliard went back empty-handed—also full of alarm at Napoleon's 'suffering, dejected air, his features sunk, his dull look; giving his orders languidly'. To Ney the Imperial message was singularly unhelpful: 'Nothing is yet sufficiently unravelled: before committing my reserves I need to see more clearly what is left upon the chessboard. What if I have to fight again tomorrow?' It was a message guaranteed to lose France the war.

Upon Ney, embattled at the very foot of the big redoubt, its effect is well-known: since it provoked his first and most celebrated outburst against the Emperor. His longstanding idol had turned into a figure of clay! 'Have we come all this distance to be satisfied with one battlefield?' he yelled at Murat. 'What's His Majesty doing *behind* his army? He doesn't see any of our successes there—only our reserves. Since he no longer makes war himself, and isn't the general any more, but wants to play the Emperor everywhere, why doesn't he go back to the Tuileries and let us be the generals for him?'[28]

For once though the hot-headed Murat was the calmer of the two. He recollected having seen Napoleon the day before, while observing the enemy's line, halt several times, dismount, and with his head resting upon the cannon, remain there for some time in an attitude of suffering.

> The King guessed that fatigue and the first attacks of the equinox had shaken his weakened frame, and that in short, at that critical moment the action of his genius was in a manner chained down by his body; which had sunk under the triple load of fatigue, of fever, and of a malady which, probably more than any other, prostrates the moral and physical strength of its victim.[29]

When he explained this Ney too calmed down and side by side, facing the worst of the gale as well as the enemy's fire, they pushed on with their conquest of the redoubt.

Soon 'a dull cheering told the Russians that they had burst over the rampart and were going to work with the bayonet'.[30] At the same time the best riders of the Cavalry Corps scrambled up the breastworks, while others forced their way through the embrasures. 'The cramped interior of the redoubt was quickly filled with a terrible press of cavalry and infantry, thrown pell-mell together and doing their best to throttle and mangle one another.' (Von Meerheim's account) Within minutes there were dead and mutilated men and horses lying six or seven deep. It was the most severe test the *Grande Armée* had ever undergone in a frontal-assault upon a set position, but they came through it magnificently.

Once in full possession of the redoubt it then became Prince Eugène's task to collect all the remaining French cavalry and hurl them against Barclay's own cavalry and his battered infantry divisions on the plain behind. This both protected Ney and Murat in the redoubt from a Russian counterstroke and was intended to cut off Barclay's line of retreat. The opposing horsemen swirled around and struck at one another for the best part of two hours. At times it was impossible to distinguish friend from foe in the clouds of dust. But Barclay's infantry got away—and the French were too exhausted to follow.

Kutuzov, 'drinking champagne and surrounded by a knot of rich shirkers' at first refused to believe that Barclay was in retreat. To the messenger, Wölzogen, he exclaimed: 'You must have been getting drunk with some flea-bitten sutler to bring a report like that! We have victoriously repulsed the

French attacks along the whole length of our front, and tomorrow I shall place myself at the head of the army and drive the enemy without more ado from the sacred soil of Russia.'[31] By 7.00 p.m. he was back inside his coach, fleeing towards Moscow.

And so, in the most gory fashion Napoleon's entry into the revered capital of his enemy had been assured. The French had suffered at least 30,000 dead and wounded, the Russians more than 50,000. But the victors took few prisoners: according to Ségur 'from seven to eight hundred . . . and twenty broken cannon were all the trophies of this imperfect victory'. Over 90,000 Russians escaped to fight another day. Had the Emperor answered Ney's call to employ the Guard against them this would never have happened.

In turn though the Marshal found himself singled out as the hero of the battle. In recognition of his prodigious efforts he was given a new and more illustrious title: *Prince of the Moscowa.*

[1] Ney's third son, Eugène Michel, had been born in 1808. His fourth was christened Napoleon Henri Edgar.

[2] General Count de Ségur: *History of the Expedition to Russia, undertaken by the Emperor Napoleon in the Year 1812,* London, 1825.

[3] Col. H. C. B. Rogers: *Napoleon's Army,* Shepperton, England, 1974.

[4] *History of the Expedition to Russia.*

[5] H. A. Vossler: *With Napoleon in Russia, 1812,* trans. Walter Wallich, London, 1969.

[6] Ibid.

[7] Ibid.

[8] *History of the Expedition to Russia.*

[9] Ibid.

[10] *The Campaigns of Napoleon.*

[11] *History of the Expedition to Russia.*

[12] Ibid.

[13] Ibid. From Russia, Junot was appointed to the 'safe' governorship of Venice, where he went mad: on one occasion wandering about the streets naked apart from his epaulettes and sword. His death in 1813 was self-inflicted. After slashing his body with a sword, he then threw himself from a high window.

[14] Ibid.

[15] Ibid.

[16] Ibid.

[17] Christopher Duffy: *Borodino and the War of 1812,* London, 1972.

[18] Ibid.

[19] Ibid.

[20] Ibid.

[21] *History of the Expedition to Russia.*

[22] *Borodino and the War of 1812.*

[23] *History of the Expedition to Russia.*

[24] Ibid.

[25] Ibid.

[26] Col. A. A. Strokov: *Istoria Voennogo Iskusstva,* Moscow, 1865.

[27] *History of the Expedition to Russia.*

[28] Ibid.

29 Napoleon, whose indecision through sickness (dysuresis) had caused the victory of Borodino to be incomplete, recalled after the battle his words in Italy fifteen years before: 'Good health is indispensable in war, and nothing can take its place'. Also his words at Austerlitz: 'To command is to wear out! There is only one age for war. I am good for six years more. After that, I myself shall have to stop!'

30 *Borodino and the War of 1812.*

31 Ibid.

CHAPTER 7

Saviour of the Eagles

The retreat from Moscow

Ney. A proud name even amongst the many which adorn the history of France. His daring courage will be for ever the admiration of all Peoples who still preserve any national sentiment for the self-sacrificing soldier who counts his life as dross in comparison with the upholding of his country's honour. As we read of Ney's chivalrous conduct throughout this campaign we cannot help feeling what poor creatures many of Homer's fabulous heroes were when compared with him.

FIELD-MARSHAL VISCOUNT WOLSELEY
The Decline and Fall of Napoleon

'Apart from this everything goes well: we have not seen a woman since the postmistresses of Warsaw, but by way of compensation we are great connoisseurs of fires.' So Henri Beyle wrote to his patroness, the Countess Daru, on 16 October 1812.[1] The future 'Stendhal' was employed as a quartermaster with the *Grande Armée,* and as such would require all of his naturally sunny and optimistic disposition to cope with the privations ahead. But his 'black humour' on this occasion does not exaggerate the facts. As witnesses Stendhal and Ségur are like Dante and Virgil on a journey through their own national Inferno.

Fire and snow. The two forces of nature which combined to defeat the French army after the Russians had failed. It was fire that came first though . . . driving the invaders away from their glittering prize, into the deadly grip of a winter they were ill-equipped to face. Every field, farm, village, town and city: burned. No fodder left. No fresh food. Nothing at all that they might be able to use. And now Moscow had gone the same way: callously consigned to the flames.

In the aftermath to Borodino, Kutuzov promptly removed himself and the last of his troops to positions south and east of the capital, determined not to fight again unless he had overwhelming superiority. (Which meant drawing in the northern and southern defensive armies of Wittgenstein and Admiral Tchichagov.) The Czar removed himself to St.-Petersburg. Doubly irrational in his panic, his final order upon leaving was for 'a prodigious balloon' to be

131

constructed; 'not far from Moscow, under the direction of a German artificer.' The destination of this winged machine was to hover over the French army, to single out its chief , and destroy him with a shower of bails and fire. Several attempts were made to raise it, but without success, 'the spring by which the wings were to be worked having always broken. . .'[2]

Upon Alexander's departure both the executive power and the ultimate responsibility for the city were transferred to its governor, the sinister Count Rostopchin. It was he who would prepare Moscow for burning: a sacrificial offering to the ideal that Russia must never sign a peace of shame. It meant the destruction of the finest and richest of his own palaces, the great houses of his friends and the homes, goods and wealth of all the capital's merchants and traders. But Rostopchin was made of the stuff to do it. After the bad news from Borodino he ordered compulsory evacuation, removal or spoiling of everything edible and the gathering of 'a large quantity of squibs and combustibles'.

The French came within sight of Moscow on 14 September. It was two o'clock in the afternoon as they viewed it from the Mont du Salut (the Hill of Salvation) and the sun 'caused the city to glisten and twinkle with a thousand marvellous colours'. Halting for a moment with astonishment, they then hurried on in growing disorder, 'clapping their hands and repeating with transports of joy *Moscow! Moscow!* Just as sailors shout *Land! Land!* at the conclusion of a long and toilsome voyage'.[3] Even Napoleon could not repress his joy. 'So here at last is the famous city,' he exclaimed, drawing rein to acknowledge the congratulations of his marshals, 'well, it was high time!'

His eyes, fixed on the capital, also expressed impatience, Ségur tells us. 'In it he beheld in imagination the whole Russian Empire. Those walls enclosed all his hope—peace, the expenses of the war, immortal glory: his eager looks therefore intently watched all its outlets. When would its gates open? When would the deputation come forth, placing its wealth, population, senate and the principals of the Russian nobility at France's disposal?' But he waited in vain. There was no deputation. And not a word from the Czar. Moreover his soldiers, hastening forward, found the gates already open: but upon a city totally devoid of life. The silence unnerved them, so they began shouting and cheering just to fill the void. 'No Pillage!' the Emperor insisted to Marshal Mortier in appointing him the new governor. 'I shall hold you personally accountable. . .' Then, overcoming his disappointment, he expressed a wish to go and take up quarters in the Kremlin, the half-Gothic, half-modern palace under its cross of Ivan the Great. Given a few more days, he reasoned, and Alexander would be sure to come to terms.

Mortier did his best to stop the pillaging. However, how exactly do you billet 100,000 men in an unattended Aladdin's cave and then prevent them from helping themselves? There was a terrible shortage of food, but enormous stores of wine and spirits, and thus fortified the soldiers could not keep their eyes and hands off the furs, jewels, silver cutlery, hand-painted dinner services, gilded mirrors, ornate furnishings and other valuables which surrounded them on every side. Often bivouacking inside the grandest houses, these things were theirs for the taking—and soon the famous *Grande Armée* discipline, already undermined in Spain, showed serious signs of falling apart. When confronted by so many temptations even Ney's known severity towards looters failed to act as a deterrent. III Corps were billeted in the western suburb of Vladimir, well away from the richest pickings; yet he found his men were sneaking into central Moscow at night, then returning in the early hours 'dressed in silks, laden with booty and laughing and singing as they drank from bottles of vodka and brandy'. The deserted Oriental bazaars, the spires, the onion-shaped domes and above all the contents of the big houses lured them back time and again. Ney bowed before the inevitable. He could not imprison the whole corps. Also, he realised that as the fires started getting a hold Mortier couldn't possibly police the streets.

There were fires even on the first night of their occupation. For Moscow had not been entirely evacuated. At the last moment Rostopchin opened up the capital's prisons, provided the inmates with large quantities of strong drink and instructed them to put the city to the torch. Mortier succeeded in extinguishing their attempts around the Kremlin and the bazaars, but he was perplexed to come upon houses 'covered with iron . . . closely shut up, untouched and uninjured outside, yet with black smoke already issuing from them'.

At this stage, according to Sergeant Bourgogne who served as a member of the fire-fighting squad, 'deliberate incendiarism was not suspected'.[4] But the convicts quickly repeated their performance, and as time went on they grew more daring. Stendhal refers to 'an incendiary mania', while Ségur describes his first glimpse of those responsible:

> Hideous-looking men in rags, and women resembling furies, jumping about in the flames and completing a frightful image of Hell. These wretches, intoxicated with wine and the success of their crimes, were no longer at pains to conceal themselves: they raced in triumph through the blazing streets. When we caught those with torches, our men actually had to cut off their hands with sabres to make them drop them.

Orders were immediately issued to shoot all incendiarists on the spot. But by now the flames were spreading, fanned by persistent breezes which carried showers of sparks and burning fragments from one roof-top to the next. Stendhal reports changing houses three times in a single night; also having to sit and eat his dinner in the courtyard while watching the sky for falling embers.[5] And Ségur goes on to tell us how 'we breathed nothing save smoke and ashes. For we were encircled by a sea of fire.' He helped save the Emperor from being killed by red-hot roofing tiles in a new conflagration around the Kremlin, and afterwards was present at the historic lunch with the marshals during which Napoleon finally made up his mind to leave. 'I've had my fill of heroics! We've done far too much already for glory. The time has come for us to turn all our thoughts to saving the remains of the army. . . .'

Clearly Alexander had not the least intention of seeking peace. He was now the creature of his die-hards. Moreover on 13 October, riding back to his quarters, Marshal Ney had felt the first flakes of snow against his cheek. With the capital half-destroyed by fire and the troops hungry there was no longer any question of Russia providing them with winter-quarters. Their hollow victory must somehow be redeemed. They would have to extricate themselves as best they could; and the sooner the better. Upon receiving these pessimistic reports from his individual commanders the Emperor set 19 October as the date for their moving off. . .

He first of all opted for a southerly route; because there was intelligence of sizeable stores held at Kaluga. However Kutuzov had blocked the passes and Napoleon no longer had men to spare for an all-out attack. Without more ado he faced about: heading back upon the main road leading towards Smolensk. Marshal Davout, commanding his rearguard, was already subject to attacks from Platov's marauding Cossacks. On the other hand the weather continued good—with frost at night, but sunshine and a minimum of wind during the day.

III Corps left Vladimir with their knapsacks full of jam, candied fruits, cake, liqueurs and only a very little flour. They carried their share of 'trophies' as a matter of course, although they were far from being the worst of the army's crows. At the start they were forced to make their way across bizarre 'fields of plenty' where a typically European riff-raff had set up an exchange and mart. Ségur's account is most vivid here:

> Enormous camp-fires had been lit in the thick, cold mud and were being fed with mahogany furniture and gilt-edged windows and doors. Around these, on litters of damp straw and a few boards, soldiers and their officers, mud-stained and smoke-blackened, sat in splendid armchairs or lay on silk sofas. At their feet were heaped or spread out cashmere shawls, the rarest of Siberian furs, cloth of gold from Persia, and silver dishes off which they were eating coarse black bread, baked in the ashes, and half-cooked, bloody horseflesh. What a strange combination of abundance and famine, wealth and filth, luxury and poverty![6]

Once through this human dustbin (soon to be pounced on by the Cossacks) they found the road was straight and their pace increased. Ney himself was supremely fit, give or take a few rheumatic twinges. The weeks of comparative inactivity in Moscow had done nothing to harm him physically. If he had changed it was due to his knowledge that the Emperor was no longer infallible; an awareness that after the major mistake at Borodino there were liable to be others. The Marshal remained completely loyal. But he was quiet, watchful and above all realistic. He viewed the retreat as a question of endurance, being constantly alert and improvising tactics whenever the Russians chose to strike.

After the Kaluga project was abandoned the corps swung round and marched directly for Mojaisk. Ney was on foot and so were most of his cavalrymen. Their remaining horses were kept for the couriers. There had been a vicious engagement at Malo-Yaroslavets on 25 October where Delzons was killed; but it convinced Kutuzov that the French were still too strong for him and so their next few days of marching continued without interruption. Mojaisk was reached on 28 October and there—quite suddenly—the winter began for real, with big, thick snowflakes covering the ground and a sharply intensified cold. Immediately it became a great problem to move the guns, while on all sides soldiers were throwing away their rich booty and cursing themselves for not bringing more flour.

Setting off the following morning Ney complained to his principal *aide* Colonel de Fézensac that there was an oppressive odour. 'It is the battlefield of Borodino, *Monsieur le Maréchal.* We are now quite close to it.' 'Ah, yes, Borodino. I should have known!' And sure enough the scene of their near-triumph soon came into view. It was now a place to be dreaded. Despite the blanket of snow and ice the stench grew almost overpowering. For the area was still littered with thousands of corpses, either rotting or half-eaten by wolves and surrounded by the implements of their destruction: gun-carriages tumbled into ravines, rusting sabres and muskets, fragments of helmets and cuirasses, broken drums, gun-stocks, tatters of uniforms and standards dyed with blood. Ney shuddered at the sight but indicated they must keep moving.

At the monastery of Kolotskoi the sights were even more gruesome, for here badly wounded men were being tended by a handful of junior surgeons who lacked everything in the way of medicine and hygiene. The amputations alone accounted for many deaths. Just a mouthful of vodka and a bullet between the teeth served as anaesthetic. At the town of Gjatsz, Ney and Fézensac discovered wounded lying unattended beside corpses three weeks old. A young man made crazy by a sabre-blow on the head was calling himself Napoleon and commanding a row of dead to get up and present arms. The French did what they could to make these poor unfortunates more comfortable, but there were few fit to travel and in the end a large majority of them perished.

Within the next three or four days the cold became excruciating. Painful frostbites were now added to their list of difficulties. Already the weak were falling out, equipment lay abandoned by the side of the road and their columns no longer marched but trudged along 'any old style'. On 31 October they reached Viasma, or as the *Grande Armée* had nicknamed it 'Ville au Schnapps' after the quantities of spirits they had found in it during the advance. As the cold got worse most of the remaining horses died, so that horse-blood broth became the daily menu. The 'grumblers' marched with red-stained beards, dipping their fingers into the regimental-kettles as they plodded along westwards.

On 3 November Ney was sent back from Viasma to rescue Davout, cut off by General Miloradovich near the town of Fiodorovskoi. With General Razout to the fore he struck into the Russians encircling I Corps for a total of five hours and eventually they were forced to give way, allowing Davout a narrow corridor to pass but with the loss of his guns and casualties. It was an important moment, because as a result of Davout's exhaustion Ney now found himself in command of the rearguard: a position he would hold for the duration of the retreat. He had with him some 10,000 men of mixed nationality, not all of whom had been with him at Borodino 'but good troops nevertheless'. *Le Rougeaud* seemed determined either to write them into the history-books or die in the attempt. 'From Viasma he began to protect France's retreat, mortal to so many others, but immortal for himself.'[7]

Once on the road again, Ségur continues, and as far as Dorogobouje 'III Corps was molested by bands of Cossacks, troublesome insects, attracted by our dying and our forsaken carriages, flying away the moment a hand was lifted, but harassing by their continual return.' However these did not even merit mention in Ney's report to the Emperor. What worried him now was the deterioration of the French army itself. 'On approaching Dorogobouje he had met . . . the disorder which prevailed in the corps that preceded him, and which it was not in his power to efface. So far he had made up his mind to leave the baggage to the enemy; but he blushed with shame at the sight of the first pieces of cannon abandoned . . .'

He ordered a halt therefore, and sat down to put his thoughts into writing. 'During an awful night in which snow, wind and famine drove most of his men from their fires, followed by a dawn which brought him a tempest, still more enemy and the prospect of an almost general defection.' In vain he had just fought in person at the head of what men and officers were left. They were now obliged to retreat precipitately behind the Dnieper. Consequently he didn't mince his words. His messenger was a Colonel Dalbignac and the Emperor quickly knew the worst:

> The first movement from Malo-Yaroslavets, for soldiers who had never before run away . . . dispirited the army. The affair near Viasma has shaken its firmness; and lastly the deluge of snow and the increased cold is completing its disintegration. Having lost everything . . . platoons, battalions, regiments and even divisions have become roving masses: generals, colonels and officers of all ranks are seen mingling with privates and marching at random, sometimes with one column sometimes with another. Their example is seducing even the veteran regiments, men who have served throughout the wars of the Revolution.

Why should the rearguard be used to defend cowards and stragglers, the Marshal asked in conclusion. Why? Surely it was better to preserve the good regiments, the best men, and push on faster with these!

Napoleon replied that he should defend himself long enough to allow everyone some stay at Smolensk, where the army could eat, rest and be to a certain extent reorganised. At which point 'Ney saw that a victim was required, and that he had been chosen. He resigned himself to the task, without question accepting a danger to which his courage was the equal.'[8] As if to test him, that same afternoon (6 November) a large contingent of Russians advanced through the nearby woods and set upon his men, half of whom 'with stiffened fingers, got discouraged'. But the Marshal rushed in amongst them:

> snatched one of their weapons and led them back to the fire, which he was the first to renew; exposing his life like a private soldier, with a musket in his hand, as when he was neither husband nor father, neither possessed of wealth, nor power, nor consideration: in short, as if he had still everything to gain, when in fact he had everything to lose. At the same time that he again turned soldier, he ceased not to be the general—but

took advantage of the ground, supported himself against a height and took cover against a palisaded house. His generals and his colonels, among whom he himself remarked Fézensac, strenuously seconded him; and the enemy, who expected to be pursuing, was obliged to retreat.[9]

By this vigorous action Ney gave the vanguard a start of twenty-four hours. Napoleon profited by it to reach Smolensk, where he found the supply situation grim. Even Stendhal began to speculate that 'our goose appears to be cooked' and the bird he referred to was sheer wishful thinking.

However, from the next day, and on every one of the seven succeeding days, Ney fought in similar fashion to prolong the army's breathing space. The Russian winter was beginning to crush them. The Marshal's beard had all but covered his face, and at night his eyebrows froze; yet he continued to defend the struggling columns to the best of his ability. On 13 November the skies were blue-black with snow, 'promising terrible storms with a further drop in temperature',[10] and the following day there were no less than twenty-two degrees of frost. Sergeant Bourgogne recorded how 'our lips were solid ice, our brains too. There was a fearful wind and the snow fell in enormous flakes. We lost sight not only of the sky but of the men in front of us.' Putting to flight still another batch of Cossacks, Ney pushed them on ever nearer to Smolensk. Sometimes it meant forming square twenty times a day, with the Marshal outside, jeering at Platov's men and encouraging them to come a little closer. Then on intervening days they marched backwards for hours on end, their leader last of all 'discharging musket after musket against the pursuers'. His rearguard reached the city on 14 November.

The Marshal entered the first house he came to and lay down on the floor, fast asleep within seconds. Of the ten thousand who had started out with him, there were now something like six, together with twelve guns pulled by horses hardly able to walk. Nor did Smolensk offer them much comfort—for the vanguard and centre had plundered it from end to end, leaving little to eat and virtually no ammunition. It was like spending three days in a ghost town. The so-called *Grande Armée* had 'departed reduced to one thick, unwieldy column', with St.-Cyr and Macdonald withdrawing from the north and Schwarzenberg's Austrians (who had not fired a single shot in the campaign) marching for home on the southern side. Napoleon and the Imperial Guard left the day before Ney entered the city. The rearguard was still very much the Marshal's responsibility.

But it was upon quitting Smolensk that his problems really began. Napoleon was pushing hard to reach the Beresina and secure its crossings before Wittgenstein and Tchichagov met there to cut him off. Only three days of marching now separated these Russian forces, totalling over 90,000 men. Also Kutuzov was cautiously approaching the French flank with another 60,000 plus the Cossacks. At this juncture the Emperor had hoped to wait for Ney to come up, but in the end he instructed Davout to wait 'on his own . . . at least until the danger of being separated with I Corps from the main army becomes too great'.

Even now Cossacks were raiding the western outlying parts of Smolensk; and at Krasnoi, where the road runs through a deep gully called Lasmina,

General Miloradovitch had blocked their way with infantry and cannon. A messenger got through from Davout, urging Ney to give up his unequal struggle 'and think about saving himself'. But the Marshal retorted he 'was prepared to take on every soldier in Russia if needs be'...and he promptly attacked Miloradovitch: full frontal. He was thrown back—but attacked again and again; the French were down to six pieces of artillery, yet when the Russians offered him honourable terms Ney answered with scorn. 'A Marshal of France never surrenders. One does not parley under the fire of the enemy!'[11] He would find a way through somehow.

The Russians continued to shell them. Ney was leading four thousand effective combatants plus fifteen hundred wounded. He summoned his remaining staff officers and announced: 'We march back towards Smolensk!' Naturally they were astonished; but the Marshal stood for no arguments. 'If necessary I march alone,' he told them. Trusting him, they therefore set about collecting up the wounded.

Immediately it was dark he ordered scores of camp-fires to be lit ... to deceive the Russians, and then issued instructions to begin retreating: silently, with no stragglers. 'We are in bad state,' he whispered to General Razout, 'but I have figured a way out.' 'What do you propose to do, sir?' the General asked him. 'Get to the other side of the Dnieper, of course.' 'But where's the road?' 'Don't worry, we'll go across country.' 'But what if the river isn't frozen over, sir?' '*Sacrebleu*, I tell you IT WILL BE!'

They did not have to march very far along the road back to Smolensk. After a couple of miles they found a stream, and after breaking the ice and noting the direction of the current, Ney decided it must flow into the Dnieper. They followed it over a series of fields, moving in total darkness. But then at nine o'clock they had a bit of luck. Razout encountered a lame peasant, the first human being they had met since giving the Russians the slip, who told them they were near the Dnieper. He also told them although the ice was thawing, they were near a sharp bend where floating ice had piled up and it was safe to cross. 'Ney was desperate, but he was not to be hurried. He ordered three hours' rest before the crossing was attempted and then, wrapping himself in his cloak, lay down and slept on the bank. *Like a child*, commented someone who saw him sleeping.'[12]

They began to go over at midnight, single file, stepping from one floe to the next, the stronger ones helping the weaker. The infantry and walking wounded all got across safely; then their few horse-drawn wagons ventured out upon the ice. One, loaded with wounded, cracked the fragile surface and overturned into the freezing water. A single man was left clinging to the edge of the ice. Ney crawled out across the floes to save him. 'Ah ... de Briqueville,' he noted, matter-of-factly, 'I'm glad we got you out!'

Obviously the rest of the wagons had to be abandoned. Still, at least they were across. And the Russians as yet had no idea what had happened to them. 'Quick! No time to waste!' the Marshal told everyone. It was four days since their losing contact with the other retreating forces. They could only rely now on Ney's inspired guesswork, heading in a wide semi-circle towards Orcha, fifty miles downstream from Smolensk. The ground was fairly open here, and after two days without seeing any Russians they began to suffer raids from

Platov's Cossacks again. They formed square and kept marching, a difficult manoeuvre even on the parade-ground let alone in snow. But it cost them a number of their wounded. Also they were begining to run out of ammunition.

On 20 November when they had almost reached Orcha, the Cossacks suddenly appeared in force, occupying a long stretch of the road and driving the French back into some nearby woods. By this time Ney's gallant band was down to less than two thousand, and he now suffered a further five hundred casualties: because with only small arms to protect them the Cossacks could select their targets more or less at will. Nevertheless it didn't deter the Marshal from leading at least one spirited bayonnet charge, crying '*Tambours! La Charge!*' just as if Platov was the man in a spot. And he successfully got away a Polish officer, Pchebendowski, to Orcha for help.

He was taking a chance on the French still being there; but once again his luck held. Moreover what a sensation his news caused! After so long without any word the Emperor had secretly given him up for lost. Therefore the message that he had almost battled his way through, Caulaincourt states, was received at headquarters 'like that of a famous victory!' Prince Eugène wept, and disputed the honour of going out to his assistance with Mortier. Davout, deeply remorseful for not helping him out earlier, also wanted to go. As for Napoleon, several miles west of Orcha with the Guard, the news was the best yet. 'At last, I have saved my eagles!' he cried. 'I have three hundred millions in francs at the Tuileries. I'd give up the lot to save Ney. What a soldier! The army of France is full of brave men, but Michel Ney is truly *the bravest of the brave!*'

However the Marshal was in desperate need of assistance and Eugène had recognised the fact. He went around the billets himself, rousing his men and putting together a force 'of four thousand: all ready to march at the news of Ney's dangers, even if it was their last effort'. The following day (21 November), he broke through to III Corps, smashing the ring of Cossacks and at long last uniting the rearguard with their fellow countrymen. 'Where is Michel Ney?' he demanded. Then someone pointed out a gaunt, bearded man in a huge, shabby cloak and the two recognised one another. '*Monsieur le Maréchal!*' '*Monsieur le Vice-roi!*' They fell into each other's arms, tears streaming down their faces. Eugène 'was delighted . . . elevated by the warlike heroism which his own chivalry had just helped to save!' Ney, 'still heated from the combat, had a few harsh words to say about Davout for not coming to his aid previously.' But even these were quickly forgotten. At a horse-steak dinner that night, when Davout again 'offered profuse apologies', Ney slapped his old comrade on the back and said '*Monsieur le Maréchal*, please, no self-reproach. I have no more reproaches to make to you. God is our witness and will be our individual judge. . .'[13]

III Corps, at the moment when Eugène arrived to help them, had been reduced to 925 men. Tolstoy refers to their efforts as 'a game of blindman's buff'. But then *War and Peace* was written by an archpatriot sitting in a heated room and with a full stomach. Count Tolstoy was born too late to command a unit through the Russian snows that year.

Napoleon promptly reinforced Ney to a strength of about 4,000 but kept him in the rearguard position. The next major task for the French was to cross the Beresina: under normal circumstances not a difficult river, but now in a state of being half-frozen, half-melting–while the Russians were determined to catch their enemy there.

On 23 November a Brigadier-General Corbineau received information that although the Russians were holding the bridges at Borisov there was a ford some eight miles to the north at Studienka, where the water was only three and a half feet deep. Corbineau crossed the river himself, found this to be true and immediately reported it to Oudinot who told the Emperor. By late on 24 November the French were already beginning to congregate at Studienka. The river would need to be bridged, of course, and it meant luring Wittgenstein and Kutuzov away from their watchdog positions. However General Eblé said his engineers were ready for the former, even if it meant pulling down every house in the area for timber; and the Emperor despatched Victor and Davout to create time-winning feints against the Russians to north and south.

Speed was now of the utmost importance. Eblé had just two field forges, two wagons of charcoal and 'a single great barrel of nails'. But his men were prepared for the ultimate sacrifice, working in the freezing river until one by one they sank back unconscious and drowned: almost 400 of them. At the same time a party of Oudinot's *cuirassiers* deceived Tchichagov into believing that the crossings were to be made many miles downstream. 'Oudinot's activities served to confirm the impression that the main French effort would indeed be south of Borisov. Accordingly, needless of the protests of many of his officers, the Admiral ordered his army to draw off . . . abandoning the positions opposite Studienka. . . .'[14]

Ney's first task at the Beresina was to patrol the eastern bank while the other units used the two makeshift bridges. (The weather, incidentally, had become atrocious again: a swirling white hell for officers and men alike.) But then on 28 November he was suddenly switched to the west bank. Tchichagov had finally discovered what was happening to the north and as a result he attacked the French in strength. Oudinot received yet another battle wound—a bullet, fired from below, entered his body at an angle of 45 degrees—and was carried from the field. Michel Ney had to push his way through the columns on the infantry bridge in order to take his place.

The Marshal advanced upon the Russians sword in hand, rallying Oudinot's men as he went and ignoring a powerful artillery barrage. It is even recorded of him on this occasion that he informed a fellow-officer of his belief in Trappist fatalism, saying Trappists always stood on the edge of their own graves repeating to one another, 'One must die, brother; one must die.' Nevertheless his counterattack proved a great success. Doumerc's *cuirassiers* cut the Russian assault-force in two, taking four cannon, routing no less than six Russian infantry units and driving the remainder all the way back to Stakavo. *Again Ney was buying the Emperor time.*

Unfortunately though the following day (29 November) developed into one of panic at the bridges. Something similar had nearly occurred on the 28th when the Russians slipped several batteries behind Victor's IX Corps and began to shell the artillery bridge. 'In the middle of this disorder,' Ségur says, 'the bridge burst and part of it fell into the river. Those on it attempted to turnback...but those who came behind, unaware of the calamity, and not hearing the cries of the men ahead, pushed them into the river and then were pushed over in their turn....' *Sauve-qui-peut* ('save oneself') now became the general attitude and on several occasions Eblé had to threaten the packed masses with a firing-squad if they didn't sort out their columns and stop 'pulling and clawing at one another in their efforts to cross'. Napoleon personally organised a huge battery on the west bank, blasted the Russian gunnery teams to smithereens and so made it possible for Eblé to repair the damage, meanwhile keeping the rickety infantry bridge open and the tired troops on the move.

By dawn on the 29th all the combat troops were over. This left just the stragglers: several thousands of whom were camped in the snow along the eastern side, united in their apathy and sloth until the word got around that Eblé was about to blow the bridges. Then they became like mad devils—writhing, screaming and fighting each other in a desperate bid to get across. Many rushed out upon the collapsing timbers and were crushed to death; others tried to swim the river or jump on ice floes and in their efforts drowned; the rest were eventually killed by Cossacks or died of cold and starvation. The disaster of France's 1812 campaign, Bourgogne states, 'had reached its utmost bounds'.

To the survivors all that mattered was pressing on homewards: still, it seemed to them, 'a million miles away'. Of eighty thousand men at the Beresina, Napoleon had succeeded in saving almost sixty-five thousand. Their route now led across marshlands; and they burned every bridge as they went to hold up the Cossacks. At night they made miserable bivouacs around fires that scorched them on one side while they froze on the other. But every morning when they struggled to their feet they left a circle of dead. It was early December and the Russian winter bore down on them relentlessly.

The days were as bad as the nights. They trudged along 'pell-mell—cavalry, infantry, artillery, French and Germans; there was no longer either wing or centre. The artillery and carriages drove on through this disorderly crowd, with no instructions other than of proceeding as quickly as possible.' No wonder the Emperor wrote to his Foreign Minister, the Duke of Bassano: *I am particularly anxious that there should be no foreign agents at Vilna. The army is not for exhibition at the moment!*

The temperature had dropped to thirty below freezing-point and the wind possessed a cutting edge that 'drove through flesh, muscle and bone'. Again, always hovering and ready to pounce, there was Platov. Ney and Victor, sharing the rearguard duties, fought engagements with his men sometimes three and four times a day. One of Ney's regiments was reduced to just four men and their Eagle. Fézensac gives us some indication of the French plight by this stage:

Let one imagine ground covered with snow from horizon to horizon . . . and across this sombre landscape a column of men, overwhelmed with misery, marching in no particular formation, half of them without arms. As they go forward, one or other falls on the icy track, between the remains of horses and the bodies of his companions. Their eyes are dull, lifeless, their faces drawn and distorted, black with filth and smoke. Their boots are fragments of sheepskin or cloth. Their heads are muffled in rags, their shoulders covered with horse-blankets, women's petticoats, half-burnt skins of animals. When a man drops from exhaustion, his comrades want to strip him before he has died.[15]

He then quotes a Russian officer, who saw men crawling along with bare feet, desperate to escape the Cossacks' lances. Some had lost the power of speech, others had gone quite mad and were prepared to set fire to any hut they came to as a means of warmth.

But he pays tribute to Ney's abilities to rouse them. 'The temptation to lie down in the snow and end their terrible suffering, the hunger, the cold and the despair, was resisted only because the leader they loved and trusted never for one moment showed weakness, indecision or even discouragement. His strong body and his strong soul seemed to be unassailable. . .'[16] He concludes that not only had Ney saved the rearguard but the whole army.

One other thing kept them going: the prospect of arriving in Vilna, where the *Grande Armée* quartermasters had supposedly built up a full magazine and large quantities of food and drink. . .

At Malodechno there was a blazing row between Ney and Victor, the Commander of IX Corps, because Ney had asked him to take sole charge of the rearguard for a few hours while he regrouped. Victor refused point-blank, and Ney rode after him to try to persuade him to change his mind. But still Victor refused, whereupon Ney cursed him to his face and galloped back to resume the position he would never afterwards relinquish. ('Veterans who witnessed or heard about their dispute were not surprised by Marshal Victor's subsequent behaviour when the Empire had fallen and the Bourbons employed him to hunt down old comrades who had served under Napoleon during the Hundred Days.'—R. F. Delderfield.)

By 4 December as the column approached Smorgoni the weather grew even worse. Sergeant Bourgogne describes it as the most terrible day of the whole retreat. More heavy snow had fallen, the temperature went to thirty-four below and the wind drove even the strongest 'half-crazy'. Only the Imperial Guard kept their formation: 'Like a 100-gun ship amongst a fleet of fishing boats'. But the rest floundered about beyond belief, Bourgogne himself staggering three miles through snow drifts away from the Vilna road. He spent the night sitting shivering on his knapsack and sharing a bottle of vile gin with another soldier who had lost his way.

At Malodechno Napoleon had dictated his last standing orders of the campaign, and at Smorgoni, on the morning of 5 December, he held a final conference with the Marshals: Murat, Berthier, Lefebvre, Bessières, Victor, Davout, Mortier and Ney—together with Prince Eugène. The time had come, he told them, for him to leave the *Grande Armée* and return by fast-sledge to Paris where the aftermath to the Malet Conspiracy and other urgent matters required his presence. He informed them he would be accompanied by only Caulaincourt, Duroc, General Lobau, his Mameluk valet Roustam and a Polish interpreter.[17] But also he would be recruiting new troops to send to their aid.

The news of his leaving must be kept secret; at least to begin with; but after a few days an Imperial decree was to be published, giving in detail the new command-structure. 'The King of Naples is nominated Lieutenant-General and will take charge of the *Grande Armée* in my absence. I hope you will yield him the same obedience as you would to myself, and that the greatest harmony will prevail among you.'[18] Sadly, he added: 'If I had been born on the throne, if I were a Bourbon, it would have been easy for me not to have committed any errors . . .' Then he took each marshal aside and spoke of personal things. 'His manner was kindly and flattering to all . . . he complimented them upon their noble actions during the campaign.'

Marshal Ney's task, he said finally:

> was to proceed towards Vilna, there to reorganise the army. General Rapp would second him, and afterwards go to Danzig, Lauriston to Warsaw and Narbonne-Lara to Berlin; but that it would be necessary to strike a blow at Vilna and stop the enemy there. In Vilna they would find Generals Loison, Wrede, reinforcements, provisions, and ammunition of all sorts; afterwards they could go into winter-quarters on the other side of the Niemen. The Russians would never dare pass the Vistula before his return.

After this he rose, squeezed their hands affectionately, embraced them and departed to rescue France from the prospect of ruin.

On 6 December, the very day after his departure:

> the sky exhibited a still more dreadful appearance. You might see icy particles floating in the air; the birds fell from it quite still and frozen. The atmosphere was motionless and silent; it seemed as if everything which possessed life or movement in nature, even the wind itself, had been seized, chained, and as it were frozen by an immense universal death. Not the least word or murmur was then heard; nothing but the gloomy void of despair, and the tears which proclaimed it.[19]

When Ney at last arrived in Vilna he found the supplies gone and the hospitals and all the public buildings there filled to overflowing with five thousand dead and dying men. They had strained every fibre to reach the city and now most of them could go no further. Murat hadn't even bothered to stop, fearing his defeated cavalry would break down entirely. He felt himself 'no longer master of the army', and so he pushed on, leaving the rearguard to its fate.

Clearly Vilna was no more a place of refuge than Smolensk or Orcha had been, but Ney saw it as another opportunity to win time. He therefore defended it for almost two days, often with only twenty or thirty men helping him to shore up the weak spots. He set up HQ 'in a very primitive house, assisted by a dozen or so gendarmes who had been wounded at Smolensk but now reported back for duty'. 'Who cares about Cossacks any more?' he joked with them. 'It's the winter that's killing us!'

They moved out on 10 December *en route* for Kovno and fighting doggedly all the way. By now the Marshal had perfected his technique of retreating:

> Towards the end of a day's march he would find a defensive position, a knoll or wood or small ravine, close his ranks and light fires. Such food as was available would be cooked and the men given five hours rest. About ten he would follow the army's route under cover of darkness, halting again at first light and repeating the process all over again. He was now commanding his fourth rearguard since Krasnoi and it was melting away like its predecessors, from 2,000 to 500, from 500 to fifty.[20]

Just ahead of him there were as many as 10,000 fresh stragglers, 'men with frost-bitten feet, with half-healed wounds, with double hernias'; but somehow, by a superhuman effort he managed both to protect and keep them on the move. If he paused to eat a single strip of burnt horse flesh it was a feast; sometimes it was boiled in melted-down snow from a grenadier's shako, with perhaps the contents of a cartridge or two added in lieu of salt. Frequently he kept the stragglers going 'by dint of cries, of entreaties, even by blows'. And yet Berthier continued to bombard him with instructions—just as if he were in command of thirty-five thousand troops!

One night the Marshal lay down beside General Wrede for a few hours sleep, and upon waking called out for the men to pick up their guns and resume the march. Not a sound. 'Their soldiers had deserted them, as well as their arms, which they saw shining and piled up together close to the abandoned camp-fires.' Ney and Wrede tramped off through the darkness. At this moment the rearguard of the *Grande Armée* consisted of just two men. . .

The next day they reached Kovno—only to find the shops plundered and the whole town full of drunken men. Some of the soldiers had gorged themselves sick on pastries; others had poured raw rum into their empty stomachs and fallen insensible upon the sidewalks. Moreover this was the moment when the King of Naples chose to let him down again. Since taking over from Napoleon at Smorgoni, Murat's operational contribution had been practically nil. He seemed more worried about political events in Naples than what happened to the remnants of the expedition. But now, just when a last stand against the Russians was urgently needed he compounded his faulty leadership with dereliction of duty: heading off towards Gumbinnen and Königsberg, and leaving Michel Ney to defend Kovno with the sick and the dying.

The Marshal had entered the town with forty men. In the course of the next day (13 December), and using every available soldier who could still walk and shoot, he brought the defending forces up to 700; including two artillery teams, three hundred Bavarians who formed the original garrison and four hundred tatterdemalion French commanded by General Marchand. If they could hold Kovno for two days there was every chance for the remainder of the *Grande Armée* to cross the frontier into the safety of East Prussia. Ney therefore placed his guns on either side of the Vilna road and slipped into a derelict house for a few hours' sleep. 'At Kovno, the same as after the

1. NEY WITH HIS STAFF by Meissonier, *Trosset, Paris*

Le dix est Née et a Été Baptisé par moi
Soussigné Michel, fils Legitime de Pierre
Ney, Maître tonnelier, et de Marguerite
graffin. le parain a Été Michel Winter.
Cousin de L'enfant de la paroisse d'enstroff
et la maraine Eve Renard, fille d'andré
Renard, mé tonnelier qui ont Signé et
marqué Michel Winter
Justin Bickelberger Recoller de la marine

2. BAPTISMAL CERTIFICATE OF MICHEL NEY

Saarlouis

3. GENERAL KLÉBER

4. GENERAL MOREAU

5. GENERAL HOCHE

6. NAPOLEON by Vernet, *Coll. P. d'Harville*

7. JOSEPHINE DE BEAUHARNAIS after an engraving by Isabey, *Malmaison-Laverton*

8. AGLAÉ NEY by Gérard

9. JOSEPH BONAPARTE by Gérard

10. MARSHAL SOULT

Photograph Bulloz, Paris

11. TALLEYRAND by Godefroy, *Giraudon, Paris*

12. MARSHAL MASSÉNA Musée de l'Armée, Les Invalides, *Larousse, Paris*

13. NEY IN THE RETREAT FROM MOSCOW by Yvon, Musée de Versailles

Photograph Bulloz, Paris

14. THE FIRST ABDICATION, FONTAINEBLEAU, 1814
Napoleon with Marshals Ney, Lefebvre, Oudinot, Macdonald and Charles de Flahaut
by Berne-Bellecour

Photograph Bulloz, Paris

15. WATERLOO: AFTER THE BATTLE

16. ARTHUR WELLESLEY, DUKE OF WELLINGTON by Goya, *Apsley House*

17. LOUIS XVIII, KING OF FRANCE by Gros, *Coll. Duke of Treviso*

18. NEY IMPRISONED IN THE CONCIERGERIE

19. JOACHIM MURAT by Gérard, *Musée de Versailles*

20. THE EXECUTION, DECEMBER 7, 1815

21. THE GRAVE AT PÈRE LACHAISE

22. DETAIL FROM PÈRE LACHAISE

Photograph by Robert Clarson-Leach

disasters of Viasma, of Smolensk, of the Beresina, and of Vilna, it was to him that the honour of our arms and all the peril of the last steps of our retreat were again confided.'[21]

On the 14th, at day break, the Russians commenced their attack. Not just with Cossacks but using units of Kutuzov's regulars who had now come up. 'One of their columns made a hasty advance from the Vilna road . . . another crossed the Niemen on the ice above the town, landing on the Prussians' territory. Proud of being the first over, they then marched to the bridge of Kovno . . . to close that outlet upon Ney, and completely cut off his retreat.'

Awakened by the sound of cannon-fire, the Marshal ran to the Vilna gate. 'He found his own cannon had been spiked, and that the artillerymen had fled! Enraged, he darted forward, and elevating his sword, would have killed the officer who commanded them had it not been for his *aide-de-camp* (Fézensac), who warded off the blow and allowed the miserable fellow to escape.

> Ney then summoned his infantry, but only one of the two feeble battalions of which it was composed had taken up arms; these were the three hundred Germans. He drew them up, encouraged them, and as the enemy was approaching, was just about to give them the order to fire when a Russian cannon ball, grazing the palisade, came in and broke the thigh of their commanding officer. He fell, and without the least hesitation, finding that his wound was mortal, he coolly drew his pistols and blew out his brains before his troops. Terrified at this act of despair, his soldiers were completely scared. All of them at once threw down their arms and fled in disorder.
>
> Ney, abandoned by all, neither deserted himself nor his post. After vain efforts to detain these fugitives, he collected their muskets, which were still loaded, became once more a common soldier, and with only a few others kept facing thousands of the Russians. His very audacity stopped them; it made some of his artillerymen ashamed, who then returned to join their Marshal; and it gave time to another *aide-de-camp*, Heymès, and to General Marchand to assemble thirty soldiers and bring forward two or three light pieces. Meanwhile Marchand went to collect the only battalion which remained intact.[22]

At about 2.00 p.m. the second Russian attack began from the other side of the Niemen, although still directed against the bridge. Obviously the last desperate action of the 1812 campaign was now approaching its climax. Ney sent Marchand and his four hundred men forward to secure the bridge.

> As to himself, without giving way, or disquieting himself further as to what was happening at the rear, he kept on fighting at the head of his thirty men and maintained himself until night at the Vilna gate. He then traversed the town and crossed the Niemen, constantly fighting, retreating but never flying, marching after all the others, supporting to the last moment the honour of Napoleon's arms, and for the hundredth time during the last forty days and forty nights, putting his life and liberty in jeopardy just to save a few more Frenchmen.

Reportage tends to become confused about these final scenes of hand-to-hand combat: largely because the final battle itself was of a haphazard nature and took place as darkness and the snow fell together. Coignet records watching Ney and Férard defend one of the last palisades as late as nine o'clock that night, 'having destroyed all that remained of our artillery and ammunition. It must be said in praise of Marshal Ney that he kept the Russians at bay entirely by his own bravery. I saw him grab a musket and face our enemy as if he had a whole army at his back. France should be grateful for such men.' Fézensac too noted how 'officers and men obeyed (the Marshal) without a murmur . . . I, who so admired his (Ney's) heroic fortitude, congratulated myself that I had the honour of trying to support his last efforts.' Ségur goes on to claim that once the fighting stopped Ney plunged into the woods and disappeared to make his own way back. But Fézensac insists he watched the Marshal supervising their late-night river crossings. 'His absence,' he says, 'would have destroyed us. His presence alone was enough to retrieve the situation. He seized yet another musket, rallied the men and kept the Russians out of Kovno until it was possible for the withdrawals to continue uninterrupted. The success of our action was entirely due to his personal intervention.'

Even after passing over the Niemen, he adds, '. . . we saw Russian troops who had crossed on the ice barring our way. They opened fire. Advance seemed out of the question; and yet to stay where we were meant falling into the hands of the dreaded Cossacks.' Then all of a sudden Ney joined them, 'showing not the slightest anxiety' for their terrible predicament. He led them along the banks of the Niemen—'through woods and by crossroads'—and put them on the road to Königsberg: which they made with less than two hundred fit combatants, a dozen senior officers and the chronically sick on sledges. They were the very last of the *Grande Armée*, and the Prince of the Moscowa was the last Frenchman of all to leave Russian soil.

At Gumbinnen in East Prussia on 15 December General Matthieu Dumas was just sitting down to his first decent breakfast in months when someone kicked the door open. 'There stood before him a man in a ragged brown coat, with a long beard, dishevelled and with his face darkened as if it had been burned, his eyes red-rimmed and glaring. Underneath his coat he wore the rags and shreds of a discoloured and filthy uniform.'

'Here I am then,' the newcomer exclaimed.

'But who are you?' the general cried, alarmed.

'What! Don't you recognise me? I am Marshal Ney: the rearguard of the *Grande Armée!* I have fired the last shot on the bridge at Kovno. I have thrown the last of our muskets into the Niemen. I have made my way here across a hundred fields of snow. Also I'm damnably hungry. Get someone to bring me a plate of soup . . .'

Once fed, bathed and with his uniform stitched up he hurried off to Königsberg and a traumatic conference with the King of Naples. . .

When Murat reached Gumbinnen he was exceedingly surprised to learn that Ney was already catching him up . . . for since it had left Kovno the rearguard was marching with greater speed than the van. Also—happily—

the pursuit of the Russians once they had reconquered their own country was slackening. They seemed to become nervous along the Prussian frontier, not knowing whether they should enter it as allies or long-standing enemies.[23]

However by Königsberg Murat had formed his own decision; and to Ney, Davout and Berthier he didn't hesitate to lay it on the line:

Peace is necessary. It's no longer possible to serve a madman. There's no hope of success in his cause. No one has confidence in the word of the Emperor and no one is willing to treat with him . . . whereas I could have made peace with the English long ago. And why not? I am King of Naples as the Emperor of Austria is the Emperor of Austria. I can act as I wish![24]

'Sire,' Ney replied with the utmost politeness, but nevertheless staggered, 'surely as a French prince you ought to act as the Emperor himself wishes. . .'

Murat pouted and turned his back on them; but then Davout weighed in, heavily, rudely. 'The King of Prussia, the Emperor of Austria are monarchs by the grace of God, of time and the custom of nations. As to you though, you are only a king by the grace of Napoleon and the blood of many Frenchmen. You cannot remain so except through Napoleon's wish, and by continuing in union with France. Obviously you are led away by the blackest ingratitude.'[25] And he declared that he would immediately denounce this act of treachery to the Emperor.

Murat was put out of countenance. He felt himself guilty. But at the first mention of treachery all were strangely silent. The marshalate, they individually decided, like the French army seemed on the point of collapse: and if this was so upon whom could the Emperor rely in the future?

[1] *To the Happy Few: Selected Letters of Stendhal,* trans. Norman Cameron, New York, 1952.
[2] *History of the Expedition to Russia.*
[3] Ibid.
[4] *Memoirs of Sergeant Bourgogne,* trans. J. W. Fortescue, London, 1899.
[5] *The Private Diaries of Stendhal,* trans. R. Sage, New York, 1954.
[6] *History of the Expedition to Russia.*
[7] Ibid.
[8] Ibid.
[9] Ibid.
[10] R. F. Delderfield: *The Retreat from Moscow,* London, 1967.
[11] Ibid.
[12] Ibid.
[13] *History of the Expedition to Russia.*
[14] *The Campaigns of Napoleon.*
[15] General R. A. P. J. de Fézensac: *Souvenirs Militaires.*
[16] Ibid.
[17] *The Campaigns of Napoleon.*

18   *History of the Expedition to Russia.*
19   Ibid.
20   *The Retreat from Moscow.*
21   *History of the Expedition to Russia.*
22   Ibid.
23   Ibid.
24   J. Bear: *Caroline Murat,* London, 1972.
25   *History of the Expedition to Russia.*

CHAPTER 8

Honours

Together with more action and another wound (Leipzig, 1813)

When Ney reached Paris at the beginning of January 1813 the Emperor had already taken the most vigorous measures to provide France with a new army. Never a man to admit having suffered defeat, Napoleon appeared to be in full possession of his original abilities as an organiser and sufficiently pragmatic to recognise that speed and the quality of recruitment were now of the utmost importance for his Empire to stand any chance of survival.

The *débâcle* in Russia had involved an army of 650,000 men (if one counts the garrisons and support troops). Of these about 90,000 managed to slog their way home, and perhaps under 50,000 were strong enough to fight another battle. Behind them they had left a Russian army more or less intact and determined to commence a push westwards which the Czar referred to as 'a war of liberation', but which—as Napoleon knew only too well—was really intended to 'let slip the dogs of war' on French soil and compel his own abdication. At present, apart from Platov's roving Cossacks, these Russian forces were being held up, thrown into confusion and generally frustrated by a combination of stubborn French garrisoning of the German cities and a numerically weak, but very brave and extremely mobile body of sharp-shooters and cavalry led by Prince Eugène: who in the next few weeks performed the most illustrious feats in his whole loyal career,[1] entirely justifying Ney's high opinion of his improvising talents in the field.

However, left to a mere task force the situation was at best a makeshift one. Directly the off-balance Russian masses were joined by the regiments of France's defecting allies, notably Prussia and Austria, plus the fresh troops promised by Sweden under the command of Crown Prince Bernadotte, then not even Eugène's ingenious screen would be able to hold them up for more 'than a few months.

To add to the Emperor's problems there was also an increasing cause for concern on the far side of the Pyrénées. Ever since Ney's recall from the Iberian Peninsula Wellington's activities there had succeeded in tying up increasing numbers of good French veterans in what was widely regarded as a war of attrition. Each time the cautious Duke seemed about to take advantage of one of his well-planned encounters so his ill-disciplined soldiery could be

149

relied upon to throw the move away in a prolonged orgy of drunken looting, rape and killings. After the capture of Badajoz, for instance, they went on the rampage for a whole forty-eight hours, in the course of which Wellington himself narrowly escaped death by a reveller's bullet. When the city fell the Duke was seen to be in tears, allegedly for his 'nearly 5,000 dead'[2]; but one begins to wonder if these tears were not out of sheer frustration at the behaviour of 'his living'! The men had become as great an anxiety for him in victory as they had been for Sir John Moore in retreat. Often it took him a week to sober them up: by which time the French had withdrawn, regrouped and chosen the best positions for another fight. Consequently when Napoleon set out for Russia, although the battles for Spain still dragged on, he felt reasonably sure that the country could, and was being held.

But on 22 July 1812 Wellington inflicted a most severe defeat upon Marshal Marmont's forces near Salamanca. This was a genuine field victory as distinct from some of his previous calculated strikes, and apart from giving him Madrid it gained him a first footing on the vital roads leading via Burgos and Vitoria towards France's back door. Napoleon, still advancing on 'the far side of the civilised world', took the news of Salamanca and of Marmont being wounded there quite calmly. He ordered reserve troops into Spain, called up Marshal Soult from the south to take overall command and Wellington was denied access to France for another year. Yet taken together with the disasters which subsequently occurred in Russia it did begin to look as if a vast, continental trap had now been laid for Napoleon. A trap ready to spring to the moment he made his next mistake; seizing his Empire in a military grip from which there was no escape.

At least a part of this trap was laid right in the heart of Paris: where suddenly, after the first bad reports from the east, almost everyone gave way to expressions of gloom and despondency. Half of those who had benefited most under the Empire, including many high officials in the administration, were prepared to turn their coats if the military situation was not resolved. Others (inevitably centred round Talleyrand) were actively intriguing to do so—with even former revolutionaries conversing in whispers about their prospects of office if they helped to bring back the Bourbons.

Ney was disgusted by the prevailing mood in the capital. After speeding across Europe—by whatever means he could commandeer, sleigh, coach or plain horseback—he was there for a few days before leaving for a period of much-needed rest at Coudreaux. But the satisfaction he felt at still being alive and the pleasures of a reunion with Aglaé and his sons were considerably reduced by the defeatist talk and the rumours of plots he now heard on all sides. He was tired, gaunt and undernourished: a man just coming up to forty-four who looked his age. He had gone through what no French general had ever done before in defending the retreat from Moscow. But he was also for the same reasons very short-tempered; and the pessimists and his old adversaries, the self-seeking politicians and courtiers, were soon the recipients of a typical outburst.

To Ida St.-Elme he apologised for upbraiding her so furiously when she'd appeared during the retreat near Marienwerder, but once their talk turned to political intrigue and a possible restoration nobody was spared. 'Yes,' I have listened to all the rumours,' Ney said, scowling. 'Fouché is said to be the instigator this time. The Emperor should have had him shot: it would have been a wise precaution. Many of those he thinks are still his friends will betray him too. They're no better than Fouché and Talleyrand. Fortunately though there is a new army being formed. The sound of cannon always frightens off birds of prey!' Then he thundered on about the courtiers and sycophants. 'A lot of use they were to him when things started to go wrong, all those windbags of the Senate, the Council of State and the Assembly! He ought to have clung to the people—they were his real strength—and to the soldiers, his truest friends!'[3] After this he set off for a ride in the Bois de Boulogne to cool down. Aglaé was much happier once she got him away to Coudreaux.

If Napoleon entertained similar ideas then he kept them strictly to himself. He had police spies trailing Talleyrand, but otherwise concentrated his efforts upon the army. In December and January he was scarcely seen outside the Tuileries; and the only real evidence of his activities was the coming and going of officers and clerks. But he certainly made good use of his time, because from this point on the new army took shape much more rapidly than the Czar and his allies could ever believe.

To begin with, all his veterans from the Russian campaign who remained fit were divided into cadres which the Emperor judged would make most use of their experience. The new men were then organised around them. The Emperor was lucky in that the Class of 1813 had just completed their training: 140,000 conscripts altogether. They were mostly fresh-faced lads, but French training methods were still the best in Europe, and on 11 January 1813 they were augmented by a further 80,000 men of the National Guard. At the same time the Class of 1814 was called forward to February. Given a 'crash-course' of training, they meant another 150,000 support troops by April. The rest came largely by demand or improvisation; particularly in the matter of specialists. Italy was called on to send 30,000 men, the Swiss to increase their regiments from four to six and even old soldiers retired on half-pay were called up. Most of the required 12,000 gunners came from the French navy, which also supplied 24 battalions of sailors to fight as infantry. And to start off a new cavalry corps 3,000 officers and NCOs were drafted from the gendarmerie.

Last, but never least in Napoleon's army, there was a major reconstruction of the Imperial Guard. After Russia:

> The Guard as such no longer existed. But there were still the Paris depots; the Veterans and four battalions of *pupilles* in Versailles, and four more in coastal garrisons; four instruction battalions at Fontainebleau; the artillery and cavalry and reserves of material at Mainz and elsewhere. In Spain there were the 3rd *Voltigeurs* and 3rd *Tirailleurs*, and elements of the 1st; the National Guard regiment, some cavalry and ten guns. The Guard tradition lived on in the generals and cadres.[4]

It was a promising start, and the revival of the Guard went so well that by March the Emperor could write: 'An officer or NCO may not be admitted into the Old Guard until he has served twelve years and fought in several campaigns . . .'[5], while General Lefebvre-Desnoëttes was told to 'sort out the newcomers and hold on to the men of good physique and conduct, exchanging the rest for good specimens in the Line'. The Guard's senior officers were undoubtedly good, particularly Roguet, Drouot, d'Ornano and of course Lefebvre-Desnoëttes himself. The Duke of Istria (Marshal Bessières) remained its overall commander under Napoleon, but Marshal Mortier was also a Guard commander and had supervised much of its organisation and retraining programme.

Altogether by April, France had half a million men in the field to replace the Army of Russia; in addition to the 200,000 in Spain and another 100,000 employed in garrisons. And on the whole the new recruits would prove themselves good troops in battle, certainly man-for-man the equal of the Russians and the regiments from Austria. Only the cavalry and horse-artillery remained weak, for a variety of reasons, but including the fact that many of the best horse-producing areas in Europe were lost to Napoleon with the defection of Prussia and the German princelings. A final reason too was the non-appearance from his Neapolitan kingdom of Joachim Murat when he could have done most to revitalise the cavalry's training and field exercises. In the end Napoleon severely reprimanded him ('I presume you are not one of those who believe the lion is dead . . . if you make this calculation, you are very much mistaken. . .'). But he did not rejoin the *Grande Armée* until August, by which time the cavalry's insufficiencies had already placed the campaign at risk.

Ney was given exactly one month to recuperate. Early in February he received a typically polite summons from Berthier to 'rejoin His Majesty in the city of Paris with all possible speed'. His services were required for the army obviously: just the mention of his name was sufficient to stiffen the veterans' resolve. But beforehand he was to be made great use of in Paris itself. At last able to turn to other, non-military affairs, Napoleon made up his mind to dispel the citizens' pessimism and boost morale with an elaborate showing of the Empire's more positive achievements; from its advances in health and education to the overall glory of its arms. And who better to illustrate the latter than the real hero of the Russian campaign? ('The Russians have *never* defeated us,' the Emperor reminded everyone, 'it was only their winter which forced us back!)

Accordingly when Michel and Aglaé shut up Coudreaux and returned to Paris they found themselves at the centre of a formidable public-relations exercise. Every day there were crowds outside their house in the Rue de Lille to cheer their appearances; and no important reception was complete without them. People even took to jumping on their carriage hoping to shake the Marshal's hand. With an almost Svengali-like pleasure Napoleon opted for a back seat role whenever the Prince of the Moscowa was present: the 'good effect our brave Ney is having' struck him as 'better than a *Champ de Mai*. . .'

Again, although his title of prince had been conferred on the day following Borodino, the Emperor decided it would be proclaimed afresh at a special ceremony in the Tuileries to which he gave the maximum publicity. And at the same time he announced that 'a grateful nation' had awarded Ney the huge gratuity of 800,000 francs: to be paid jointly by Rome, the Mont de Milan, Westphalia and Hanover. Nor was Aglaé overlooked. As a princess of the Empire she was being very warmly received by Marie Louise, while Napoleon singled her out with the flattering compliment that she was 'a person loved and respected by both his wives. . .'

The adulation continued through February and well into March. But then it was time for Ney to rejoin the army: at present in its final stages of assembly on the far side of the Rhine, near Mainz. As a result he was not around to console his wife for the tragic news which quickly put an end to her social triumphs.

For a year, possibly even longer, Aglaé had seen less of Hortense de Beauharnais in order to stay more often with Josephine at Malmaison; and her place as the former's confidante had been taken over by her sister Adèle, the widow of General du Broc. Hortense's letters never ceased praising her:

> Since her husband's death and my return from Holland, Adèle . . . has remained constantly with me. She devotes herself to her duties as friend and comforter. Often too I have seen her take off a brilliant court dress and abandon pleasures that would have seemed to many to be altogether absorbing in order to take alms to beggars living in the most miserable hovels. . .[6]

However the two had gone to take the waters at Aix-les-Bains—and one day they set out on foot to visit a picturesque waterfall. Hortense crossed the stream first on an unsteady plank, then '. . . as I turned, what a tragic spectacle met my eyes! Great God, could it be true? My friend, swept away by the current, vanished beneath my very eyes. I succeeded only later in recovering her inanimate body. . .'[7]

With a heavy heart the Queen of Holland made arrangements for the burial; and afterwards there was a distraught reunion in Paris with Aglaé, Antoinette (Madame Gamot), Josephine and Madame Campan. Poor Aglaé! Her life story is criss-crossed by the deaths of loved ones, and on this occasion it took all of Josephine's gentleness and understanding to dry her tears. The former Empress also knew the pains of losing someone. Ney's own condolences were tender, but brief. He apologised for sending scribbled notes again. But the Emperor had taken command of the army on 16 April; and he was clearly in no mood to wait for the Allies to come at him.

Prince Eugène's tactics had bought France the time. And the Allied generals were said to be quarrelling. But in fact the Russians and their new Prussian allies already had 80,000 men in Saxony under the joint command of Wittgenstein and Blücher. Once their quarrels were patched up it would not be long before these troops were reinforced. Napoleon's number one priority, therefore, was to throw them back into Bohemia; cutting their links with the safe fortresses along the River Elbe before the Allies woke up to his speed and direction. If he controlled the theatre of operations between his existing forward positions along the Lower Saale and the Elbe itself, then he stood a very good chance of making his own strategy work: which was to march on

Berlin. 'If circumstances permitted he would afterwards relieve the besieged fortresses along the Vistula—Danzig, Thorn and Modlin', thereby allowing a further 50,000 veterans to rejoin his army.[8] Even so, the most immediate hope was 'that his second conquest of Berlin' would not only knock Prussia out of the firing line but also 'so disorganise the Coalition that all the princes of Germany would confirm their fidelity and alliances with France'.[9] The final objective would be the re-occupation of Stettin and all points south along the River Oder, cutting off the Russians from their bases and returning Poland to the French fold.

The French began their advance as two distinct army groups. To the north the Army of the Elbe (58,000 men) was concentrating around Merseburg, with Davout's detached I Corps (20,000) hovering nearby and Sébastiani de la Porta's 14,000 cavalry travelling towards it from the Hamburg direction. The Elbe group also included Roguet's division of the Guard, plus Latour-Maubourg's cavalry, and its first task was to act as a defensive hinge along the River Saale while the more southerly Army of the Main crossed the river and then swung north-east to capture Leipzig. This Main group, now moving very fast indeed, consisted of Ney's III Corps (45,000 men) travelling from Meiningen-Gotha, Marmont's VI Corps (25,000) from Eisenach and a further 36,000 men divided between Bertrand's IV Corps and Oudinot's new XII Corps coming from Coburg. To their rear marched the real élite of the Imperial Guard (15,000 men). Also they were very strong in artillery; even if this didn't fully compensate for the poor cavalry.

Ney made his crossing of the Saale on 29 April near Weissenfels and almost immediately was spotted by the Allies' reconnaissance patrols. The following day the Russians launched a brigade of cavalry against his lead division—only to be bloodily repulsed. 'These lads are heroes,' Ney exclaimed to the Emperor as they watched the young conscripts stand their ground, 'they fight with a bravery I did not suspect!' Napoleon nodded, and ordered him to push on without delay towards Lützen.

The next day (1 May) there was more fighting to effect a crossing of the Rippach, which although narrow runs through a steep ravine. Again though the French were successful; and once over, an elated General Souham (the commander of Ney's lead division) reported being able to see the round tower and belfry of Lützen across the open fields. Such is war; but it was a success almost at once overshadowed by the death of Marshal Bessières. At 12.45 p.m. Ney had been conferring with him about how to make the best use of the Guard's limited cavalry. Then at 12.55 'a cannon ball hit the wall enclosing a neighbouring field, ricocheted, and struck the Marshal, hurling his mutilated body under his horse's hoofs.'[10] The ball severed his wrist and pierced his chest, killing him instantly.

At this point in his *Mémoires* Marbot is somewhat unkind to Bessières, stating '. . . The Emperor regretted him more than the army, which had never forgotten that it was advice given by the Marshal on the evening of the Battle of the Moscowa that prevented Napoleon from completing his victory by sending in his Guard.'[11] Ney, in spite of being the main sufferer at Borodino, was more forgiving. 'This is our fate,' he remarked sadly after riding up to view the body.

That same night, and visibly distressed, the Emperor called Ney, Marmont and Duroc to a meeting at Poserna (where the innkeeper scratched the names of his illustrious guests on a windowpane with his diamond-ring). Napoleon said he might replace Bessières with Soult. Also he gave them very specific instructions for the following day. It now seemed safe, he announced, for the Elbe army's commanders (Lauriston and Macdonald) to advance directly upon Leipzig. Marmont was to close up on Lützen, at present being occupied by Ney's men; and the latter would instead provide a new right flank based on the four villages of Kaya, Rahna and Klein and Gross Görschen. He said he was not expecting an attack from this direction, more likely the enemy would retreat towards Dresden—but he wanted to take every precaution. Finally Ney was 'to send out two strong reconnaissance forces, one towards Zwenkau, the other towards Pegau'.[12] They drank their wine, and because Napoleon was still brooding upon Bessières' death Ney did not argue. But it is no mean feat to occupy four country villages on a pitch-black night (Souham was still busily establishing himself in Gross Görschen at 11.45 a.m. the next day when Blücher attacked). And how do you send out active, far-reaching reconnaissance patrols when you have no proper cavalry?

On the morning of 2 May Napoleon suggested that Ney accompany him 'to witness the events at Leipzig'. They arrived just in time to see Lauriston easily defeat Kleist and his men pouring across the bridges into the city. But then they heard the first, sinister rumblings of a cannonade to the south-west. It could mean only one thing. No wonder Kleist was so weak at Leipzig . . . obviously Wittgenstein and Blücher were attacking Ney's positions in force! There was not a moment to lose; and the Marshal spurred away with Napoleon's reassurances: Macdonald would be detached to follow him, also the Guard. Nevertheless when he reached the area (shortly before one o'clock) he found the battle for Lützen and the villages well on the way to being lost. General Berthèzene, who was on detachment from the Guard and had first sighted the enemy army 'which seemed immense' from a steeple-top in Lützen, quickly gave him a situation report. Marmont was still holding the Starsiedel ridge but in the face of a terrible bombardment from what seemed like 400 guns. At the same time Souham had been forced to evacuate Gross Görschen, while Bonnet's infantry and Campan's marines were being put to the bayonet in Klein Görschen and Rahna. Only Kaya of the villages remained completely in French hands (although that too would be hotly disputed before the fighting ended).

Ney decided upon an immediate counter attack. He could not afford to wait for Macdonald. He called up his only reserves, the three divisions from Lützen, threw Marchand's against Blücher's right flank and drawing his sword personally led the other two into the fray. Perhaps the most desperate hand-to-hand combat of the entire campaign now took place. The Emperor himself arrived at 2.30 p.m. and also joined in: proving a definite inspiration to the younger soldiers, some of whom had been on the point of breaking. 'This was undoubtedly the day, of his whole career, on which Napoleon incurred the greatest personal danger on the field of battle,' Marmont later recalled. 'He exposed himself constantly, leading the defeated men of III Corps back to the charge.'[13] By 4.00 p.m. the pace had still not slackened.

Blücher was wounded and his command passed to General Yorck, but Wittgenstein chose this moment to throw in his own reserves, again vigorously attacking the village of Kaya.

Ney's men held their line only with great difficulty, stiffened by a contingent of the Young Guard, and it was not until 5.30 p.m. that things started to turn in their favour. Then suddenly there was Macdonald on the left, and Betrand to relieve Marmont: just what Ney required! Slashing right and left with his big sword he plunged down the slope in the direction of Klein Görschen which his men eventually recaptured. Meanwhile Napoleon had assumed overall command. Drouot's 70 cannon were rushed up and blasted the Russian centre at almost point-blank range, leaving the remainder of the Young Guard to deliver the *coup de grâce*. As the latter advanced in four columns the Allied troops broke and fled—estimated to have lost nearly 20,000 dead and wounded. The French too had lost heavily; but at least they could announce a resounding victory. It had been one snatched from the very edge of defeat and as well as demonstrating the renewed power in France's army it gave the Czar and his allies a nasty shock.

Afterwards Napoleon expressed great alarm at the sight of Ney. 'My dear Cousin! But you are covered in blood!' The Marshal looked down at his gory uniform. 'It isn't mine, Sire,' he replied calmly; '. . . except where that damned bullet passed through my leg!'

Unfortunately, due to the scarcity of good cavalry, there was no question of a pursuit. And more or less the same thing happened after the battle of Baützen on 21 May. Although another French victory, it was not one of Ney's best performances. He had been given twenty-four hours to overhaul his battered corps and then begin a march north on Wittenberg to relieve the French garrison there. He was called to Baützen, east of Dresden, only at the last moment and was ordered to complete an encircling movement from north to south-east, leaving Bertrand, Marmont and Macdonald with Oudinot to hold the front. However, the orders from HQ were muddled, communications almost non-existent, and against the advice of Baron Jomini, Ney stopped to attack the strongly-fortified village of Preititz when he ought to have bypassed it. Napoleon had ordered him to take the village, but the Marshal should have used his own initiative. In the end it cost the French dearly in men and resources and gave Blücher time to extricate his forces. Again there was no proper cavalry to pursue them. The Russians took the worst hammering at Baützen though (Wittgenstein spoke of resigning over what he described as the Czar's 'incompetence') and it wasn't long before the Allies were seeking an armistice. Napoleon welcomed it. He had lost his great favourite General Duroc at Baützen; also the army was tired.

In all, this armistice lasted for fifty-six days; during which time Michel ate well, hunted and even went fishing! But Metternich, who had been brought in as an intermediary, already had an offer in his pocket of half a million pounds from England if Austria added her 150,000 troops to the Allied cause. Moreover Sweden was at last ready to join in. Accordingly, at the Congress of Prague, Metternich demanded that Napoleon give up Holland, Belgium, Spain and Italy, agree to a free Confederation of the Rhine and allow Russia to annex Poland. After two French victories it was intolerable!

Hostilities were resumed in August. By which time Ney had lost the services of Baron Jomini. Historians have generally assumed this was caused by their argument at Baützen, while Marbot and others clearly regarded him as a traitor (Marbot states he deserted with some of Napoleon's battle-plans). But there is rather more evidence that Jomini had been involved in a dispute with Berthier over promotion; also one must remember that he was not French but Swiss, and a professional theorist of war rather than a patriotic fighting general; finally, and with Napoleon's permission, he had held a commission with the Russian army since Tilsit. Ney never afterwards spoke of him disparagingly or unkindly.

Another surprise was the belated re-appearance of Joachim Murat. Having grudgingly left the sun of Naples, he was just in time to command the right-wing at Dresden (26-27 August 1813), where he did fairly well. Considering the acrimony at their last meeting in Königsberg when Murat had abandoned the *Grande Armée*, Ney exercised great self-restraint in dealing with him now. He greeted him with affected politeness; but he still did not trust him. Furthermore Murat had arrived too late to do much about improving the cavalry.

At Dresden Ney had temporary command of two divisions of the Old Guard and fought with marked success, especially on the first day, when he trounced Schwarzenberg's Russo-Austrian centre and re-took all the positions previously lost by St. Cyr.[14] But then the campaign started to go wrong. Ordered to take over Oudinot's bogged-down march upon Berlin on 2 September, and to reach the city 'within a week or eight days at the most', Ney set off only to discover within twenty-four hours that all his reserves (25,000 men) had been sent to reinforce Macdonald near Baützen. Then on 6 September, again largely due to lack of horse reconnaissance, the Marshal found himself trapped by the entire Swedish army near Dennewitz.

Sword in hand, Ney fought his way out, ably supported by Generals Bertrand and Reynier. But Oudinot, still smarting at being placed under Ney's command, failed to provide adequate support and in the end the French suffered 10,000 casualties to Bernadotte's 7,000. It meant abandoning the march on Berlin. On 10 September Napoleon ordered Ney 'to take up a position on the right bank of the Elbe near Torgau',[15] where he hoped to reinforce him. However, events were now taking place elsewhere which forced Napoleon to abandon the plan. The whole of Germany—except Saxony—had joined the Allies. The French army was starving; and had lost 300 of its vital artillery pieces. All of which led to a contraction of the front and then, beginning on 16 October, to a major defeat of France's army and Napoleon's revived hopes at Leipzig.

Ney's part in 'The Battle of the Nations' as it came to be known was limited, for he was badly wounded on the third day, whereupon Napoleon ordered the entire sector under his command to be abandoned. It proved to be an extremely ill-fated battle, not only for the Marshal physically, but also with everything it caused to descend upon the French nation.

Blücher was the first Allied commander to strike towards Leipzig, bringing a reluctant Bernadotte in his wake and subsequently drawing the Russians and Austrians there as well. More than half a million men took part in the four days of fighting which ensued. Berthier issued the Emperor's instructions for a general concentration in the Leipzig area on 14 October. Ney was at Dessau, where two days earlier his leading division (Souham's again) had frustrated Bernadotte's efforts to cross the River Mulde. Also on the 14th Murat's cavalry was engaged in a sharp, but inconclusive action to the south of Leipzig, which ultimately determined the positions Napoleon's army would take up there. By the evening of the next day some 200,000 French troops were occupying these positions. 'Facing them to the south there were massing 203,000 soldiers of Schwarzenberg's Army of Bohemia; a few miles to the north-west Blücher was driving forward the 54,000 men of the Army of Silesia; a good many miles to his rear were the 85,000 men of the Army of the North (*Bernadotte's Swedes*)—still near Halle.'[16]

The Emperor was ill: a recurrence of the troubles which had caused his similar lack of resolution at Borodino. And Ney had already written warning him that the army was in bad shape. 'The morale . . . is shaken. To command under such conditions is to be only half a commander, and I would rather be a grenadier. I am ready to shed my last drop of blood, but I want it to be shed for some useful purpose.' At a face-to-face meeting which followed he returned to the same theme, but Napoleon was both uncertain and committed—and as a result silenced him with a glare.

Much of the fighting took place in woods and marshland. All of the troops north of Leipzig were placed under Ney's command, amounting to four army corps. On 16 October there was a lot of bitter fighting in this sector, but Ney's position was constantly being weakened by Napoleon's detaching first one and then another regiment in order to stiffen Macdonald and Poniatowski further south. The Marshal coped with this drain on his resources until the sudden mass defection of his Saxon units on the afternoon of the third day: three squadrons of cavalry, eleven battalions of infantry and three batteries. 'They simply marched away from the French with their bands playing patriotic German tunes.' That same morning Ney had been in fierce fights against Von Bülow and Langeron to retain the villages of Paunsdorf and Schönefeld. Two horses were killed under him, and at Schönefeld he received a musket-ball in the shoulder.

The Saxons' going had left him outnumbered by three to one. When later on that same afternoon Napoleon decided upon a strategic withdrawal Ney had lost much blood and was hardly able to sit upright in the saddle. On the southern front the battle was lost: although not due to lack of bravery but because ammunition was running out.

Early on 19 October the French began to pull back. Orders were given for 30,000 troops to hold Leipzig while the main army crossed the River Elster to safety. By mid-morning the withdrawal had been proceeding along model lines for a good five hours without the Allies realising what was happening. But then, due to an error in communications which originated with Napoleon himself, a panicky corporal fired the demolition charges on the line of retreat (the way out of Leipzig was a tortuous system of causeways and bridges). It left the whole of the French rearguard trapped. Ney, despite his wound, managed to ride his horse across the river; so did Oudinot. But the dashing Poniatowski—a marshal for a day and Napoleon's penultimate creation—

perished. Cut off from the main army with only a handful of skirmishers, he charged the enemy's units again and again. Eventually, wounded four times and faint from loss of blood he plunged into the deepest part of the river rather than accept capture. His last words before he drowned were 'Poland' and 'Honour'.

Ney by this time was already speeding back towards France, having eluded the 43,000 Bavarians under General Wrede sent to cut him off. ('Wrede!' the Marshal growled, gritting his teeth with pain. 'And to think that he fought by my side in Russia. What a strange thing war is!')

'The Battle of the Nations', although inconclusive as regards the actual fighting, in fact was a disaster for France. Her actual casualties were 40,000 whereas the Allies suffered almost 60,000. But the capture of the French rearguard represented a further 30,000 men and this included six generals dead, twelve wounded and thirty-six captured. Napoleon might be able to raise more troops, but general officers were harder to come by. Also, he had lost the King of Saxony, his last remaining European ally. The French had discharged over 220,000 rounds of ammunition and were forced to abandon over 300 guns, their transport trains and most of their military stores.

As for Ney, when he arrived back in Paris all he could say to Aglaé was: 'I feel tired. My wound is hurting, but most of all I feel tired. More tired than I have ever felt in my life. . .'

---

[1] *Napoleon's Viceroy.* Eugène fought in Germany as if following a manual of Ney's tactics in Russia; also Napoleon paid a glowing tribute to his stepson in his *Mémoires*—'Eugène was one person who never caused me the slightest anxiety'.

[2] *Wellington, The Years of the Sword.*

[3] *Les Mémoires d'Une Contemporaine.*

[4] *The Anatomy of Glory.*

[5] Ibid.

[6] *Mémoires de la Reine Hortense.*

[7] Ibid.

[8] C. de Montholon: *Mémoires de Napoleon*, Paris, 1823.

[9] Ibid.

[10] *The Anatomy of Glory.*

[11] Marbot: *Mémoires.*

[12] Marmont: *Mémoires.*

[13] Ibid.

[14] General Moreau was killed in the closing stages of this battle. Having returned from America, he was serving as the Czar's personal military adviser and voicing his criticisms of Schwarzenberg's tactics when a French cannon ball killed him. The Czar ordered his remains to be buried in St.-Petersburg with the honours due to a Russian general.

[15] Napoleon: *Correspondence.*

[16] *The Campaigns of Napoleon.*

CHAPTER 9

Some Battles for France

From Brienne (January, 1814) to the first abdication . . .

Ney's leg-wound at Lützen was superficial and healed up almost immediately; but the wound he received in the shoulder at Leipzig proved more serious. After the ball had been extracted it took the best part of two months before he could ride and swing his sword-arm again. By which time both the Emperor and France had urgent need of his services.

Following their near-collapse at Leipzig the French had retreated to their national frontiers; and already one marshal (Macdonald) was pressing Napoleon to negotiate. 'Sire, you have no army left, only a few poor souls dying of hunger . . . you have lost, and your only thought now must be of peace.'[1] But the Emperor dismissed the idea as absurd. Would the Allies let him keep his throne? Never! Very well, they must fight on. . . .

But fight with what? Lord Wellington had crossed the Pyrénées, the Austrians were pouring into Switzerland, Von Bülow's Prussians driving down through the Netherlands and Blücher crossing the Rhine with 60,000 more of them. Behind these was Schwarzenberg's huge army: 200,000 men, including the bands of Cossacks who were guaranteed to terrorise the French countryside. Napoleon, desperate, entrusted the immediate defence and his remaining units to three men: 'In the north to Macdonald who was dynamic and honest, but unlucky; in the centre to Marmont who was not entirely loyal; and in the south to Victor, a good soldier but with few other talents.'[2] Meanwhile he sent frequent messages to Marshal Ney to rejoin the army as soon as possible and he set about trying to raise fresh troops.

There was very little time. Admittedly the Allies were still squabbling—largely because they could not agree on a political solution. Metternich wanted a regency on behalf of the King of Rome since the child had an Austrian mother. The Czar thought on the whole he favoured Bernadotte as Head of State. England's Lord Castlereagh was in agreement with Talleyrand: that a return of the Bourbons was most desirable. Nevertheless they were agreed on one thing: that their armies must converge on Paris and force the Emperor's surrender and abdication. Again, therefore, it became a question of what forces Napoleon could gather to defeat their plans —and how soon.

At present about 80,000 exhausted and half-beaten men were guarding the frontiers. Another sixty-thousand Frenchmen were cut off in Germany and Poland. The Imperial Guard was still intact; and the cavalry, although small, showed signs of improvement: thanks to a major effort at the depots. But the main problem was new recruitment. The Leipzig campaign had drained the country of its manpower to such an extent that there were now few suitable conscripts available. Napoleon sent out decrees calling up three-quarters of a million men. The response was 120,000—and many of these arrived too late to fight.

In the end it was the reduced size of the army which determined the style and extent of the campaign to hold France together under the leadership of its Emperor. Instead of being at the head of a *Grande Armée*, capable of delivering overwhelming hammer-blows against the enemy, Napoleon was forced to become a guerrilla fighter: his tactics being to strike and retreat, hoping to damage the enemy, perhaps even throw him off balance, but above all to confuse him, keeping his forces divided and as a result away from Paris. It meant a lot of hard riding and marching, to say nothing of the frequent night-raids and sharp 'commando' type attacks. Also it was a form of warfare to which he brought hardly any previous experience (unlike the Rhine Army-trained Ney). And yet the more I have read about these engagements at the beginning of 1814, the more convinced I've become that they inspired Napoleon's most brilliant generalship since the great days of Austerlitz and Friedland. How so few, and such badly-equipped French units could hold up the Allied advance for so long is a continuing source of fascination. Often they were outnumbered by ten to one. Yet for weeks, then months, they stalled and harassed the Allies so effectively that even highly professional commanders like Blücher and Von Bülow began to look decidedly second-rate. Obviously this chapter must be concerned with the part played by Ney in the campaign. But it should not obscure the fact that the Emperor was in overall control; and that although he was merely delaying the inevitable, not even the Revolution had caused France to be defended with so much spirit.

This last, late flourish of Napoleon's abilities as a general began on 2 January when General Curial of the Guards' depot at Metz received orders 'to send Meunier's division from Saarlouis to Épinal, and to alert those still being formed by Decouz at Thionville and Michel in Luxembourg'.[3] It was the correct move, because only the day before, while the Emperor was dictating these orders, Blücher had forced his way across the Rhine at three points including Mannheim. By this time Ney had returned to the army, capable of light duties and ignoring Aglaé's protests that he had allowed his body to be 'turned into a pin-cushion' for Allied bayonets. 'It's usually bullets that get me!' was his typical rejoinder. Napoleon immediately entrusted him with overseeing the work by Decouz and Michel to form two more Young Guard divisions. They were expected to defend the Vosges.

At first Blücher made remarkable progress. By far the boldest of the Allied commanders, on 17 January he was over the Meuse and by the 22nd he had seized a bridgehead over the Marne at Joinville: in all an advance of 75 miles. Schwarzenberg, upset by Mortier's fighting withdrawal, had only just reached Bar-sur-Aube: at least two days of straight marching for a link-up with Blücher. And from this point onwards Napoleon's tactics began to prove effective. At Bar-sur-Aube, Mortier cut up the Austrians' advance units and slipped away under the cover of a thick fog, ready to fight them again another day. (He also left behind him a Schwarzenberg more cautious than ever.)

On the same day Ney was advancing towards Brienne—to which Blücher, after also pushing Victor out of St.-Dizier, had advanced with 25,000 men and 40 cannon.

For Brienne—the scene of his schoolboy training—Napoleon issued orders that the troops were to form up on either side of the road to St.-Dizier, with Victor nearest Vitry and Marmont a mile beyond. 'Ney, with Meunier, and Decouz, supported by Lefebvre-Desnoëttes, Rottembourg and Oudinot' would be posted a mile beyond Marmont. The Marshal was instructed 'to spread the news of the offensive, send back the baggage, and issue bread and wine. If there is no wine but bottled champagne, use that. It had better be drunk by us than the enemy. . .'[4] Unfortunately, a copy of the orders fell into the hands of some Cossacks, who passed them on to Blücher. But this did not prevent the French from falling upon the Prussians' flank. Napoleon attacked with Grouchy's cavalry on the morning of the 29th; then at three o'clock in the afternoon Ney whipped his two divisions into Brienne itself while Victor seized the castle. In the dark, twisting streets below the castle Ney's *voltigeurs* caught the Russian battalions serving under Blücher in a devastating cross-fire. Decouz was killed, but by the evening Blücher had had enough. He retreated under the cover of darkness, leaving behind him over 4,000 wounded for the French to look after.

At 9.00 a.m. 1 February Napoleon began a well-disguised pursuit. He moved to La Rothière, just north of where Blücher had camped. Marshal Ney had orders to go back to Lesmont bridge; although it meant marching in a howling gale and there were the first signs of snow. At noon Victor reported the enemy were marching in three great columns back upon Brienne. Evidently 'Old Forwards' had recovered his nerve; although in fact the Prussian commander had been reinforced by Schwarzenberg and Wrede to a strength of nearly 100,000 men. Ney's first set of orders were cancelled, therefore, and his divisions of the Young Guard were called back towards Brienne, where by 8.00 p.m. that night Napoleon had fought Blücher to another bewildered standstill. The French lost more heavily on this occasion, but the Allies suffered a thoroughly disruptive check and Ney's Young Guard were left to retreat in good order towards Troyes.

The Emperor now predicted (correctly) that the Allies would attempt a two-pronged drive on Paris: by Blücher through the valley of the Marne and by Schwarzenberg along the Seine Valley route. Accordingly he decided to hinder them next by splitting his forces, sending Marmont to defend Château-Thierry and Meaux against Blücher while he himself attacked Schwarzenberg with forces including Ney's towards Bar-sur-Seine. But then he discovered that Schwarzenberg had stolen a march on him, so he hurriedly changed his plans.

On 6 February the Emperor 'faded back to Nogent behind a screen manipulated by Mortier and the Old Guard,'[5] where he had a conference-style breakfast with Ney, Berthier, Oudinot and Grouchy. Napoleon also learned that Murat had joined the Allies. 'Murat, to whom I have given my sister! Murat, to whom I have given a throne! It doesn't seem possible . . .'[6] But Ney expressed little surprise; and none whatsoever when the Emperor rejected the Allies' latest peace terms: that France must return to her pre-1792 boundaries.

Schwarzenberg had slowed up again. Blücher was still the faster of the two, and Napoleon attacked him in swift succession at Champaubert (10 February), Montmirail (11 February) and Vauchamps (14 February): all French victories thanks to the Emperor's surprise tactics. At Montmirail, Ney at the head of six battalions of the Guard *élite* launched an attack which crushed in the Prussian commander's left flank and it began to look as if the French might roll up their enemies and destroy them piecemeal. Sacken's Russians fled at the mere sight of Ney, throwing down their guns and leaving soup still bubbling in their kettles. The Prince of the Moscowa tasted it and pulled a face. Cabbage again!

Meanwhile, due to Blücher's soaking up all the punishment, Schwarzenberg's separate move towards the capital was at last beginning to make some progress. His Army of Bohemia was west of the River Yonne and his Cossack reconnaissance-patrols had reached the Loire. Clearly Napoleon needed to do something substantial over in that direction, and he therefore decided to leave Marmont and Mortier shadowing Blücher while he force-marched his remaining troops towards Chalmes: a mere 18 miles from Paris. A further move took him to Montereau and a hard-fought, but decisive victory over Schwarzenberg on the 16th. For the moment his capital was safe, the Russians and Austrians were in retreat.

More battles followed; some lost, some won. On 25 February Ney was ordered to Arcis-sur-Aube with his Young Guard: until Napoleon discovered that Blücher had lunged northwards, when they were ordered towards Sommesous. Such was the pattern of this war! The Marshal saw action at Craonne, Laon, Reims, Châlons-sur-Marne and finally back at Arcis itself before another month had passed. And the French could not afford to lose any more men.

Ney expressed this opinion to the Emperor shortly before they went into action together at Arcis. 'The enemy can afford losses. In our case even our victories bring us nearer to defeat.' Furthermore he had an alternative suggestion. Why not let him take the remnants of the Young Guard into Alsace and Lorraine and bring about a widespread insurrection by the populace? On 15 March, Ney had captured Châlons from Blücher, and he knew how angry the people there were at the Prussians' indiscriminate looting. With so many local revolts across their lines of communication the invaders would be forced to pull back. However Napoleon rejected the idea.' He was now fighting with 'an abandon the like of which he has not shown in two decades', according to Berthier, and so he gambled on a series of strikes against Schwarzenberg, Wrede and Barclay de Tolly along the line of the Aube. Leaving Marmont and Mortier to cover Paris he set off early on 20 March for Arcis, thought to be held by Wrede, but weakly.

In the battle which subsequently developed on 20 March he would find Schwarzenberg and Barclay facing him as well, *in strength*. Sending Ney and

Sébastiani across the river, the Emperor shadowed them from the north bank —and by 11.00 a.m. Arcis was safely in their hands. Napoleon crossed the river at one o'clock and joined Ney near the village of Torcy-le-Grand. 'He accepted without question the assurance of a single staff-officer that no more than 1,000 Cossacks were in the vicinity.'[7] Yet within the hour the French were being attacked by the enemy's massed cavalry. They held their own: Ney being especially active in providing support for Sébastiani's beleaguered horse-troopers. But then during the night Schwarzenberg successfully deployed no less than 80,000 men with 370 guns in a vast arc around the French. Once Napoleon saw these on the morning of 21 March he knew the game was up. Ordering an immediate retreat, he began to make plans for a stand nearer Paris and a new HQ, possibly at Fontainebleau. . .

Marshal Ney watched grimly as the retreat got under way. He had crossed the Aube still fighting, and behind him Oudinot's sappers were just about to blow the bridge under cover from Rottembourg's *tirailleurs*. Some of the troops and staff were marching off towards the capital; others appeared to be straggling away in the general direction of Champagne. There was a lamentable lack of clearly discernible orders. Had Berthier lost his touch?

If Berthier had lost his touch, then the Emperor himself was running out of luck. And Ney was beginning to accept the fact. Napoleon thought he had merely lost Arcis. His Marshal believed they were in the final stages of losing France.

On 23 March Tettenborn's Cossacks intercepted a packet of despatches from the Duke of Rovigo (Savary) in Paris, and brought it to the Czar. Savary wrote: 'There are influential personages in Paris, hostile to the Emperor, whom we have reason to fear if the enemy approaches the capital . . .' This confirmed a letter to Nesselrode, the Russian diplomat accompanying the Allied forces, composed by Talleyrand and incorporating a postscript in invisible ink: 'You are walking on crutches . . . use your legs . . . march straight to Paris . . . they are waiting for you'.[8] In effect, it meant that 'the gracious cripple in powder' (Lord Acton's description of Talleyrand) had all but triumphed in his campaign to depose Napoleon. Also that he was engaged in multifarious plots to hand Paris over to the Allies without a struggle.

Because of the interception the Emperor heard nothing more from Paris for two days. But then he was informed that the capital was in uproar. Its defences were incomplete. And Savary (at the head of police) succeeded in getting his second message through about the extent of Talleyrand's activities. Napoleon woke up to the situation immediately. He ordered a speed-up of the retreat towards Fontainebleau, and on the morning of 27 March conferred with Berthier, Lefebvre, Ney, Caulaincourt and Maret. 'His army must reach the capital in order to utilise its large resources before the enemy,' he said.[9] At dawn on the 28th he left with the Guard, the artillery and Ney. The cavalry were to follow, using Macdonald and Oudinot as their rearguard.

On the 29th the Emperor detached Ney and sent him back to defend Troyes; the French must try to delay the enemy there. He had received a further message from Lavalette, his loyal postmaster-general: 'Sire, your presence is imperative if you wish to prevent your capital being handed over to the enemy. There is not a moment to lose.'[10] Also at this time the Czar—fortified by the message from Talleyrand—was bullying the hesitant Schwarzenberg into resuming his march upon Paris. By contrast the gin-swilling Blücher needed no second bidding. He hated the French sufficiently to enjoy the prospect of a march to the Tuileries and another looting session. Napoleon's last words to Ney were to look after the army. 'Arrange for bread, provisions and supplies. The soldiers are famished. Also they have marched clean through the soles of their shoes!'

On 31 March Ney was still seeing to Troyes' defences when he received the startling news that Paris had capitulated—although as yet he had no knowledge that this was largely the result of Marmont's treason. The Duke of Ragusa (his title from this day forward passed into the French language as *raguser*, a verb meaning to *betray*) had been putting up a good show in front of the city together with Mortier. But then Talleyrand suddenly appeared bringing attractive offers from the Bourbons, and Marmont in a remarkable fit of self-importance decided 'that the destinies of France, of Napoleon, of all Europe, perhaps, were in his hands alone'.[11] Selected Allied units were allowed into Paris the next day (30 March), and on 5 April, while negotiations were still in progress, Marmont would additionally 'betray his own corps, 12,000 strong, into the hands of the Allies'.[12] With this he sealed the fate of the Empire.

Michel Ney's role over the next few days became crucial.

Napoleon had now reached Fontainebleau; and from there sent out orders for all his senior commanders to bring their forces to join him. After hearing the news from Paris he was already considering Orléans as a new provisional capital. 'By 1 April he had 36,000 men near Fontainebleau; two days later he could count 60,000.'[13] But what were these against the Allies' half-million? Meanwhile, in the real capital, Talleyrand was declaring him to be deposed, and playing host to the Czar while wild-looking Cossacks rode up and down the streets. Marie Louise and the King of Rome were at Blois on the Loire: sent there on the advice of Joseph Bonaparte according to Charles Doris.[14] Marshal Mortier had ordered the remains of the Guard to Fontainebleau where they prepared for a fight to the death if need be in defence of the Emperor's person. The whole situation was rapidly deteriorating into chaos.

Ney arrived at Fontainebleau on 3 April, and the next day, Palm Sunday, he attended a special parade of the Guard. At noon he appeared on the terrace of Le Cheval Blanc with Berthier, Moncey, Bertrand and Drouot and from there listened to Napoleon's rousing proclamation: 'Despite the success of the Coalition Army, over which it will not have cause to gloat for long, do not let yourselves be downcast . . . the Emperor is watching over the safety of all . . .'[15] Also, the Emperor announced, they would soon be marching towards a big, set-piece battle for Paris. It was the first time Ney had heard of this, and he judged it to be sheer folly. The 'grumblers' of the Guard noticed the long faces of the Emperor's suite, especially the Duke of Vicenza's (Caulaincourt, who thought the best they could hope for now was a Regency). Nevertheless the soldiers broke into shouts of *Vive l'Empereur! A Paris, à Paris!*

Ney remained silent during the rest of the parade; although he knew very well that any march on Paris with the troops in their present pitiful state was out of the question. But as the men began to disperse he made an important statement. He told his colleagues that whatever happened the Emperor must be prevented from further military action. 'And I will tell him so,' he said firmly; whereupon Lefebvre and Moncey agreed to go with him to the Emperor's study.

Entering, and bowing respectfully, they found Napoleon in conference with Berthier, Maret, Caulaincourt and Bertrand. He looked up at the newcomers but said nothing.

'Is there any news from Paris?' Ney asked him politely.[16]

'None whatsoever,' Napoleon replied, telling a deliberate lie.

'Well, I have heard bad news, Sire,' Ney informed him, still very politely. 'The Senate has declared against you. It has sided with Talleyrand and announced you deposed.'

Then the Emperor tried to bluster. 'The Senate has no power to do this, since the Empire is based on the votes of the whole nation! Only the people can demand the abdication of their Emperor! And as for the Allies, I am going to crush them before Paris!'

Ney immediately protested that further military operations would avail them nothing. That the state of the army forbade it. He pointed to the example of his own so-called two divisions, now reduced in size to less than a single regiment. And he spoke of the dangers of civil war; with Marshal Lefebvre supporting him. 'A pity,' he concluded, 'that peace was not made sooner. Now the only thing left for you is abdication . . .'

Napoleon noticeably winced at the mere mention of this dread word coming from one of his own officers—although he had used it himself only a few seconds before. Did they and their colleagues want to live under the Bourbons, he demanded of the marshals?

'Sire,' Ney replied calmly: 'I know that my family and myself will be only too happy to serve your son if he rules in your place.'

But this only caused Napoleon to revert to the idea of a military solution. He ran his finger over a large map stuck with pins. 'The enemy is in a position which is easily assailable,' he said, his voice rising. 'Look! In a few days we can cut their lines of communication. Then the people will rise against the foreigner . . .' He seemed to be returning, when it was too late, to Ney's idea of a popular rising in Alsace and Lorraine.

Meanwhile Berthier said nothing; and Caulaincourt was busily committing to memory everything that transpired.

At this point the arguments were interrupted by the arrival of Marshals Macdonald and Oudinot. 'You see, I have another fifty thousand men,' the Emperor added, referring to what he thought (erroneously) was the size of their joint corps. 'They will march with me to Paris. When we have driven the enemy from our gates, it will be time for the people to decide . . .' But even as he was speaking Ney was exchanging hurried whispers with Macdonald: 'Nothing short of abdication can get us out of this mess. The situation is desperate! He must be made to realise it . . .'; and when the latter spoke there was a notable absence of *politesse*.

'I must declare to you that we do not mean to expose Paris to the fate of

Moscow,' Macdonald announced with brutal frankness. 'We have reached our decision, and we are resolved to make a speedy end to all this.'[17]

Still Napoleon would not listen. Still he argued for a march on Paris. 'I don't care about the throne,' he told them. 'Born a soldier, I can become an ordinary citizen again without regret. I wanted France to be great and powerful . . . I wanted her especially to be happy. It is not for my crown I am fighting, but to prove that Frenchmen were not born to be ruled by Cossacks. If I lose this battle the poor French must submit to their rule. For myself I want nothing . . .'[18] He noticed the marshals shaking their heads. 'You do not agree?' he stormed at them. 'Then I shall march upon Paris without your agreement!'

'But, Sire, the army will not march on Paris,' Ney interrupted to remind him.

'The army will obey me!' Napoleon banged his fist upon the table, and looked Ney squarely in the face.[19]

The Marshal could contain himself no longer. The fervent patriot had become detached from the loyal subordinate. And when he replied it was with an awful finality. '*Sire*,' he announced in a hard, utterly ruthless-sounding voice which Napoleon had never heard from him before: '*The army will obey its commanders!*'[20]

He had deliberately forced himself to speak in this way because he felt it was the *only* way to convince Napoleon. However, the effect upon the Emperor was terrible to behold. Looking stunned, deeply hurt, physically disgusted, he slumped into a chair. So even Ney was ungrateful enough to desert him! He murmured for the marshals to retire, leaving him to consult with Caulaincourt. And later that afternoon he wrote out a single-sentence, carefully-worded offer of abdication:

> Since the foreign Powers have declared that the Emperor Napoleon is the only obstacle to the re-establishment of peace in Europe, the Emperor Napoleon faithful to his oath, declares himself ready to descend from the throne, to leave France, and even to lay down his life for the welfare of his country, which is inseparable from the rights of his son, those of the regency of the Empress, and the maintenance of the laws of the Empire. Given at our palace of Fontainebleau: 4 April 1814 . . .[21]

Napoleon signed the paper and instructed Caulaincourt to take it to Paris. Ney and Macdonald accompanied him; as the army's commanders they would have to negotiate the terms for its surrender with the Czar Alexander, dominating influence upon the Coalition (if one discounts Talleyrand, that is). At Essones they were joined by Marmont: who, for some reason they could not fully understand, appeared greatly embarrassed. (His actual dealings with the Allies were still very much a secret.) He confessed to having been in communication with Alexander and Schwarzenberg but spoke only in the vaguest terms about the present disposition of his troops.

The Czar had taken up residence in Talleyrand's house in the Rue St.-Florentin: an ominous sign. Nevertheless, when they arrived at midnight he was still up and received them with every courtesy, listening with studied interest as Caulaincourt read aloud the abdication note. Ney then begged him to use his influence towards obtaining its acceptance by the Allies, leaving France free to direct her own destiny. He stressed that the people did not want the Bourbons back, and that the peace of Europe would be better served if there was a regency under Marie Louise. Alexander in reply promised to consult the other Allied leaders and indicated that he might support the regency idea. After all, he pointed out, in Germany he had promised the countries free institutions following their liberation. Why not also in France? With high hopes that they had won their case the delegation then withdrew.

By breakfast time though the situation was altogether different. Talleyrand had used the intervening hours to press his own separate arguments upon the Czar, gradually convincing him that a restoration of the Bourbons promised greater stability than a regency. Alexander:

> found in Talleyrand no reflection of his own uncertainty. With over-whelming arguments and with irresistible logic the impression which the emissaries of Napoleon had produced on him was effaced. Whatever regency was set up, Talleyrand pointed out, Napoleon would in reality be the power behind it, within a year he would be once more openly in control, the army would once more be in the field, and all that had been accomplished would be undone.[22]

So, the views of the wily politician at this juncture overturned Ney's good intentions. Also the news came through of Augereau's abandoning Lyon and the reduced chances of Soult holding Toulouse against Wellington. Suddenly the Allies were in a stronger bargaining position than ever; and when Marmont officially disclosed the surrender of his troops they agreed there was no longer any need to offer Napoleon terms.

When Marmont informed Ney that General Souham had led his units over to the Allies the latter swore at him. 'God!' he exclaimed. 'It is the end of the Emperor and his family.' Marmont either pretended or showed genuine distress. 'I would have given my right arm to prevent this happening,' he said. 'Your arm,' Ney replied. 'Not even your head would be enough!'[23]

At the Marshal's next interview with Alexander the King of Prussia was also present: another ominous sign. They had decided against a regency, they informed him. Napoleon must surrender unconditionally. His dynasty was to go into liquidation—or rather, it would be moved. They were prepared to be generous. He might retain the title of *Emperor*, and they would give him the island of Elba, off Italy, together with a suitable income. Marie Louise would receive the Duchy of Parma, with reversion rights to the King of Rome. As regards the Bourbons, however, they were adamant. *Come what may, Louis XVIII would be restored to the French throne.*

Ney saw clearly that all further argument was useless. In any case, he had no troops to back up his words and therefore no separate source of power. Accordingly, although extremely brought down, he agreed to convey their intentions to Napoleon at Fontainebleau; and himself submitted to the Provisional Government which Talleyrand had been forming. He said he hoped it would represent the majority of Frenchmen.

'I don't envy you carrying this latest news to Fontainebleau,' Talleyrand smirked, obviously enjoying his long-worked-for moment of triumph. Somehow Ney managed to conceal his distaste for the little man with the elaborate *coiffure* who had outwitted them all. He promised to send him a report of events: another mistake, since Talleyrand would publish the report in the *Moniteur* of 7 April and extract the utmost propaganda value from it.

Briefly, what happened between the Marshal and Napoleon is as follows . . .

When Ney reached Fontainebleau it was only to discover that the Emperor was actually fortifying the place. And as he passed on the facts about Marmont's defection, together with the Czar's changed, harsher attitude, it seemed as if his words were falling on deaf ears. For suddenly Napoleon was announcing a new campaign! He would retreat upon the Loire, he said, together with the forces 'of Augereau, Soult, Suchet, Eugène, the depots, and the people. The Guard was to leave the next morning at 6.00 a.m.' 'But, *Sire*,' Ney expostulated: 'This is madness! You have no army; just a few disorganised units. Augereau is with the Allies, Soult is beaten by Wellington and Eugène is still in Italy. How can you possibly consider further action? The soldiers will not support preparations for a civil war . . .' Ignoring him, Napoleon carried on in the same vein for several minutes like a man deranged. So much so that Ney, Macdonald and Caulaincourt jointly forbade his orders to be transmitted to the remaining troops.

Not until the following morning (6 April) did the Emperor finally give in. After one more attempt to win over the marshals, shouting at each in turn and finding that not one would budge, he pulled a small mahogany table towards him. 'Very well,' he said angrily: 'You wanted a rest—go and take it.' And he sat down and wrote out his definitive, unconditional abdication, declaring that 'There is no sacrifice, including my life that I would not be willing to make for France.' Also he composed a second document for the benefit of the Commissioners (Ney, Macdonald and Caulaincourt) granting them powers 'to negotiate, conclude, and sign such articles, treaty, or convention as they see fit, and to stipulate all arrangements concerning our interests and those of our family, the army, ministers, Councillors of State, and other of our subjects who have followed in our path.'[24]

Afterwards, with officers and *aides* quitting Fontainebleau for Paris, Napoleon said to Ney: 'So, I have no more generals, nor friends, nor comrades-in-arms'. He had calmed down; but still spoke with ill-grace.

'*Sire*, we are all your friends,' the Marshal protested.

'Ah—but Caesar's friends were also his murderers,' Napoleon concluded savagely.[25]

[1] Marshal Jacques-Étienne-Joseph-Alexandre Macdonald, Duke of Taranto: *Recollections*, London, 1893.
[2] *The Anatomy of Glory.*
[3] Napoleon: *Correspondence.*
[4] *The Anatomy of Glory.*
[5] Ibid.
[6] Bourrienne: *Mémoires.*

7   *The Campaigns of Napoleon.*
8   *The Anatomy of Glory;* also in Duff Cooper's *Talleyrand.*
9   Ibid.
10   Antoine-Marie Chamans, Comte de Lavalette: *Mémoires et Souvenirs,* Paris, 1831.
11   William Milligan Sloane: *The Life of Napoleon Bonaparte,* New York, 1896.
12   *Napoleon's Marshals.*
13   *The Campaigns of Napoleon.*
14   Charles Doris: *Secret Memoirs of Napoleon,* first pub. Paris, 1815.
15   *The Anatomy of Glory.*
16   *Les Mémoires du Général de Caulaincourt,* Paris, 1933. Also C. Doris.
17   Macdonald: *Recollections.* Also Caulaincourt.
18   *The Anatomy of Glory.*
19   Macdonald: *Recollections.*
20   Ibid. Also Caulaincourt.
21   Napoleon: *Correspondence.*
22   Duff Cooper: *Talleyrand.*
23   Caulaincourt, who calls this section of his *Mémoires: No Peace with Napoleon.* Also Emil Ludwig's *Napoleon.*
24   Napoleon: *Correspondence.*
25   *The Anatomy of Glory.*

CHAPTER 10
A Soldier Under the Bourbons
1814-1815

But the new *régime* was not for him—the swarm of returning aristocrats had no place for the *Bravest of the Brave*, and only furnished the ludicrous spectacle of pretentiousness patronising genius.

HUBERT RICHARDSON
A Dictionary of Napoleon

Although Michel Ney was a soldier of quite remarkable talents one would hardly call him a military genius. The only living Frenchman with claims to that particular distinction was now swatting mosquitoes on Elba: lord of a Ruritanian-style court consisting of ardent Bonapartists and die-hard ex-Guardsmen, his boredom only partly alleviated by a visiting Marie Walewska and their natural son. However, if Ney was no genius, then the antics of the restored Bourbons soon began to make him look like one to the ordinary French ranker and his junior officers. Historians have rightly criticised the *émigrés* for their greed and superior airs; but a lot has still to be written about their blundering behaviour in dealing with the remnants of France's former *Grande Armée*.

Ney was certainly the best soldier the Bourbons had to call upon: more skilled than Victor, a better fighter than Marmont and more popular with the army than Soult or Macdonald. (Davout, arguably his superior as regards strategy if not tactics, was in bad odour: having refused to surrender Hamburg until many weeks after Napoleon's abdication.) For the new regime to succeed, therefore, it had far more than just a temporary need of the Marshal. To begin with King Louis XVIII relied upon the Allies to buttress him; but eventually he would have to depend on France's own forces proving loyal to a revised concept of monarchy. And good relations with Ney were an essential part of this.

Unfortunately it was a fact the returning Royals failed to realise, probably because they had no revised concept of monarchy. In years of exile they had grown older, but in no way wiser; and now they hadn't the slightest understanding of how the average Frenchman thought or felt. They came back autocratic and unforgiving. Louis wanted to ignore the Revolution and date his reign from the death of Louis XVI's son. His brother, the Count of Artois, was even more extreme. Leader of those *émigrés* soon to be named 'The Ultras', he demanded aristocratic privileges which would have turned the clock back to the days of the Sun King at Versailles. Their conciliatory gestures towards the moderates in France were quickly forgotten. Before long it was all noisy clamourings for the seizure of old property and a witchhunt against former revolutionaries. *Needless to say*: it didn't take very much longer for these same noises, plus the rumbles of discontent growing up around them, to reach the sensitive ears of the man on Elba.

On 28 April 1814 the Emperor and his suite had 'embarked on the British brig-of-war HMS *Undaunted* at St. Raphaël and set sail for Elba. Two days later, the newly restored government of King Louis XVIII signed the Treaty of Paris with the Allies.'[1] By this treaty the French agreed to return to their pre-November 1792 boundaries, while at the same time all decisions on the future of Europe passed into the hands of the four victorious Powers, shortly to go into continuous conference at Vienna The treaty also gave the Allies' representatives and limited numbers of their troops access to France's capital.

Marshal Ney first set eyes on his new sovereign at Compiègne, some eighty kilometres to the north-east of Paris. In sailing from England the Bourbons had decided to stage their triumphant re-entry into the capital, not from exile, but from one of their former royal palaces. And so the Provisional Government, still headed by Talleyrand, organised an interim welcoming ceremony at Compiègne for 29 April. Most of the Imperial marshals attended and were given prominent positions to the right and left of the main entrance. Swiss Guards and a contingent of National Guardsmen wearing white sashes formed the official Guard of Honour.

Louis and his party arrived at five in the afternoon, their coach drawn by six horses making a wide sweep of the courtyard before stopping close to where the marshals, Czar Alexander of Russia, Talleyrand and other dignitaries stood waiting. Their actual appearance did little to inspire the receiving body. Following years of overeating and inactivity the king's watery brown eyes and pouchy cheeks gave his face the look of a superannuated guinea-pig. He walked with obvious difficulty, leaning heavily on the arm of his niece, the ugly, unsmiling Duchess of Angoulême, daughter of Louis XVI and Marie Antoinette. ('A typical housekeeper,' the Czar commented later.) Behind these two tottered an aged death's-head, the Prince of Condé, assisted by his son the Duke of Bourbon: grandfather and father to the late Duke of Enghien whom Bonapartist colonels had kidnapped, tried and shot. 'What a bunch!' Talleyrand murmured.

For better or for worse though Talleyrand was now committed; and inside the palace he would be forced to sit through a number of upsetting audiences, his nimble disloyalty for once immobilised by the lack of any alternative. 'Sit' incidentally being the operative word: for etiquette was renewed with a vengeance, and as Louis graciously pointed out, Talleyrand's family background determined that 'he might be seated in the royal presence'. At the same time he reminded the ex-Bishop of Autun how 'in the course of history the House of Capet had been far more successful than that of Talleyrand-Périgord'.[2]

But the King's political pronouncements were to cause the most alarm. 'The Senate's decision *freely to call to the throne Louis Stanislas Xavier, brother to the late Louis XVI* on the basis of a Constitutional Charter, was communicated to him by the Marquis of Maisonfort who ended his speech with the words, *Sire, you are King of France. Have I ever ceased being that?* replied Louis coldly.'[3] To Czar Alexander he insisted 'absolutely' that he was King of France by divine right, and that he 'the legitimate King, would be pleased to give the Constitution to his people, as an act of Royal Benevolence. The Charter therefore was to be regarded not as the will of the Senate, but as the expression of a wish which the Sovereign would translate into an act of Government.'[4] And he would date this concession as having been given 'in the nineteenth year of my reign'. Small wonder that Talleyrand limped away to note: 'I have formed my impression of the man's character: selfish, insensitive, Epicurean, ungrateful.'[5] While the Czar remarked: 'He seems to think he is Louis XIV!'[6]

Later that evening, however, Louis received the marshals, an audience at which he made both a conscious effort and a more favourable impression. In reality he had a softer disposition than the other returning Bourbons; nor did he lack intelligence—although this last quality was frequently distorted by bitter memories and an exalted sense of his own position. Berthier introduced the marshals one at a time and the King found a nice thing to say to each. At the end, standing to indicate the audience was over, he waved aside his courtiers, announcing that on the whole he preferred two of the marshals to support him. 'If I could campaign,' he said, 'then I would feel proud to march in your company. It is on you, *Messieurs les Maréchaux*, that I desire to lean. You have always been good Frenchmen!' Furthermore, at a banquet which followed, the tables were so arranged to give the marshals seats of exact equality with the *émigré* nobility.

The next morning, at the first of what were to become known as the King's *levées*, Ney found himself singled out for special treatment. Moreover extremely complimentary treatment. Louis revealed quite an extensive knowledge of the Marshal's career, wounds—even his life-style. 'During my years in England,' he said flatteringly, 'at Hartwell in Buckinghamshire: I followed all of your campaigns with the greatest interest.' The two conversed amicably for several minutes; at the end of which the King addressed Ney as '*mon ami, Le Prince de la Moscowa*' and asked whether he had any particular advice to give about making his reign truly secure. 'I can answer that very easily,' the Marshal replied. 'Sire: change the name of the Imperial Guard to the Royal Guard and your throne will last forever.'[7] The implication being that this fine body of men, founded upon personal loyalty, if once won away from their attachments to Napoleon would serve Louis with identical devotion. The King rose, and beamed. 'That sounds like good advice,' he said.

3 May. Louis XVIII made his solemn entry into Paris, preceded by a general distribution of hand outs which called upon the French to 'demonstrate their love for the true King in return for his safe-guarding their individual and public liberties'. He would also 'bestow on them' a new Constitution, the result 'of our work with a Commission'.[8]

Curiosity if nothing else brought the crowds out. The King's carriage this time was pulled by eight horses. But even skilful tailoring—a blue greatcoat with golden epaulettes—failed to disguise his immensity and age from the capital's citizens; while the expression on the face of the Duchess of Angoulême made it perfectly clear to those lining the route that she viewed *all* of them as the same monsters who had guillotined her parents. Again the King was accompanied by 'death's-head' Condé and his son, the Duke of Bourbon. Behind their carriage rode other leading extremists: Charles of Artois, posing in the uniform of a National Guardsman, his sons the Dukes of Angoulême and Berry, and finally the odious Blacas and the Duke of Duras, Louis' favourite adviser and 'grand master' of etiquette respectively. Following these came the marshals, wearing white cockades but otherwise in their Imperial dress-uniforms. Berthier led them—and came in for the most heckling. 'Go to Elba, Berthier!' the Bonapartists called out.[9]

Marshal Ney rode along with a very straight face. Vitrolles refers to his 'withdrawn look',[10] but really there is no point in supposing that this was anything more than a dignified military bearing. His aggravations during the new reign had yet to begin. For the moment he seemed to have the King's confidence, was about to receive various honours and therefore had no cause for complaint. No, the more immediate problems that day concerned the attitude of the Imperial Guard.

Five hundred grenadiers of the Guard had been ordered up from Fontainebleau to escort the King into his capital, and on to Notre Dame for the solemn *Te Deum*. Others were lined up on the Pont Neuf, where the royal procession paused—quite deliberately—before the statue of Henri IV: the first Bourbon King of France.

> This was an honour, to be sure; but the day began badly. On the 'grumblers' arrival at the Napoleon barracks General Friant deluged them with advice, enumerating everything he expected of them—good discipline, impeccable uniforms, etc. But first, they must remove from their bonnets the plaques with the cuckoo, since it was no longer in season. In vain had they brushed their bearskins and attached white cockades—the *Tondu's* insignia was still there, making their disguise incomplete.[11]

As the hours of waiting dragged on their mood changed from bad to worse. For the first time in its existence, the Guard was not at the head of the line. This post was reserved for the Russian guards. 'Was Louis Stanislas Xavier really King of France? The veterans of Tilsit ground their teeth. And here was Friant wanting them to cry *Vive le Roi*!'[12] They also recognised Napoleon's green carriages being used. The arms on each of the hammer-cloths had been covered over with a paper fleur-de-lis—which it amused the street-urchins to peel off; meanwhile half the grooms still wore Empire livery. 'This parade seems more like a funeral,' one grenadier muttered. Again, they demanded to know: 'Where are our bands? Why must we have Russian, Austrian, even Prussian music?' But the final insult was delivered by the Duke of Angoulême. He arrived to take his place in the procession dressed as an

English general! Just tactless behaviour? To an elite body of men, reared on hard-fighting, love of country and love of its commanders, nothing could possibly have occurred to anger them more . . .

By the time Louis' carriage halted on the Pont Neuf the guardsmen were seething. Chateaubriand's recollections are the most famous:

> I do not believe that human faces have ever worn such threatening and terrible expressions. These grenadiers covered with scars, these conquerors of Europe were forced to salute an old king, a veteran of years and not of war, while an army of Russians, Austrians, and Prussians looked on in the conquered capital of Napoleon. Some puckered their foreheads into a scowl that pulled their enormous fur bonnets down over their eyes, as if they did not want to look; others drew the corners of their mouths into grimaces of scorn and rage. When they presented arms they did it furiously. It must be admitted that no men were ever put to such a test, nor made to suffer greater torture . . .[13]

If this description sounds somewhat lurid, then there is the additional evidence of Madame de Boigne—who watched the procession in the Rue St.-Denis:

> All I have to say is that its (*the Imperial Guard's*) aspect was imposing but it froze us. It marched quickly, silent and gloomy. It stopped, by a look, our outbursts of affection; the shouts of *Vive le Roi* died on our lips as it strode past. And as it passed, the silence became general, so that nothing could be heard but the monotonous tramp, tramp of the quick step striking into our very hearts.[14]

On the Pont Neuf Louis XVIII was subjected to the full impact of this silent hostility. Visibly alarmed, he signalled for the carriages to drive on; and once the various important personages had disappeared into Notre Dame, General Friant used his head. Although a Bonapartist himself (who would be wounded on Napoleon's behalf at Waterloo), nevertheless he marched the grumblers straight back to barracks—and kept them there. Having clearly upset the King, he didn't want the additional scandal of their brawling with visiting Russian and German troops when the festive spirits began to flow. By this time however the long-term damage was done. For as Louis struggled to his feet inside the cathedral, and as the rich, massive chords of the *Te Deum* reverberated all around him, he was already (mentally) sealing the Guard's fate. Marshal Ney's advice would be set aside. Instead of seeking to win their affections, he resolved to disband, disperse, *remove them from any further place in France's army!* Yes; he must break the decision to Ney gently though . . .

It was probably with this in mind that Louis' cultivation of the marshals now proceeded at a rapid rate. 'They were confirmed in their ranks and Imperial titles, and fourteen of them became peers in the new Upper House which was to be convoked as soon as the business of Constitution-making was over.'[15] Again Ney was shown preferment, ahead of Berthier and even those two 'zealous Royalists', Victor and Marmont.

He became a chevalier of the Order of Saint-Louis, received the posts of honorary colonel to a number of regiments (each of which meant an increase on his salary) and as Duke of Elchingen and Prince of the Moscowa was created a Peer of France. (Although the King requested him not to mention the last title too loudly whenever the Czar was present!) On a purely military level he became Commander of the VI Military District (which also made him Governor at Besançon) and was given special responsibilities for overhauling the French cavalry. Taken altogether it meant a promotional bonanza. Certainly enough to keep most men contented.

But Michel Ney was not a venal man; and in this Louis had miscalculated. He cared about France, the army and his own immediate family: three things which would influence the remainder of his brief service under the Bourbons.

France, he hoped, was being taken care of in the series of discussions begun between the Royalists, Talleyrand, members from the two Chambers and the inevitable teams of lawyers. These followed the King's giving the nation a Constitution on 4 June and were designed to reconcile the Royalists with all former adherents to the Revolution. Also on 4 June Ney swore a peer's oath of 'inviolable fidelity' to the monarchy, so there is no reason to suppose that he was unhappy with the attempts being made to improve France's government. But the army and his family concerned him more directly, and here his initial confidence in the new regime would very quickly ebb away.

Because, regrettably, General Bonnal's great work on Ney only takes us up to March 1812 no research of his day-to-day activities as a soldier for Louis XVIII was carried out until Commandant Henry Lachouque's *The Anatomy of Glory*. This last book is essentially a history of the Imperial Guard, but it contains enough references to Ney for one to piece together what he was doing. Other historians have expressed surprise that he did not spend more time at Besançon, a governorship he delegated to the King's commissioner there. The truth is that his job with the cavalry kept him almost at full stretch, beset by problems which his Bourbon masters merely exacerbated.

On 12 May a Royal *Ordonnance* announced that the army was to be cut back from half a million to just over 200,000 men. France was at peace and the country needed to economise. But Dupont at the War Ministry carried out the programme with indecent haste. (Louis had wanted Ney as War Minister, only to be talked out of it by Blacas and Artois.) Naturally it led to hardship and ill-feeling. Veterans, former heroes, were turned into civilians without any particular artisan-skills and often without their back pay. Many were forced to beg: a sight which did little for the morale of the troops still serving. Also on 12 May it was announced that the Imperial Guard would be replaced by *La Maison Militaire du Roi*.

This came as a profound shock to Ney, but there was nothing he could do to stop it. In the *Maison Militaire* all appointments lay within the power of the King, and all officers must be nobly born. It was a return to the days of privilege; the Revolutionary army as Michel Ney knew it would cease to exist. Instead of serving the nation these new soldiers became the exclusive property of the King. Meanwhile the Guard was to remain at Fontainebleau until its members could be formed into new regiments or posted to garrison duties around the country. The King's decree stated:

The infantry of the Old Guard will form two regiments of three battalions, the first to be known as the Royal Corps of Grenadiers of France; the second, as the Royal Corps of Foot Chasseurs of France. The cavalry of the Old Guard will form four regiments called the Royal Corps of Cuirassiers of France, Royal Corps of Dragoons of France, Royal Corps of *Chasseurs à Cheval* of France, and Royal Corps of Light-horse Lancers of France. . .[16]

Marshal Oudinot was entrusted with the infantry and created 'Colonel-General of the Royal Corps of Grenadiers and Chasseurs of France'. The mounted were handed over to Ney's new cavalry pool.

However, if the latter could do nothing to preserve the Guard in name, then he fought a skilful and successful campaign to keep the horse part of it intact; and generally to improve the French cavalry's equipment and efficiency in the face of stringent budgeting from Paris. He had inherited an unenviable task. In Napoleon's last, complicated sequence of battles across France the cavalry had taken a good deal of punishment. It was now characterised by broken harness, threadbare uniforms and underfed mounts. But Ney tackled the job with his usual abundant energy.

He began by paying the men's arrears out of Napoleonic funds. This was money which Government ministers badly wanted to get their hands on, but Ney beat them to it.[17] Then he experienced a host of difficulties because the quartermasters had been allowed to store clothing reserves with private individuals; while many rankers had deserted with their arms and equipment. As well as lambasting the War Office for its slipshod methods he also delivered a sharp rebuke to his cavalry colonels. 'I want proper inventories of men, horses and stores', he informed them. '*And fast!*'[18]

A white silk standard blazoned with the arms of France was presented to each corps commander except the dragoons who were given a *guidon* (broad-pennant). All but the *cuirassiers* retained their old uniforms. The latter wore blue coats without reveres, like the Line, trimmed with scarlet, with white cuff flaps, yellow grenades on the skirt flaps, and fleur-de-lis buttons. Marshal Ney ordered the officers to wear moustaches.[19]

He had good senior officers. General Guyot as Colonel-in-Chief of the *cuirassiers*, with General Jamin as vice. Generals Lefebvre-Desnoëttes and Lion held these positions in the chasseurs, and Ornano and Letort in the dragoons. Colbert headed the lancers, and from this body Ney expelled a large number of homesick Dutch, replacing them with trained Frenchmen and Mameluks. 'I should like to know whether I shall be retained,' Colbert put to him in a memorandum. He was—and in justifying this Ney described him to the War Office as 'a consummate officer of the greatest distinction'. His other letters to the War Office were more blunt:

There are not enough NCOs, the horses are mediocre, the clothing worn and motley, including bits and scraps of uniform of the guards of honour, scouts, lancers of Berg, etc. The tack and equipment are delapidated. Musketoons, lances and more than 300 pistols are missing. The accounts are chaotic, the magazines as empty as the treasury. The quarters at Orléans are poorly ventilated and the stables inadequate.

This was only the first of a long stream of complaints which the Marshal dashed off to Dupont; but other commanders were taking it easy and, gradually, the War Minister gave in if only to keep him quiet. Within two months the cavalry's 'spirit was good, its discipline strict, and the bread and forage of good quality.'[20]

Ney was particularly interested in the dragoons. 'They are the ones you fight with when in most difficulty,' he once remarked. The new dragoons were presented to him by General Ornano, the *seigneur* of Aubenas—and like Napoleon, of Corsican descent. 'The men are very handsome, the horses fine, strong and well cared for,' the Marshal noted, 'the officers, NCOs and soldiers are animated by an excellent spirit, perfect discipline, and have a splendid appearance. The barracks are not much to look at, but well kept. The stables are ample and sanitary.' His sharp eye missing nothing, and Dupont didn't escape another nagging. 'The clothing and equipment need repair and the troopers are lacking two pairs of grey overalls,' the Marshal reminded him.[21]

'I beg you, *Monsieur le Comte,* to garrison the Chasseurs of France at Saumur,' Ney wrote to the Minister on 30 July. He seemed to be continually on the move, inspecting one regiment here, another there. But at least his efforts were now proving themselves and a degree of real optimism had crept into his despatches:

> The spirit of the Chasseurs is excellent. The soldiers know their business ... and are handsome in build and appearance. Their clothing is practically new and perfectly maintained. The tack and equipment are complete, in good condition and very well kept. The barracks and stables are excellent. Only the beds are mediocre. The horses are beautiful and of a good type.[22]

The regiment still owed the horse-dealers at Caen 131,000 francs, dating back to 1813; but Ney had used his good offices to obtain extended credit for them, while at the same time prising money for the new beds out of a minister who by now was considerably in awe of him. 'Thank God there is only one Marshal Ney in France,' Dupont exclaimed to his Paris hostess upon receipt of yet another demand for funds.

The cavalry remained staunchly Bonapartist. At a review in June three-quarters of the chasseurs refused to cry *Vive le Roi.* 'Had he (Ney) recommended Saumur as their garrison so these subversive utterances would not reach the royal ear?'[23] When the Duke of Angoulême came to review the same chasseurs a month later, and raised their pay by 5 sous, the troopers spent this extra money toasting the Emperor with cannon balls of strong red wine. But if he didn't make them Royalists then at least Ney gave them back faith in their own capabilities. Slowly but surely, helped by the money he secured for new equipment and horses, France's cavalrymen became a powerful fighting force again.

'Last come the *gods*, the supreme reserve, whose privileged position the whole army—and even the Guard—had envied.' This was the body of 'Royal' Cuirassiers. Ney completed their reorganisation in time for a big review at Blois on 23 July. His 'praise of their spirit and conduct was unstinting.'[24] By autumn all of the mounted regiments were up to strength. And at the end of the year, when Marshal Soult replaced Dupont at the War Ministry, he was so impressed by the revived cavalry that he sent its units off for combined manoeuvres with Mortier's troops in the northern (XIV) Military District, 'where they would be appreciated'. But they must not stay for too long in any single place. 'The prefects of police had warned against leaving these regiments in one garrison for fear of subverting the inhabitants.'[25] So much for the Bourbons' popularity in France! Ney was aware that certain officers and NCOs had a regular correspondence going with Elba. He put it down to nostalgia though; and he didn't blame them. After all, who wasn't missing Napoleon's care and expertise by this time?

Although still loyal to his peer's oath, the Marshal had other, more provocative reasons for feeling discontented. Apart from having to battle for every last franc he was now subjected to constant interference by persons without the least qualification for doing so. At the War Office he was dealing with fellow-officers. But the *émigrés*, and especially the Ultras, seemed to think they had an automatic right to dabble in France's military affairs.

The Count of Artois, for instance, after proving himself a totally incompetent officer when endeavouring to combat the Revolution, took it upon himself to speak on the army's behalf at Royal Council meetings. His achievements in this capacity varied from deluging Ney (who was excluded from the Council) with idiotic advice to having the Duke of Havré, past seventy and bent over like a hairpin with rheumatism, created a new marshal. He was also fond of displaying himself in gorgeous uniforms, which resulted in the cancellation of vital training-sessions so that he could have troops on the parade ground for narcissistic reviews.

An even greater nuisance was his second son, the Duke of Berry. For here Ney had to put up with a young man who fancied himself a born general. He insisted on being the Marshal's admirer, though this was largely an excuse to have Ney accompany him during parades and inspections.

> He made a promising début. In the early days Ney was with him during an inspection when one of the men was heard to cry *Vive l'Empereur!* The Duke spotted this dangerous agitator and asked him quite reasonably why he should love the Emperor so much. *Because he led us to victory*, came the reply. *Is that so surprising*, said the Duke mildly, *since he commanded men like yourself?*[26]

But very soon this little, roly-poly creature began to imagine that he could replace Napoleon in the army's eyes! On 25 July, again with the Marshal present, he reviewed both Royal Grenadiers and the Chasseurs of France at Fontainebleau and presented their colours.

Had this military amateur been more tactful he would have avoided aping the Emperor's mannerisms. He expressed a genuine passion for the army and said he wished it to love him; but to address the grenadiers as *thou*, pull their ears, taste their soup, and question them brusquely in the manner of the Little Corporal was the royal road to failure. The response of the troops was glacial.

The newly revived Order of Saint-Louis was awarded to the generals, colonels and other officers present . . . and the Order of the Lily was presented to all. Then, obviously infuriated by his reception and venting his spleen under his breath, His Royal Highness beat a hasty retreat without eliciting any response from the Grumblers beyond a few surreptitious smiles.[27]

Sensing he was making a fool of himself, Berry's attempts at *bonhomie* broke down. On one parade Ney witnessed this at close-quarters. 'When a Captain of Infantry . . . contradicted His Royal Highness, he had his epaulettes torn off and stamped on by the enraged Prince.'[28]

It's hardly surprising that the Marshal lost confidence in his new masters. What future was there for his beloved army if it had such men as these to lead it? Certainly men of his own background, who had risen on their abilities, were not intended to lead it. For the *École Militaire* was now re-established, announced a Royal edict, so that 'the nobility of our Kingdom may enjoy the advantages that were accorded to it by the edict of our ancestor given in January 1751'.[29] But that means my own sons will be barred from entering, Ney remarked gloomily to Davout, with whom he was openly associating again. 'Our ancestor?' Lefebvre exclaimed: 'Why, I am an ancestor myself!'

However, what finally turned Michel Ney's discontent to genuine anger, and then led to several explosions in his old, celebrated style, had its origins away from the parade ground or even establishments like the *École Militaire*. Characteristically though it involved mostly the same group of people, specifically due to their treatment of both himself and his wife in the capital.

Ney, on the evidence of his previous career, could never be described as a social animal. And in any case his Herculean labours with the cavalry kept him almost constantly on the move, dashing first to Nevers, then on to Orléans, then up to Fontainebleau, back to Saumur, north again to Angers. But he was jealous of his military titles and he did love Aglaé: and for each of these reasons the Bourbons' behaviour was guaranteed not only to infuriate him, but also—ultimately—to lose them his support.

Despite their tantrums in front of the troops (which they expected the Marshal to condone) Artois and Berry always refused to let him travel in their carriages; or even to sit down at the same table. Yet on his occasional visits to Paris he would discover these two and their coterie fawning upon visiting dignitaries, several of whom had played important parts in bringing down France's Empire. Ney put up with this for a time; but already the lid on his temper was beginning to rattle.

When the first explosion occurred it concerned, oddly enough, the Duke of Wellington. He was in Paris as Great Britain's official ambassador to the Bourbon court and quickly gained a reputation for splendid party-giving. He 'met Ney almost at once, out hunting in the Bois', and later on was introduced to 'his handsome wife, Aglaé'.[30] These encounters were perfectly friendly. Following Bussaco we know that Ney maintained a high regard for Wellington's abilities; nor was it ever recorded that he bore a grudge against any of the few people who had defeated him in battle. But one September evening at the Tuileries, when he saw how the Duke was being lionised by the Ultras, he suddenly let fly at the courtiers standing nearby. 'That man,' he exclaimed, 'did well in Spain thanks to Napoleon's mistakes, not because we had bad generals. Just let us meet him one day when luck is not all in his favour! Then the world will see him for what he is. And now to see him flattered in this manner—in the presence of Marshals of France—he, our country's worst enemy!'[31] He wasn't alone in such feelings. For here is General Foy writing in his diary:

> Lord Wellington . . . is held in horror by everybody. Our telegraph is at his disposal two hours daily . . . he signs his letters from his headquarters in Paris; he has an air of saying to us: if you sit on the fence it is me you will be up against. We see him coming away from the King *en frac* and in boots. The Princes go and dine with him after manoeuvres. O Napoleon, where are you?[32]

No wonder Wellington was soon writing home: 'I think we are getting a little unpopular. . .'[33]

Ney's next outburst was aimed in another direction though. 'His blood pressure rose even higher when he heard that the Emperor of Austria was talking of removing the bronze Column made of Austrian guns. *So!* he roared, in a most terrible voice. *Does he want us to capture enough to make a second column?*[34] Blindly, stupidly, when the Ultras heard about these things he was uttering they went out of their way to alienate him further, turning each chance to annoy him into a kind of sport. 'Knowing how sensitive he was of the honour he felt due to all the marshals, another of the courtiers threw him into a violent temper by pretending not to recognise one of Ney's famous colleagues and remarking, either of Lefebvre or Macdonald: *Now who is that fellow over there?*[35]

In spring, when the violets bloomed and the first swallow flew in, they would come to regret their misplaced humour: dashing away north by whatever means they could find for a second time.

But it was Aglaé Ney who bore the brunt of the Bourbons' snobbery and bad manners; for she had to stay on in Paris while her husband visited the troops. Also the verbal thrusts and snubs of the different Ultra menfolk were like so many gnat-bites compared to how their women treated the marshals' and other First Empire wives. It was this treatment which in the end drew from Ney his most virulent attack of all upon the Royalists.

Aglaé at first had appeared to be enjoying the Restoration. Suddenly, after the dark, desperate days of the Empire's passing, typified by its unsmiling faces at Napoleon's court, everything in Paris became brightness and receptions. Her great friend Hortense de Beauharnais was back in the capital, a favourite of the Czar and with Charles de Flahaut as her lover. She,

Aglaé Ney and the Duchess of Abrantès (now a widow, Junot having committed suicide in 1813) took turns at giving glittering parties. At their salons one could meet generals, writers and philosophers, musicians like Cherubini, painters as varied as Prud'hon, the fashionable Barons Gros and Gérard or the young Ingres, even some of the more enlightened *émigrés*: all people promising to give France, thanks to the peace, a whole new social character and charm. If Ney protested from time to time about the expense Aglaé would kiss him and tell him 'not to be such an old grouser'.

Also at this time she was receiving the pleasant attentions of a young Englishman, Michael Bruce. She'd met him at Hortense's and expressed delight upon hearing how he had managed to avoid fighting France during the recent wars. He gained acceptance as Aglaé's official cavalier-servant and he was soon a frequent visitor at the Neys' Paris home. 'For the first, and last, time since her marriage she permitted herself the pleasures of a flirtation in which sentiment, an unexpected romantic side to her being, was gradually awakened. She was a woman of tact and the situation was, especially in Restoration Paris, made respectable by convention.'[36] This harmless relationship, with no reported misdemeanours, was also accepted by Ney who grew to like the young man. Bruce for his part became devoted to the Marshal, and at a later date would reappear, quite dramatically, in a bid to save him.

Happy days; but unfortunately for Aglaé they were of limited duration. A sensitive woman, gentle and considerate, she was by temperament totally unprepared for the unkindnesses about to be showered upon her at court.

It took the *émigrés*, after hurrying back to Paris 'in carts and wagons and wicker salad-baskets', some three or four months to establish a new court ceremonial. But once they did so then the King, more moderate in other things, became their avid partisan and caused it to be made inviolable. Czar Alexander's comment—'He seems to think he is Louis XIV'—was not far short of what Louis hoped to be in this respect. The Sun King at Versailles had organised his court on the basis of exclusiveness, precedence due to one's rank and a most elaborate ritual of movement. It is all fully described by Saint-Simon, a pitiless, but brilliant observer. On the exclusiveness: 'Those who had the *entrée* entered the private apartments by the mirror-door that gave on to the gallery and was kept shut. It was only opened when one scratched at it and was closed again immediately.'[37] The precedence and ritual are perhaps best summed up by this snippet about the King's hatmanship: 'For ladies he took his hat quite off, but more or less far, as occasion demanded. For noblemen he would half remove it, holding it in the air or against his ear, for a few moments, or longer. For landed gentlemen he only touched his hat. Princes of the Blood he greeted in the same way as the ladies.'[38] But the Sun King's Versailles had its own political logic. After the disastrous rebellions known as the *Fronde* (literally 'sling' or 'catapult') it kept a pack of troublesome nobles locked up around his immediate person, while the real government of France was carried out from Paris by a group of able, middle class ministers. In contrast, Louis XVIII wanted a ceremonial which would 'demonstrate the lofty heights on which the Kings of France traditionally dwelt.'[39]

Harold Kurtz gives us a very detailed picture of how the Restoration ceremonial actually worked. The King had entrusted it to the Duke of Duras —still wearing his pre-Revolutionary clothes—and to the Marquis of Dreux-Brézé, 'the son of Louis XVI's grand master of ceremonies whose conduct on 23 June 1789, had caused the Third Estate to declare itself a National Assembly'.[40] Duras, in daily consultation with Louis, created the details of etiquette; Dreux-Brézé supervised their application.

> The result was that the ceremonial once in use at Versailles was now being enforced in the far less spacious conditions of the Tuileries; and it was restored with a rigidity and self-consciousness that would have driven Louis XV away in horror. It was not merely that long-forgotten gradations of privilege and precedent were revived with merciless rigidity, and old Court charges brought back in antiquarian devotion. The trouble was that these matters, in the conditions of 1814, quickly assumed the character of political weapons to defend the King's position against the onrush of the liberal age. The ludicrous discomfort which this revival inflicted was only equalled by the offensiveness to which it gave rise: for what was the good of being born a Duchess if you could not show your superiority over one created by Bonaparte? It was the innate defensiveness of the whole sad process of artificial respiration which led to all the trouble.[41]

Aglaé Ney, although a Princess of France, now found herself reduced to being a lady of second rank because the *tabouret* (stool) was re-introduced.

> *A Duchesse à tabouret* was allowed to be seated at the King's table. Duras decided that those who had this privilege should be the first to enter the throne room on the great reception days while lesser ladies should line up in the ill-lit, draughty and cramped Salon de la Paix and wait until their betters had left the Royal presence before being admitted themselves.[42]

Thus, Aglaé and the other marshals' wives were forced to stand about 'while dowagers of the Faubourg St.-Germain would triumphantly sweep through the folding door separating the brilliantly-lit throne room from the gloomy antechamber.'[43] Eventually an officer would signal for them to go through, 'and the ladies who had once surrounded the Emperor and Empress in these very halls, would be allowed to enter to pay their homage to the King who, depending on the state of his gout, would receive them either standing up or seated.'[44]

From here they then moved on: through the Galerie de Diane and down the stairs to the Pavillon de Flore, where the Duchess of Angoulême received them. 'The same procedure of waiting and tardy admittance would be repeated', before their ordeal terminated with a visit to 'Monsieur' (the Count of Artois) at his Pavillon de Marsan. For this they had to descend to the basement and walk along a kitchen corridor, or risk the cold air on their lightly-clad bosoms and mud on their satin slippers by crossing the road outside. For not even the *Duchesses à tabouret* were allowed to go back by the same route. 'In his deliberations with Duras, the King lavished the utmost care on . . . every aspect of Court procedure, including the costumes to be worn. No Cabinet Minister ever succeeded in rousing his attention to a similar degree.'[45]

The Duchess of Angoulême's salon was always the worst part though. After hours of waiting for presentation to the King, the marshals' wives would be herded on towards the most daunting presence in Restoration Paris. Daunting, and haughty, but by no means remote. For the Duchess and her 'ladies' proceeded to demonstrate their own ill-breeding by setting about their women visitors on every conceivable occasion. Barbed innuendoes over position; sneering references to their ancestry: the attacks were formidable. Some of the visitors gave as good as they got. Lefebvre's wife, Catherine Hubscher, now the Duchess of Danzig but better known as 'Madame Sans-Gêne' ('Rough and Ready'), countered by describing how she once scrubbed floors as if it was the most natural thing in the world for a Duchess to do.[46] Aglaé Ney, on the other hand, was absolutely crushed. One must remember that she had been brought up by Madame Campan, Marie Antoinette's close friend; and her own mother, a Royalist, had killed herself on the day of the Queen's execution. Accordingly she felt that a kind of mutual sympathy should exist between her and Marie Antoinette's daughter. Instead of which, at the first of the new-style receptions, the Duchess of Angoulême patted her on the cheek just as she might a seamstress or hairdresser. 'Our dear little Aglaé,' she drawled, 'aren't you the niece of Madame Campan, the baker's daughter?'[47]

One evening early in November, Ney returned to Paris, dusty and tired, only to be confronted by his wife in floods of tears. 'There now: what's the matter?' he demanded with his usual gruff tenderness.

'The Duchess of Abrantès and I were at the Tuileries today,' she sobbed, 'we went to see the King about the bazaar we've been planning. The Duchess of Angoulême passed by and we bowed, whereupon she stopped and looked us over from head to foot. *You are, I believe, Madame Junot: are you not?* Then she turned to me. *And you, of course, are Madame Ney . . .*'[48]

'What!' the Marshal bellowed. Then, through clenched teeth: 'Don't let that foolish woman upset you any more, *ma petite*. I'll soon settle things with her!' And off he stormed to the Tuileries—like Wellington, still wearing his riding-boots.

Brushing aside the ridiculous Dreux-Brézé—'Out of my way, *laquais!*'—he marched directly into the King's presence. Louis recoiled in alarm at the sight of his ferocious expression, but he was not the prime object of the Marshal's wrath. After paying his respects in the most cursory fashion Ney swept on to the Duchess of Angoulême. No exact account survives of all that he said. It apparently began with 'I and others were fighting for France while you sat sipping tea in English gardens', and was terminated by: 'You don't seem to know what the name Ney means, but one of these days I'll show you!' In between though the Duchess and her blue-blooded coven were treated to a very full description of their physical selves, their ancestry and what he thought about their manners. Moreover Ney—normally so polite with women—did not stint himself of that colourful barrack-room language which has served French generals with equal effectiveness from Du Guesclin to De Gaulle. Then he left: as abruptly as he'd arrived.

The impact was shattering. For once the Restoration's top women were lost for words; left rocking on their dainty little heels. When eventually the Duchess of Angoulême emerged to dress the King's sores she looked pale and was noticed to be trembling slightly. But by this time anger had replaced incredulity. Louis, timid by nature and frightened at what had gone on, proposed patching things up with Ney by creating him a Gentleman of the King's Bedchamber. She froze him with a glance. 'No!' she hissed. 'Never! Because that . . . that marshal is no gentleman!'

As for Ney, he was not the sort to appear again at Court until someone had extended an olive-branch. Soon afterwards though he bumped into Lavalette. (A Count of the Empire, formerly Napoleon's Postmaster-General.) Lavalette had heard something of the roof-raising which had gone on inside the Tuileries. It didn't surprise him, for he also knew a bit about Aglaé's problems at Court. Secretly, still being Napoleon's man, these episodes had given him great satisfaction. 'It's as well that you keep away from the Court,' the Marshal told him, 'these damned *émigrés* have become insufferable!' And he repeated that one day he would show them what the name Ney meant.[49] 'Let's pray that won't be too long,' the Bonapartist said to himself.

'Anyway, how are things with you?' Ney asked him, changing the subject.

'Excellent,' Lavalette replied, refusing to do so. 'One of our friends has just left for a certain island off Italy.'

'Really! Who?'

'Ida St.-Elme—'

'Well, you've picked a good one, Lavalette. I suppose she was wearing pants?'

'Yes.'

'That woman has courage and sense, and she is devoted to the Emperor.'[50]

Ney was aware that the Bonapartists were illegally corresponding with Elba, having chanced upon the evidence among his own troops. But he dismissed from his mind the possibility that this might lead to anything. In fact he would have been horrified to discover differently, for the King versus the Emperor struck him as a likely formula for civil war in France.

Lavalette believed the opposite. A considerable communications expert, he was already masterminding the flow of secret information to Napoleon. 'Marshal Oudinot reports that a revolt is brewing in Lorraine.' 'The men of the Revolution are closing ranks . . . and prepared to fight.' 'The garrison at Metz is Bonapartist to a man.' 'The Paris clergy are divided over the Concordat.' 'Talleyrand heaps insults upon the army, holding its glory up to ridicule.' 'The *émigrés* are taking over the sub-prefectures.' 'The proprietors of lands and businesses are ranged in opposing camps.' This is merely a selection from what he was passing on.

The truth was that the Government of Louis XVIII had already degenerated into paternal anarchy, a kind of benevolent muddling which many people found worse than Napoleon's efficient tyranny. Even Wellington wrote to Castlereagh: *There are Ministers but no Ministry*, and everyone agreed with Madame de Staël that things could not go on like this.[51]

Soldiers openly referred to the greedy Louis as the 'pig' of hearts or spades when they played cards. As for his mistresses, they wondered how he could possibly entertain them. 'Do you believe in Jesus Christ?' one asked. 'Yes; and in his Resurrection,' came the answer.[52]

Now at last Lavalette could add Marshal Ney's name to the list of those who showed their dissatisfaction with the Bourbons. And soon he was able to tell General Foy—no doubt upon the return of Ida St.-Elme—that Napoleon on Elba had become 'very gay and active'.

[1]   *The Campaigns of Napoleon.*
[2]   Talleyrand: *Mémoires*, ed. Duc de Broglie, Paris, 1891-2.
[3]   Harold Kurtz: *The Trial of Marshal Ney*, London, 1957.
[4]   Ibid.
[5]   Talleyrand: *Mémoires.*
[6]   J. C. Comte Beugnot: *Mémoires, 1783-1815*, Paris, 1866.
[7]   *The Anatomy of Glory.*
[8]   *The Trial of Marshal Ney.*
[9]   Ibid.
[10] Baron de Vitrolles: *Mémoires et Relations Politiques, 1814-1830*, ed. E. Forgues, Paris, 1884.
[11] *The Anatomy of Glory.*
[12] Ibid.
[13] Chateaubriand: *Mémoires d'outre-tombe*, Paris, 1849-50.
[14] Comtesse de Boigne: *Mémoires*, Paris, 1908.
[15] *The Trial of Marshal Ney.*
[16] *The Anatomy of Glory.*
[17] Ibid.
[18] Ibid.
[19] Ibid.
[20] Ibid.
[21] Ibid.
[22] Ibid.
[23] Ibid.
[24] Ibid.
[25] Ibid.
[26] *The Trial of Marshal Ney.*
[27] *The Anatomy of Glory.*
[28] *The Trial of Marshal Ney.*
[29] Ibid.
[30] *Wellington, The Years of the Sword.*
[31] *The Trial of Marshal Ney.*
[32] M. Girod de l'Ain: *Vie Militaire du General Foy*, Paris, 1900.
[33] *Wellington, The Years of the Sword.*
[34] R. F. Delderfield: *The March of the Twenty-Six*, London, 1962.
[35] Ibid.
[36] *The Trial of Marshal Ney.*
[37] *Saint-Simon at Versailles*, trans. Lucy Norton, London, 1958.
[38] Ibid.

39 *The Trial of Marshal Ney.*
40 Ibid.
41 Ibid.
42 Ibid.
43 Ibid.
44 Ibid.
45 Ibid.
46 *The March of the Twenty-Six.*
47 Ibid.
48 *The Bravest of the Brave.*
49 Antoine Marie Chamans, Comte de Lavalette: *Mémoires et Souvenirs*, Paris, 1831.
50 *Les Mémoires d'Une Contemporaine.*
51 *Wellington, The Years of the Sword.*
52 Henri Houssaye: *1815. La Première Restauration, Le Retour de l'île d'Elbe et Les Cent Jours*, Paris, 1893.

CHAPTER 11

"The Hundred Days"

On 7 March 1815 Prince von Metternich, Austria's Chancellor, Minister for Foreign Affairs and also something of a presiding puppet-master at the Congress in Vienna, received the following, thoroughly alarming message from his consul general at Genoa: 'The English Commissioner Campbell has just entered the harbour enquiring whether anyone had seen Napoleon at Genoa, in view of the fact that he had disappeared from the island of Elba. The answer being in the negative, the English frigate without further delay put to sea.'[1]

'Campbell' was Sir Neil Campbell, and the official title was merely a cover for his being the Allies' spy on Elba. However, on 16 February he went to Leghorn with dispatches and to take some leave (his mistress was residing on the mainland). When he sailed back to the island on 28 February he discovered all the eagles had flown! After making secret preparations for weeks Napoleon succeeded in embarking with over 1,100 of his *aides* and close supporters. They left at night in seven small ships. And ever since Campbell with an English squadron had been criss-crossing the Ligurian sea, vainly trying to catch them.

Metternich, whose politics were reactionary, now emerged as a very cool customer indeed. He had been (and remained) one of Napoleon's most implacable opponents—even behind his diplomatic smile as Austria's Ambassador to France (1806-9) and while he negotiated the marriage between 'usurper Buonaparte' and Marie Louise. Almost his first act upon receiving the bad news was to seek out Talleyrand, the Bourbons' representative in Vienna. The little lame intriguer expressed himself first and quite forcefully on the subject. 'He will land on some part of the Italian coast and then fling himself into Switzerland!' 'No, I don't agree,' Metternich replied, 'he will go straight to Paris . . .'

The Prince then had a very busy evening. As he noted in his diary:

> At a quarter past eight I was with the Emperor Alexander, who dismissed me with similar words to those of the Emperor Francis. At half past eight I received an identical declaration from King Frederick William III. At

nine o'clock I was at my home again, where I had directed Field Marshal Prince Schwarzenberg to meet me. At ten o'clock the Ministers of the four Powers came at my request. At the same hour adjutants were already on their way, in all directions, to order the armies who were returning home to halt.[2]

Thus, before retiring that night, Austria's Chancellor had achieved a unanimous agreement. *Come what may Napoleon would be stopped.* If the Bourbons and their Royal troops failed to halt him on the road to Paris, then the Powers were prepared to invade France from all sides. 'The Corsican adventurer,' they proclaimed, 'is a public enemy and a threat to the world.' Wellington, also present in Vienna, was consulted on the military situation. But he, like Talleyrand, got it wrong: imagining the ex-Emperor's landfall would be Italy. Within days Lord Castlereagh wrote to offer him the command in Belgium and the Czar was saying: 'It is for you to save the world again.'[3]

The ships sped along faster than Austria's couriers though, and the British Navy never even caught sight of them. On 1 March Napoleon's flotilla dropped anchor in the Golfe-Juan. Captain Lamouret and twenty grenadiers were the first ashore. A horrified customs officer brandished the health regulations in their faces, only to have a grenadier stick a tricolour cockade in his hat. 'It's the Emperor,' said Lamouret, 'he can issue his own regulations!'[4]

General Drouot and a Captain Bertrand (not the general) were next ashore, both in civilian dress. They set off for Antibes to urge co-operation on the garrison there. Meanwhile Lamouret sent patrols towards Cannes. Napoleon was rowed ashore at 5.00 p.m. and his standard raised. To Cambronne he said: 'You shall command the vanguard in my finest campaign. You will go on ahead—always ahead. But remember that I forbid one drop of French blood to be shed to recover my crown.'[5] When the news reached Cannes there were cries of 'Vive l'Empereur!' and a prudent mayor hastily dispatched a thousand rations of bread and meat to the landing force.

Their march, just as Metternich had predicted, would be straight towards Paris. The following day, 2 March, they were at Grasse along what has come to be known as the 'Route Napoleon'. Today this is a good road for cars which winds and hairpins through some spectacular scenery, but in that period it was very difficult: often little more than wide cart-track leading high above all kinds of fearful precipices. Nevertheless by 5 March Napoleon was being welcomed in Gap—where another astute mayor ordered the public buildings to be illuminated in his honour. The most crucial test would come when the thousand reached a major city: Grenoble, garrisoned by the 7th Military Division.

In Paris, meanwhile, the panic was on. News of Napoleon's quitting Elba had arrived there—thanks to the French telegraph system—on 4 March. King Louis XVIII's podgy fingers shook as he read the message. 'It is revolution once more,' he groaned. Many of the sycophants and Ultras around him also shook, including the two toady marshals, Victor and Marmont. Soult, the Minister for War, was much less of a lackey at the Bourbons' court, but even he showed some distress. He was a man who liked to come down on the winning side, and in this case the outcome appeared to him far from certain. His official position compelled him to take energetic measures though, which included sending 30,000 troops towards the Bonapartist stronghold of Lyon, France's second city. The Count of Artois would be in command, assisted by his sons the Dukes of Angoulême and Berry—and aided as soon as possible by three marshals, Macdonald, Gouvion Saint-Cyr and Michel Ney. (Soult's idea was that the former Imperial marshals might help to keep the troops loyal to the extremely unpopular Royal dukes.) Artois set off for Lyon on 5 March.

But where was Michel Ney as these momentous events were taking place? In fact, for days the most prestigious soldier in the Bourbons' army stayed totally unaware of them: rusticating out at Coudreaux. After forbidding Aglaé to risk any more insults at court or in Royalist society they had closed up their house in the (renamed) Rue Bourbon and prepared to stay in the country until the next field-exercises were due. The Marshal continued to smart at the treatment doled out to him by the restored monarchy; and in the evenings, playing cards with his friends among the neighbouring *petite noblesse*, he would speak with obvious regret of the Empire and its passing. He referred to Napoleon as a great man again, expressing sorrow that he had overreached himself in Russia and at the Battle of the Nations, making mistakes which led to his downfall.

In other ways though Ney became decidedly happy. After over twenty years of non-stop campaigning he was enjoying a rest. And he was also enjoying Coudreaux. To begin with he had looked upon it as a necessary status symbol. 'A Marshal of France must own a good house in Paris and then buy an estate in the country!' Yet there were long periods when visiting it became quite impossible. 'Oh, these interminable campaigns! When will they ever come to an end?' Now, however, Coudreaux offered him a very pleasant life-style; and the confirmed man of war was growing increasingly interested in his estate's management. There was a long harsh winter in northern France that year: week after week when the air never lost its bite and ice filled every ditch and stream. Ney hunted boar, tramped the frosted-over brecklands and supervised his fences being mended, often hammering in the nails himself. 'Why not? I used to grab a musket on the retreat from Moscow when my men were under pressure!' Aglaé saw more of him in these weeks than at any other time during their married-life; and often their domestics were startled to behold the 'famous Marshal of France' giving piggy-backs to his two youngest sons around the gravelled courtyard.

It was almost an idyll: occasionally brooding, but generally gay and carefree. Moreover it didn't end until 6 March when an officer pulled up at the gates of Coudreaux with a letter from Marshal Soult. Brief, and carefully polite, it requested him 'to make haste to Paris, from whence he should proceed to Besançon to take command of his troops of the Sixth Division.'[6] Never a mention of Napoleon's flight from Elba.

Ney looked searchingly at the messenger. 'This letter says little. What does it mean? What sort of mission is this I am to undertake?' Fearing an outburst the officer pretended ignorance. 'I do not know, *Monsieur le Maréchal*. The Minister for War said it was urgent, and instructed me to find you with all possible haste.' The Marshal remained suspicious. 'I should have thought Marshal Soult would tell a man what sort of wild goose-chase he is sending him upon.' Nevertheless he complied with the instructions and arrived in the capital very late the following afternoon.[7] *The same day that Napoleon and his thousand arrived outside Grenoble.* Immediately he went to the Rue Bourbon for his uniform and dress-sword; but he hardly had time to change before his notary, Monsieur Batardy, suddenly turned up. They greeted one another in a room still swathed with dust-covers.

'This is extraordinary news, *Monsieur le Maréchal,*' the notary began.

'What news?'

'Why? Have you not seen *Le Moniteur?*'

'Batardy, you exasperate me! What news? I had an order from the Duke of Dalmatia to come to Paris at once and proceed to Besançon to take command of my troops there. Good God, will no one tell me what is happening?'

'He has landed: is on his way to Paris . . .'

'Who has landed?'

'The Emperor! He is on his way to Paris, and *Monsieur* the King's brother is starting for Lyon this morning to stop him. It will be civil war . . .'

Michel Ney could scarcely believe his own ears. 'What a disaster! What a terrible business is this they've got me into! Am I to be sent to stop him then? Is that why Soult would tell me nothing in his letter?' He turned to Batardy again. 'Where did he land? What do you know about it?'

'I know little except that he landed at Cannes and is on his way to Paris. Everywhere, they say, the people are flocking to him. It looks as if there will be terrible fighting.'

'No!' the Marshal countered, in that same hard voice Napoleon had heard before the enforced abdication. 'At all costs it must be averted!'[8] 'Even so,' he added, leaning against the mantelpiece: 'If he had not known that there was discontent in the country he would never have dared set foot on French soil . . .'

Since the abdication Ney's feelings of remorse for his part in the drama at Fontainebleau had been growing. Of late, in various conversations and stung by the Ultras' barbs, he had professed affection as well as a resurgence of his original admiration for the ex-Emperor. 'After all,' he would say: 'This is the man who recognised whatever talents I have as a soldier . . . who promoted me to become a marshal of his, or rather France's Empire. By comparison with him these Bourbons are pygmies! No wonder I nearly died for him so many times in battle . . .' On the other hand, Ney's loyalty to France itself, his adopted country, had preceded and was stronger than the debt he felt he owed Napoleon—and at the time of the abdication had overridden all other considerations. He had done what he believed was right for France then; to prevent the country's disintegration in the face of foreign invasion and civil war. And here now was his former idol become the grand disturber once again, threatening to engulf the whole nation in another civil war just to satisfy his own, outsize ambition. Well, he could not stand idly by and let him do it!

Within the hour he was confronting the Duke of Berry at the Tuileries, but that arrogant young Ultra could offer him no further news. And so he went on to see Soult at the War Office. Personal relations between the two marshals had improved since their fearsome rows in Galicia and the Asturias. Nor on this occasion did Ney blame Soult for not mentioning Napoleon's escape in his letter. He simply assumed the Minister had wished to acquaint him with the facts face-to-face; so there could be no misunderstandings. 'I come for instructions, *Monsieur le Ministre*,' he said.

'*Monsieur le Maréchal*, Napoleon Bonaparte has just invaded France!'

'So I heard upon reaching Paris. It's most unfortunate. *But he must be stopped—*'

'Yes, there's going to be bloodshed,' Soult went on: his anxieties over the situation partly alleviated by Ney's display of firmness and calm. 'You are to go at once to Besançon, where you will find detailed orders in the hands of General Bourmont.'

'I'll be ready in two hours. First though I want to go by and see the King.'

'He's unwell tonight. He won't see you.' (Why was Soult so sure of this? Had he been *ordered* to omit the truth about Napoleon from his letter? Were the Bourbons so uncertain of Ney that they felt they had to get him back to Paris ahead of the news? One wonders.)

'Nevertheless I'm determined to see him. Don't try to prevent me!'

'Very well, go if you must and try to see him, but . . .' and Soult found himself addressing a set of retreating footsteps.

It was eleven o'clock by now, but after due insistence Ney gained admittance to the King, whose great bulk was spread upon a *chaise longue*, his gouty feet encased in satin bootees. The Marshal assured him that he remained 'at France's service'. Louis, much relieved at this, avoided saying anything that might upset him. Instead he appealed to the known patriotism of an officer the Bourbons (as even their Ultras now realised) quite desperately needed. 'What we most desire,' Louis said with a late-in-the-day earnestness, 'is that France shall have no civil war, shall be given an opportunity to develop in peace and contentment. War is always a terrible thing, but it is even more terrible when brother turns against brother. I appeal to you, *Monsieur le Maréchal*, to use your great abilities and your popularity with my soldiers to end this rash enterprise and prevent a recurrence of bloody strife.'[9]

'Sire,' Ney replied: 'You are right. France must not have another civil war. Bonaparte's enterprise is sheer madness. I am leaving at once for Besançon, and if need be I will bring him back to Paris in an iron cage!'[10]

He bowed and withdrew, leaving Louis deep in speculation. Was the Marshal so devoted to the Royalists' cause, or was he simply acting out a part? 'I did not ask him to say all that,' the King observed to his courtiers. But then he could not resist yet another joke at Ney's expense: now the Marshal was safely out of earshot. 'Napoleon in an iron cage, eh? Well, I would not like such a bird in my room!'[11] The courtiers duly sniggered, one of them proposing a home for it in the Jardin des Plantes (the capital's zoo) and for days 'Ney's iron cage' was an amusing phrase on the lips of everyone at court.

Other, more sincere followers of the King took a purely practical view of Ney's involvement. 'Such was the general confidence in his ability as a soldier that the spirits of the Royalists rose considerably.' Fanny Burney (a gifted diarist):

received an encouraging note from her husband who, as a member of the King's bodyguard, was unable to leave the Palace of the Tuileries during these anxious times. *We have better news*, General d'Arblay wrote. *I cannot enter into details, but set your mind at rest.* This news (says Fanny Burney) hung upon the departure of Marshal Ney to meet Bonaparte and stop his progress. The King at this time positively announced and protested that he would never abandon his throne nor quit Paris.[12]

Honest folk like Fanny Burney and her husband had need of Michel Ney's name to comfort them, because the news bulletins which now began to arrive in the capital were truly appalling—unless one happened to be a Bonapartist.

On 7 March Napoleon had approached Grenoble, where 'Generals Marchand, Mouton-Duvernet, etc. had concentrated their forces.' The ex-Emperor bared his chest and said: 'If there is any man among you who wants to kill his Emperor, here I am . . .' A captain shouted 'Fire!', but nothing happened. Then suddenly all of the rankers 'tossed their white cockades into the air . . . and shouted *Vive l'Empereur!* (Major) Lansard, in tears, offered his sword in surrender,' while Dumoulin (a glass manufacturer) galloped out from the city 'bringing 100,000 francs and his strong right arm to the Emperor'. More importantly, young Colonel La Bédoyère brought over the whole of the 7th Infantry. (La Bédoyère was promoted to general and became the returned leader's *aide-de-camp*.) 'Now all is well,' Napoleon told Drouot. 'In ten days we shall be in the Tuileries!'[13]

Despite the efforts by Marshal Macdonald, Lyon and its troops went over to Napoleon on 10 March. The city had grown rich thanks to the Emperor's previous encouragement of the silk and other trades there, and its middle classes now showered him with gifts and acclamations. As for the troops, when the Count of Artois arrived to review them they made funny faces at him. So the landing force of just over a thousand had become a small army, with Drouot already organising it into battalions.

In Paris Louis XVIII dismissed Marshal Soult. The Duke of Berry was given command of the Army of Paris, 40,000 men, with Marshal Macdonald as his Chief-of-Staff.

The hopes of the Royalists now rested on Marshal Ney at Besançon. He alone could arrest Napoleon's progress, and fortunately he was well-placed for a flank-attack. He was now marching to Lons-le-Saunier, and the capital waited anxiously for news of battle. Up and down the Paris streets young Royalists paraded, waving flags and endeavouring to muster recruits. *To arms!* they cried. *Rise against the usurper, the tyrant, who is bringing back war and despotism.*[14]

However, few volunteered to join Ney and his professionals.

At Besançon on 10 March the Marshal discovered there were no standing orders and less than a thousand depot-troops, all chronically short of ammunition and none of them showing much enthusiasm for the Bourbons. The remainder had gone off to Lyon with the Count of Artois, where they'd promptly deserted to join Napoleon. When the Duke of Maille, head of

Artois' household, rode up with this news Michel Ney had already decided on Lons-le-Saunier as the best place to collect all the troops still loyal in the eastern Departments: nine thousand perhaps against Napoleon's ten. '*N'importe! Qu'à cela ne tienne!* When he comes by . . . I shall fall upon his flank or his rear, and by God, I shall settle accounts with him!'[15] Also he dashed off a message to Marshal Suchet at Strasbourg: could the latter reinforce him? 'We are on the eve of a tremendous revolution. Only if we cut the evil at its root is there any hope. But I fear many of the troops are infected . . .'

His two principal *aides* at Besançon were Bourmont and Lecourbe: both bitter enemies of Napoleon. Bourmont had once fought with *Les Chouans* (literally 'screech owls'), the Royalist rebels in the Vendée; while Lecourbe, an old friend of Ney's in the Army of the Rhine, had been dismissed by Napoleon for suspected complicity in the plotting by Moreau. After which he'd lived in retirement on a small farm in the Jura until Louis XVIII restored him with the rank of general. With these two and the troops from Besançon, Ney set out for Lons-le-Saunier—himself riding ahead in a post chaise. Halting to change horses at Poligny, he was greeted by a sub-prefect who expressed the opinion that Napoleon could not advance much further towards Paris. 'We will stop him shortly,' the Marshal agreed, 'and then I will take him back to Paris as my prisoner.' 'It would be better to take him back dead in a tumbril!' the sub-prefect said glumly. 'No, no, that would never do,' Ney told him. 'You don't know Paris. The Parisians must see him. It will be the last act of his tragedy!'[16]

Just as he was setting off again though he received reports that first one and then another of the garrisons towards Lons had revolted against their officers. A contingent bringing up his artillery was stopped at Chalon-sur-Saône and the townspeople there tipped all eighteen pieces into the river. Outwardly, despite this, the Marshal remained positive. He called on the officers he encountered *en route* 'to remember their duty to the battle on the King's side; if necessary,' he said, 'he himself would fire the first shot.'[17]

It was past midnight when Ney entered the Hôtel de la Pomme d'Or at Lons-le-Saunier. A commercial traveller, Monsieur Boulouze, sat dozing by the fire. 'I left Lyon early this morning,' he told the Marshal:

> the Emperor was there, and the whole city in uproar. He rode through the streets with the people cheering frantically on all sides. At the Place Bellecour he reviewed the troops and delivered a long harangue. *People of Lyon*, he said: I LOVE YOU. *I shall in more tranquil moments return to occupy myself with your city and its manufacturers, for you have always occupied first place in my affections, and the elevated character by which you are distinguished merits all my esteem. At the moment of quitting your city to return to my capital, I feel that I should make known the sentiments with which you have inspired me. Soldiers, we shall go to Paris, and with our hands in our pockets. It is all arranged.*[18]

What had been arranged, or was Napoleon bluffing? Michel Ney stared into the fire and wondered. Then Monsieur Boulouze handed over a sheet of paper. 'Here is a copy of the proclamation issued by Napoleon. They are scattered everywhere in Lyon.'

The Marshal was still reading it when the prefect for the Jura, Monsieur de Vaulchier, arrived—together with the Marquis of Saurans, another of Artois' *aides-de-camp*. 'Nobody knows how to write like this nowadays,' he informed them, 'Napoleon can stir a soldier's heart. The King would do well to imitate him. In fact, His Majesty ought to be here . . . to inspire his troops—even if he appeared on a stretcher!'[19] 'Just listen to this,' he continued, reading a part aloud:

> *Frenchmen, you require the government of your choice, which alone can be legitimate. I have crossed the sea to rescue my rights, which are also yours. Soldiers, rally to the colours of your chief. His rights are yours, and those of the people. Victory will march with us at the charging step, and the eagle with the national colours will fly from steeple to steeple to the towers of Notre Dame!*

Ney strode up and down the room now, greatly agitated. 'That, *Messieurs*, will undoubtedly have a powerful effect upon the soldiers!' He turned to Saurans. 'Your master is not the equal of Napoleon. He has a greater regard for his lace sleeves and silk breeches than for the honour of his country and the welfare of his countrymen. He is no fighter, or he would not now be running away from Lyon at the mere mention of the name Bonaparte. He has blundered, badly, and you can tell him I said so!'[20] He continued to abuse them. Why had King Louis disbanded the Imperial Guard? Why were the Ultras so snobbish that he, a Marshal of France, could not ride in their carriages with them?

The two Royalists stood dumbfounded, expecting him at any moment to declare for the invader. Instead of which he suddenly sighed and dropped back into a chair:

> You think I have been won over by this proclamation, and perhaps I have given you just grounds for thinking so, but you are wrong. I was only attempting to show you the difference between the King and his counsellors and our ex-Emperor. As for Napoleon: he has no love for me, for he has never forgiven me the abdication. He would be quite willing to have my head cut off. Don't worry. His Majesty can rely on my services still. I'll take a rifle. I'll fire the first shot, and everybody will march.[21]

For the next two days the Marshal busied himself hurrying up troops, collecting ammunition and sending out an intelligence network to obtain information concerning the movements of Napoleon and the temper of the people. The reports were not good. Napoleon now had fourteen thousand men, whereas Ney could count on six or seven at the most. Nevertheless, on 13 March he was reassuring a depressed Bourmont: 'We may be short in numbers, but we'll give him a good dressing down!'[22] Minutes later, when he walked into a restaurant and heard an officer shout '*Vive l'Empereur!*' he had the man arrested. It was also on the 13th that another prefect visited Ney:

Monsieur Capelle of Bourg. His people, he said, had 'chased him out of town'. It was the Revolution all over again! Couldn't the Marshal, he asked, buy help from the Swiss? After all, for centuries they had hired themselves out to the Kings of France. Michel Ney ordered him to be thrown out. If alien soldiers put their feet upon French soil then the whole nation would declare for Bonaparte!

But that same evening, quite late, two strangers sent him a note asking if he would see them in his room at once. Upon entering they closed and locked the door. Then they took off capes and wigs to reveal themselves as junior officers in the Old Guard, both of whom Ney knew well. (After Waterloo he was charitable enough not to mention them by name.)

'*Monsieur le Maréchal*, we come to you from General Bertrand in Lyon. He requests us to present his compliments, and to give you these letters from him and also the Emperor.'

Bertrand's letter warned Ney not to view Napoleon's return as a mere bit of child's play. Everywhere, he wrote, the ordinary people were flocking to the tricolour, and if the Marshal chose to resist them he (Ney) would make himself solely responsible for the ensuing civil war. He urged him to rejoin the Emperor, and enclosed an order instructing him to report with his men at Chalon. Napoleon's note was friendly, straightforward, *but firm*: 'My Cousin —My Chief-of-Staff (Bertrand) sends you your marching orders. I don't doubt that the moment you heard of my arrival at Lyon you placed your troops under the tricolour flag. Execute Bertrand's orders and join me at Chalon. I shall receive you as I did after the Battle of the Moscowa.'[23]

Ney looked up at the messengers. 'Bertrand isn't lying to me?'

'No, *Monsieur le Maréchal*. The Emperor's return was arranged some time ago by international consent, and the English squadron was purposely withdrawn from Elba to facilitate his escape. Austria is his ally, and the Empress and the King of Rome will shortly leave Vienna to join him.' (This, knowing what we do of Metternich's activities, was definitely a lie.) 'The march on Paris is becoming a triumphant procession . . .'

'Am I expected to break my oath of loyalty and desert the King then?'

'The King, *Monsieur le Maréchal*, has deserted France! He left Paris days ago.' (Another lie, but soon to become the truth.) And they handed him a further document: a proclamation calling upon his troops to join the Emperor. Ney saw his own name together with his titles inked in at the bottom.[24]

Still he hesitated. He promised to give them an answer in the morning; together with safe conduct passes, whatever his decision. And for the rest of that night he pondered the situation. He was pledged to the King, but would his soldiers fight against Napoleon? And if they did, then it meant civil war: begun by he, Michel Ney! He thought of his lost-and-found leader again; of the Napoleon who had created him Marshal of France, Duke of Elchingen and Prince of the Moscowa . . . who had made Aglaé a Princess. After which he thought of the Bourbons, fat Louis and the dreadful Artois, who wouldn't even share a lunch with him. Finally too he considered Fontainebleau and those terrible abdication scenes. But above all he thought of his first, unswerving loyalty. What was the best for France?

There were the easy ways out, of course; and others were taking them. He could assemble his troops on the parade ground, declare that any further resistance to the Emperor was useless and say that he wished to avoid civil war; then he could keep his oath to King Louis by resigning the command. Or, like Oudinot and Moncey, he might perhaps have pretended illness. Instead he chose a strangely paradoxical course—and one which was to do him great harm at a later date.

When breakfast-time came, he sent for Bourmont and Lecourbe. He told them what he had heard the night before, but added certain refinements of his own: including the fact that he had helped to plot the Emperor's return.

> Three months ago we were all agreed on it. The King had broken his promises; the country was on the way to ruin. For a time there was talk of making the Duke of Orléans (later Louis Philippe) King, but the Bonapartists won out. It had been agreed before the Emperor left Fontainebleau that he would return, and Queen Hortense was one of the conspirators. A few months ago, a commissioner was sent down to Elba to lay the conditions before the Emperor, and even Soult himself was in on the plot. The King, I have been informed, has already fled, and no harm will be done to him. He will be aided in making his escape to America, if, indeed, he has not already reached England.

A. H. Atteridge in *Marshal Ney, The Bravest of the Brave*, Hortense de Beauharnais in her *Mémoires* and Ida St.-Elme in her *Memoirs of a Contemporary* all seem in accord that Ney made this extraordinary statement. And certainly something very like it was to be quoted by his accusers after Waterloo. But why, if it is true, did he indulge in such an elaborate, yet transparent tissue of lies? Hardly in the hopes of winning over his principal *aides*, for Bourmont and Lecourbe saw through it right away. Bourmont was convinced that Ney had invented the story. 'But *Monsieur le Maréchal*,' he said: 'You pledged your honour to defend the King whether he was in Paris or not.' While Lecourbe was even more indignant: 'How can you ask me to serve that fellow? He has never done me anything except harm, and the King has been very kind to me. I have sworn to serve the King and I have a sense of honour: I'll never serve that damned Bonaparte!'[25]

My own view is that Ney, having at last made his agonised decision to rejoin Napoleon, was merely giving two anti-Bonapartists their chance to escape. It fits with what he next said and did. 'Sense of honour?' he went on reproachfully:

> Yes, I too have a sense of honour. Do you think that I can have my wife insulted by the Duchess of Angoulême and the other parasites who hang around the court? I won't accept that sort of thing any longer, for it is plain that the King does not want us. It is only by a soldier like the Emperor that the soldiers of France will ever be treated with respect. I gave Louis his chance, for I did not want France plunged into civil war and I don't want it now. The whole country is going over to the Emperor, and to fight him would be folly, even worse. *Here is what I have written to explain my views.*

The paper was the proclamation sent to Ney by the Emperor, but he had copied it out in his own handwriting and destroyed the original.[26] After reading it, the two generals protested vehemently but finally agreed to sanction Ney's action with their presence, if nothing more. (Both slipped away after the vital parade of troops which they now helped to organise.)

10.30 a.m. on 14 March. Ney's carriage drew into the Place d'Armes at Lons. The troops were drawn up, facing inwards, and surrounded by an anxious crowd of townspeople. 'At ease!' the Marshal called out, striding to the very centre of the square. 'Soldiers!' He turned slowly on his heels to view them and then carefully, dramatically began to read the proclamation. 'The cause of the Bourbons is lost, *forever*. The legitimate dynasty, chosen by France, is about to re-ascend the throne. It is the Emperor Napoleon, our sovereign, who has the right henceforth to rule over our beautiful country . . .'[27] He got no further. His staff officers stood back in silence and shock, but then his words were drowned by the troops' thunderous 'Vive l'Empereur!' three times over. 'Soldiers,' he tried again: 'I have often led you to victory. Now I am about to lead you to the immortal phalanx which the Emperor is leading to Paris . . .' The cheering burst out again, 'caps were stuck on bayonets, and hats thrown into the air by the civilians fell inside the square and were trampled under the feet of the frenzied soldiers who had broken ranks. Junior officers and rankers threw themselves into each other's arms, and the soldiers were swarming about the Marshal, who was laughing, crying, and hugging his men, mad with the intoxication of the moment.'[28]

Not everyone was so intoxicated though. A Colonel Dubalen forced his horse through the madding crowd and shouted at Ney: '*Monsieur le Maréchal*, my oath to the King will not allow me to change sides. I give you my resignation!' 'I won't accept it,' Ney shouted back, 'but you are free to go. Just take care your own men don't give chase . . .'

Which, in fact, did happen. Dubalen spurred off with several dragoons in pursuit—and one horseman overtook him, only to be knocked out of his saddle streaming blood from a sabre-wound. Dubalen got clean away. (All the way back to Paris, where ahead of Bourmont he broke the news of Ney's changing sides.) 'Is there no more honour?' Louis XVIII sighed, while preparing to break *his* oath by fleeing to Ghent . . . Marshals Victor and Marmont going with him.

At Lons, meanwhile, order was quickly restored. The troops were re-formed and—singing and shouting—marched away to their barracks. Already they had ripped off their white cockades. General Mermet was ordered to inform Suchet that: 'I have taken the decision to join His Majesty the Emperor who is marching to Paris.' And he had instructions to collect the remaining troops from Besançon. Ney's *aide-de-camp* Clouet then departed to join the King and in doing so reproached the Marshal. 'What could I do?' he replied: 'Hold back the sea with my hands?'

That night his remaining officers gave a noisy party at the Pomme d'Or and the following morning (15 March) they set off to join the Emperor. Did Michel Ney enjoy their party? Or did he have second thoughts? In two days they reached Dijon, where a message requested them to 'report to the

Emperor as soon as possible at Auxerre': the city where Ney's brother-in-law, Monsieur Gamot, was Prefect.

'Napoleon, having continued his triumphal march through Chalon, Autun and Avallon, reached Auxerre on 17 March.'[29] Ney and a few of his officers arrived very late that same night. He was shown into Bertrand's quarters first, who received him most cordially and said: 'The Emperor is anxious to see you. Can you see him now?' But the Marshal shook his head. 'I don't wish to see him tonight. I am drawing up a paper to present to him in explanation of my conduct. No doubt he has heard much concerning me!'

Napoleon, when informed of this, didn't know whether to be angry or amused. 'What do I want with his justification? Perhaps his conscience is troubling him. Tell him to forget his writing and get a good night's sleep. I shall embrace him in the morning.'[30] However, always with a flair for spectacle, he decided that their official reconciliation would take place before the assembled troops.

18 March. 'Napoleon wore the sword of Austerlitz on the belt with a diamond buckle that his mother had given him on the day he left Elba.'[31] He was on foot, although a trooper stood nearby holding Tauris, his dapple-grey Persian. With him were Drouot, Bertrand and behind them a contingent of Polish lancers under Jerzmanowski, dismounted but still holding their crimson-and-white pennants. The remaining troops were assembled on either side as Ney was conducted towards the small, stout figure of the returned Emperor. 'The meeting was melodramatic enough to please even the most sentimental.'[32] But was it? The Marshal held a document he had composed during the night, a kind of manifesto—and before any formal greeting was exchanged he began to read it aloud: his manner more blustering than convincing. Obviously he was nervous, and in a highly-charged emotional state. 'If you continue to govern tyrannically,' he began, 'then I am your prisoner rather than your supporter. From now onwards the Emperor must govern with one object only, the happiness of the French people and the undoing of the evil which his ambition has brought upon the country . . .'[33] 'This fine fellow is going mad,' Napoleon observed in an undertone to his aides.[34] But they, like he, had noticed that there were tears in Ney's eyes. The right word, the correct gesture now, and he would be theirs totally. Napoleon was an expert in such matters. 'My dear cousin!' He moved towards the Marshal, arms outstretched. 'Haven't I promised France a constitutional monarchy? At Grenoble I explained that I was here to prepare for the rule of my son, to safeguard the peasants and the army. Of course, I will read all you have put down here.' (He eased the document from Ney's fingers.) 'But really, my cousin, you are too fine a soldier to be worrying your head about politics and statesmanship. We must get on to Paris!'[35]

Cheering broke out on all sides as the two men embraced. Napoleon had won and he knew it. 'Now tell me,' he went on gently—as the troops formed column under their officers' commands: 'How reliable are the men under your control, and what are the feelings of my people in the eastern Departments?'

19 March. Palm Sunday. As the bells rang out for early mass the Emperor set off for Paris. He moved with great speed now; also great confidence.

Ahead of him galloped an officer with a final message to Louis XVIII: 'My good brother, there is no need to send any more troops. I have enough!'[36] But Louis had already fled, and the citizens of Paris prepared to turn their coats again.

At 9.00 a.m. on the evening of 20 March the Emperor's carriage rattled at breakneck speed through the gates of the Carrousel. The troopers of his escort shouted and brandished their sabres, though there was little point in this for the crowds cheered frantically: 'The same Parisians who only last April had consigned him to the devil.'[37]

> So dense was the crowd, so eager the rush, that the horsemen were driven backwards and the postillion came to a halt 10 yards from the Pavillon de Flore. The carriage door was flung open. Napoleon! The crowd seized him, dragged him from his carriage and passed him from hand to hand into the entrance hall, where other officers bore him from them, hoisted him shoulder high and carried him to the staircase. From above the Bonapartists were pouring down the staircase, from below the tide was driving upwards. *For God's sake!* shouted Caulaincourt to Lavalette, *get in front of him!* Lavalette flung himself forward, turned his back to the crowd, and clinging to the stair-rail made a buttress of his body. Thus he pushed his way up the staircase, step by step, a pace ahead of the Emperor, murmuring all the time *it's you, it's you.* And he—he seemed to see nothing, to hear nothing. He was borne forward, his arms extended, his eyes closed, a fixed smile on his face, like a sleepwalker's.[38]

Once the doors closed behind him though his expression changed. He was back—and within minutes issuing orders with all the snappy decision of his First Consul days. The intense, infectious activity of 'The Hundred Days' (in reality one hundred and thirty-six from invasion to second exile) had begun in earnest.

There were enemies a-plenty. And not just in Vienna and London. France's Chamber of Deputies seemed likely to prove difficult; also the men of commerce and high-finance. But he had the army and now obviously the ordinary people. On the other hand, the army didn't necessarily include the marshalate—and he had to have marshals to fight against Wellington, Blücher, Barclay de Tolly and Schwarzenberg. He was sure of Davout, and within days would gain Suchet, Mortier, Brune and Jourdan (the last two surprisingly, for he had always treated them shabbily, Brune in particular). Augereau had insulted him and Napoleon struck his name from the list of marshals. He hoped to use his powers of persuasion on Macdonald, Oudinot and, above all, Berthier; but the former opted for the King and Oudinot posted him sick-notes. As for the Prince of Neuchâtel, he declared himself too war-weary to join in on either side. Instead of these three a combination of Bourbon stupidity and old, mingled loyalties had delivered over to him Marshals Soult and Ney.

Soult the Emperor had no strong feelings about; he was not tainted by 'the business at Fontainebleau' and so as an experienced fighting-marshal ought to have his uses (or misuses, as Napoleon discovered at Waterloo). But Michel

Ney was altogether different. At Auxerre Napoleon had smiled with his mouth only. Contrary to what everyone else at the meeting believed, including Ney himself, the Emperor remained totally unforgiving about the leading part the Marshal had taken during the abdication scenes. His smile and honeyed words at Auxerre, like his blandishments in the note which preceded their reunion, were simple expediencies—and no one could rival *Le Tondu* at such things. He could charm the birds down out of the trees. But now he was the Emperor for real again, the undisputed master of Paris ('and he who holds Paris has France!'). Ney had served his purpose in this restoration, but with the end achieved the means were political history. From here on the Marshal would find he was definitely *persona non grata* with France's leader; while his uses even as a soldier shrank to the 'only when it becomes absolutely necessary' point.

From Auxerre he had returned to Dijon to collect his troops, so he didn't reach Paris until 23 March. By which time Napoleon's new cabinet was already formed and at work. Maret, Duke of Bassano, as Secretary-of-State, Fouché Minister of Police again, Gaudin as Minister of Finance, Cambacérès as Minister of Justice, Caulaincourt Minister for Foreign Affairs, Carnot Minister of the Interior—and Davout as Minister for War. Michel Ney was amazed by the industry displayed by these and the people under them. Napoleon left word that he was too busy to see him; and within twenty-four hours the Marshal found himself packed off on a mission to the northern and eastern regions: from Lille to the Swiss frontier. He was one of the *commissaires extraordinaires* being sent to the provinces 'to inspire confidence and counteract the effects of the declaration signed by the Allies on the 13th.'[39] His instructions, he commented later:

> carried the express order to announce everywhere that the Emperor would not and could not resort to war, having agreed not to do so in talks concluded on the island of Elba between himself, England and Austria; that the Empress Marie Louise and the King of Rome were to remain in Vienna as hostages until he had given France a liberal constitution and fulfilled the conditions of the treaty, after which they would join him in Paris.[40]

One notes again that strange Imperial fabrication that the eagle's flight from Elba had somehow been managed by the Powers. It was a claim foisted upon Ney at Lons, and now he—presumably still half-believing it—was expected to sell the idea to ordinary Frenchmen in the north. A liberal constitution? I can imagine what Metternich would have thought of that!

Certainly in his top-level dealings Napoleon had already discarded the ruse. He was chasing both peace and war with equal enthusiasm. On 1 April he wrote to Francis I of Austria:

> My sole aim will be to consolidate the throne . . . so that I may leave it one day, standing upon unshakeable foundations, to the child whom Your Majesty has surrounded with his paternal kindness. A durable peace being an essential necessity for this deeply desired end, nothing is nearer

to my heart than the wish to maintain it with all the Powers, but above all with Your Majesty. I hope that the Empress will come by way of Strasbourg, orders having been given for her reception on this route into my realm. I know Your Majesty's principles too well not to feel every confidence that you will be most eager, whatever the trend of your policy may be, to do everything possible to accelerate the reunion of a wife with her husband, a son with his father.[41]

He also wrote to England's Prince Regent, who like the Austrian Emperor considered it beneath his personal dignity to reply. However, on 8 April Castlereagh answered a letter from Caulaincourt:

Sir. I have been honoured with two letters from Your Excellency bearing the date the 4th inst. from Paris, one of them covering a letter addressed to his Royal Highness the Prince Regent. I am to acquaint Your Excellency, that the Prince Regent has declined receiving the letter addressed to him, and has, at the same time, given me his orders to transmit the letters addressed by Your Excellency to me, to Vienna, for the information and consideration of the Allied Sovereigns and Plenipotentiaries there assembled.[42]

Vienna meant Metternich, of course, who reaffirmed the Powers' inflexible stance of 7 March: 'No peace with Bonaparte!'

On 4 April the Duke of Wellington arrived in Brussels to take command of the Anglo-Dutch forces, while the Prussian Army of the Lower Rhine began to assemble at Aix. It was just as well, Napoleon observed after the failure of his diplomacy, that France's own war-effort was at last taking shape. Both he and Davout worked every day from 6.00 a.m. until dusk. The Emperor managed to borrow 100 million francs at eight per cent from the bankers of Amsterdam (despite their British connections!). Helped by this Davout set out to galvanise and re-equip the former Bourbon Army. And—naturally—to revive the Imperial Guard.

He began with 200,000 men. Volunteers raised the number to over 300,000. 'All were French, most veterans, and their morale was higher than that of any army since at least 1809. The troops were determined to wipe out the shame of their defection the previous year, and Allied spies reported their almost frenzied enthusiasm for the Emperor.'[43] To hold the towns and cities there were 200,000 National Guards. With these, and the new conscripts— the Class of 1815—Napoleon hoped to form an army of half a million with which to face the Allies' one million plus. A lot more ordnance was being cast, and the cavalry (thanks largely to Ney's efforts during the past year) continued to be, even in Wellington's view, 'the best in Europe'. Also, this time the Emperor took care to fortify Paris. 'He drew up the plans himself, all the redoubts, couronnes and lunettes, in half an hour!'[44]

He did not allow the war-effort to occupy him completely though. Together with brother Joseph, he proved extremely successful in winning over men of ability and repute to his cause: bankers, industrialists, the press, even some of the old, revolutionary politicians. La Fayette was offered a peerage (which he refused, although agreeing to join the Chamber of Deputies). And Benjamin Constant, since celebrated for his novel *Adolphe*, expressed delight upon being invited to draft the promised constitution. 'He needed France solidly behind him,' Napoleon told the ardent Liberal, 'and France in return would demand certain liberties, in particular liberties of the press, which had been granted, then withdrawn, by the Bourbons. She shall have them again,' he declared.[45] Constant wasted no time. Within a fortnight he had drafted a constitution:

> calling for two-chamber government, the electoral colleges—a mere 15,000 under the Bourbons—to be enlarged to 100,000 as under the Empire. The Assemblies would meet in public and have the right to amend laws proposed by the Government. Other guarantees would be trial by jury and complete freedom of expression. Napoleon did not particularly like this constitution, which would make it difficult to govern, but he gave it his approval. So did the Council of State. It was promulgated on 22 April and approved by the people in a plebiscite, with 1,305,206 votes for and 4,206 against.[46]

Another thing the Emperor did was to abolish the slave-trade.

Even so, there could be no doubt about it: this was, in Edith Saunders' phrase 'The Hour of the Army'—or at least the hour of its Bonapartist *élite*. Without them Napoleon could never have maintained his position. Officers below the rank of colonel and seasoned rankers vied with one another to express the greater devotion.

> On 2 April the Imperial Guard gave a banquet to the troops who had marched with Napoleon from Grenoble to Lyon, and to those members of the National Guard who had been on duty at the Tuileries on the evening of his arrival. 15,000 soldiers and militiamen sat down at their tables in the Champ de Mars and 1,000 officers dined nearby in the École Militaire. Toasts were proposed to the health of the Emperor, the Empress and the King of Rome. Then, under the influence of free-flowing wine, the officers with unanimous inspiration jumped to their feet, crossed their swords over the tables and swore they would die for their country. Finally a voice was raised: *To the Column!* and everyone set off in procession to the Place Vendôme and the column erected to Napoleon's victories; at the head of the procession a bust of the Emperor was carried aloft like a sacred relic, while drums rolled and a band played *La Marseillaise*. Arrived at their destination the officers placed the bust in front of the column while the inhabitants of the square hastened to illuminate their windows with lanterns and candles. Those who didn't had stones thrown at their windows.[47]

But if it was the hour of the army it was not yet the hour for Marshal Ney. He arrived back in Paris on 15 April and very promptly forwarded a report to Davout on the civil and military situation he had encountered in the north. The War Minister then made an appointment for Ney to give the Emperor a verbal summary of this same report. Unfortunately, although the Marshal went with the best intentions, it proved to be a difficult meeting from the outset. He realised he was being given the cold shoulder. Apart from not having forgiven him for Fontainebleau the Emperor was now acquainted with the 'iron cage' story: drip-fed to him by certain Bonapartists who themselves had changed sides. Again, at Dijon, Ney had indulged in a mild jest with his officers—'I congratulated myself on having forced the Emperor to abdicate, now here I am serving him again!'[48]—and this too had been processed back. 'Nothing could have been more dangerous in times of political unrest; he was a man to whom all minds would turn when a scapegoat was required.'[49] Napoleon listened to him in stony silence, then made a gesture to indicate that he should withdraw. Nothing was mentioned about a new command; nor did Michel Ney expect one after so strained an encounter. Atteridge claims that, before he left Napoleon's presence, the Marshal referred to his promise to the King: 'It is true that I told him I would like to bring you back to Paris in an iron cage. But the fact is that I had already made up my mind to join you, and I thought I could not say anything better calculated to conceal my real plans.'[50] I'm afraid this doesn't ring true though. There is too much evidence that his crisis of indecision persisted right through to his first night at the Pomme d'Or in Lons.

After several days of moping about in the War Ministry's ante-room, still without a job or even the prospect of one, he decided he would return to Coudreaux—and there he stayed for the next six weeks. During these weeks the activity in Paris and various towns and cities to the north reached its peak, but for Ney it was a period of neglect comparable only to his worst times under the Bourbons. At least though he missed the latest joke going the rounds of the Bonapartist salons: *Il faut etre né(y) pour ça*—a shallow reference to his changing sides.

At last an envelope did arrive from the War Ministry, but instead of bringing his appointment to a military command he discovered it was a formal invitation to the Emperor's *Champ de Mai*, the open-air ceremonial at which Napoleon would take an oath of fealty to France's new constitution and also distribute eagles to the regiments. To complete the spectacle it was necessary to include a group of marshals, and the absence of the Prince and Princess of the Moscowa would have excited surprise. But the invitation didn't mean he was returned to favour, far from it. Neither Aglaé nor himself had been assigned a special part in the ceremony. They were merely called in order to be seen.

1 June. Everybody sensed that this would be the last big parade before the fighting started. It was certainly a glittering affair. Early on a fine, cloudless day the troops and the makings of a huge crowd began to assemble between the École Militaire and the centre of Paris. Near the school itself pavilions had been erected, and 200 eagles and 87 flags were massed before a special altar. At 10.00 a.m. the dignitaries arrived. Princes, ambassadors, members of the court and government, cardinals and bishops: Aglaé Ney could hardly remember so dazzling a throng since the Emperor's 1804 coronation. Meanwhile the Imperial Guard, the Paris garrison and the National Guard regiments were drawn up in lines on the Champ de Mars. 50,000 men altogether.

At 11.00 a.m. the Guard batteries upon the terrace of the Tuileries fired a 101-gun salute. This was echoed by five other batteries based upon the Pont d'Iéna, Les Invalides, the heights of Montmartre, the Château de Vincennes and the Champ de Mars itself. It meant the Emperor had left the palace. At noon his procession appeared near the review ground. It was led by the Red Lancers and *Chasseurs è Cheval*, then came the heralds-of-arms and nineteen coaches. The Emperor's was drawn by a team of eight horses and surrounded by the plunging mounts of four Marshals, Soult, Jourdan, Grouchy and Ney. With the first three, Napoleon was seen to lean out and exchange pleasantries, but for Michel Ney there were only glares. Behind these came the *aides-de-camp*, Savary, Dautancourt and the *élite* gendarmes, followed by the dragoons and horse grenadiers. The route was lined with infantry.

The Parisians shouted themselves hoarse as the Emperor rode on to the *Champ de Mai*, but they had expected to see him in his old green uniform and familiar *bicorne*. Instead he mounted his throne wearing 'a white satin suit and his shoes were trimmed with rosettes. From his shoulders hung a purple velvet cloak embroidered with gold and lined with ermine. On his head a black bonnet, shaded with plumes and looped with a large diamond in front.'[51] It reminded everyone of Joachim Murat in his heyday.

After Mass, which the Archbishop of Tours celebrated in his best double-forte voice against the background of more artillery salvos, delegates of the Electoral Colleges approached—one of whom read a loyal address which was followed by a *Te Deum*. Then those bearing the brass eagles and other troops:

> closed ranks in a semi-circle and the Emperor addressed them. *I entrust these eagles with the national colours to you. Will you swear to die in their defence?* Then, to the Guard: *And you, soldiers of the Imperial Guard, do you swear to surpass yourselves in the coming campaign, and die to a man rather than permit foreigners to dictate to the Fatherland?* These solemn appeals were answered by cheers and cries of *We swear!* and *Vive l'Empereur!* which increased during the parade.[52]

It was a good fête; although not one word had the Emperor addressed to Ney, and that same night Michel and Aglaé found themselves excluded from the banquet to which all the other dignitaries had been invited. Worse still, the Marshal was informed that various officers who had deserted him for the King at Lons were now being reinstated as battle-commanders. Bourmont had a division in IV Corps under General Gérard, with Clouet as his *aide-de-camp*. Colonel Dubalen held command of the new 64th Regiment, and General Mermet (who at Besançon had 'neglected' to read Ney's/Napoleon's proclamation to the soldiers) was invited by Davout to take over the VI Military District. Even Lecourbe had made his peace with Napoleon and was given command in the Jura. (He was an expert at mountain warfare and enough of a patriot to fight the Austrians.)

But the climax came two days later. Ney had gone to the palace for authority to draw thirty-seven thousand francs from the treasury: his arrears in pay together with certain expenses incurred while surveying the northern frontier. Suddenly, appearing from his own office and on his way to another,

Napoleon caught sight of him. 'You here!' he growled with obvious hostility. 'Well, well, Ney. I really thought you had emigrated!'

The Marshal's mouth fell open—then it closed with a snap. 'Sire,' he replied: 'I ought to have done so long ago.'[53] He jammed on his hat, about-turned and walked out: not only upon his former idol but also on the whole Imperial dream, *presumably forever.* 'It's finished,' he told Aglaé when he returned to the Rue Bourbon. 'All over. Tomorrow morning we leave for Coudreaux, and there, this time, we will be staying!' The romance was dead, the reality impossible to accept. 'We must think and build again . . . *Il faut tenter de vivre!*'

But was the romance over? Napoleon, as we know, had a complex habit of never quite releasing those formerly close to him. Out of favour they might be, even thoroughly disliked—or simply just past their usefulness. And yet in moments of crisis he was inclined to remember and turn to the old, familiar faces. Josephine, Jerome Bonaparte, Fouché and Marshal Brune had each of them experienced recall in this strange way; one day 'out in the cold', the next summoned back with open arms. And June 1815 promised to be the greatest crisis of all.

Napoleon appeared to take a last farewell of Paris on Sunday 4 June which had been ordained a public holiday. 'There was a further distribution of eagles in the Louvre and free wine flowed from 36 fountains in the Champs Elysées.'[54] But in the end he didn't leave until Monday 12 June. On the 10th he went to the theatre for the last time, to the Comedie Français where Talma was aptly playing *Hector.* Afterwards he spoke to the great actor. 'So Talma, Chateaubriand says that you gave me lessons in how to act the Emperor; I take his hint as a compliment, for it shows I must at least have played my part well!'[55] By this time most of France's troops were already moving north, the Imperial Guard bringing up the rear. When the Emperor himself departed, in his blue and gilt carriage, another important-looking envelope was speeding towards Marshal Ney at Coudreaux.

[1]   Edith Saunders: *The Hundred Days*, London, 1964.
[2]   Ibid.
[3]   Philip Henry Stanhope: *Notes of Conversations with the Duke of Wellington, 1831-51*, London, 1888.
[4]   *The Anatomy of Glory.*
[5]   Ibid.
[6]   *The Bravest of the Brave.*
[7]   Ibid.
[8]   Ibid.
[9]   Ibid.
[10]   Ibid.
[11]   *The Creevey Papers*, ed. John Gore, London, 1934.
[12]   *The Hundred Days.*
[13]   *The Anatomy of Glory.*
[14]   *The Hundred Days.*
[15]   *The Bravest of the Brave.*
[16]   Ibid.
[17]   *The Hundred Days.*
[18]   *Mémoires de Madame la Duchesse d'Abrantès.*

[19] *The March of The Twenty-Six.*
[20] *The Bravest of the Brave.*
[21] Ibid.
[22] Ibid.
[23] *The Hundred Days.*
[24] *The Bravest of the Brave.*
[25] Ibid.
[26] Ibid.
[27] *The Hundred Days.*
[28] Legette Blyth: *Marshal Ney, A Dual Life.*
[29] *The Hundred Days.*
[30] *The Bravest of the Brave.*
[31] *The Anatomy of Glory.*
[32] *The Hundred Days.*
[33] John Naylor: *Waterloo*, London, 1960.
[34] *The Bravest of the Brave.*
[35] Ibid.
[36] Naylor: *Waterloo.*
[37] *The Anatomy of Glory.*
[38] Houssaye: *1815—Waterloo.*
[39] *The Hundred Days.*
[40] Lt.-Col. Charras: *History of the Campaign of 1815*, London, 1857.
[41] *Correspondance de Napoleon Ier.*
[42] *The Hundred Days.*
[43] Cronin: *Napoleon.*
[44] Ibid.
[45] Ibid.
[46] Ibid.
[47] *The Hundred Days*; also Houssaye: *1815.*
[48] Houssaye: *1815.*
[49] *The Hundred Days.*
[50] *The Bravest of the Brave.*
[51] *The Hundred Days.*
[52] *The Anatomy of Glory.*
[53] Houssaye: *1815.*
[54] *The Hundred Days.*
[55] Cronin: *Napoleon.*

CHAPTER 12

The Set-Pieces

Quatre Bras and Waterloo

The greatest favourites of destiny make mistakes.

VICTOR HUGO

Breaking the seals, he found it was a letter from Davout, but incorporating a message from the Emperor. 'Send for Marshal Ney and tell him that if he wishes to be present for the first battles, he ought to be at Avesnes on the fourteenth.' (Dated 11 June 1815)

It must have set Ney's thoughts teeming like a mill-race! All through the intense activity of 'The Hundred Days' Napoleon had kept him—quite deliberately—at far more than an arm's length. Their meetings were few, invariably cool and once or twice decidedly-abrasive. The old, comradely ease based upon mutual military respect had gone, apparently for good; and while the Emperor prepared to defend his self-propelled restoration against the Allied attacks he offered Ney no new command. Davout was the most relied-upon marshal now, with Soult drafted to replace Berthier as Chief-of-Staff and Grouchy, at long last created a marshal, the rising star. Mortier, another who had rallied to Napoleon, was given command of the Old Guard, Suchet had the Army of the Alps and Brune was asked to keep order in the difficult (largely Royalist) area of Provençe. Obviously the Emperor felt very differently towards Michel Ney following the traumatic scenes which surrounded his abdication. He had judged the Marshal hard-hearted and ungrateful then—and despite Ney's subsequently bringing the army over to his side doubts about the Prince of the Moscowa being trustworthy tended to linger.

As for Ney, he was still a tangle of conflicting ideas and emotions. He knew that his changing sides back to Napoleon had hopelessly compromised his relationship with the Bourbons. Should the Emperor fall again it meant his own personal fate was sealed. And yet the Emperor refused to employ him! He, who had even more at stake, because now his actual life depended upon the outcome! Many things had combined to restore his beliefs in the Empire. As a practical soldier, always close to those troops placed under his command, he was fully aware of the army's mood: in the face of everything defiantly nationalistic and anti-Bourbon. Moreover the social pin-pricks escalating to snubs and open insults which he—and especially his wife—had received from the returned Royalist 'Ultras' were sufficient to convince him that merit no longer counted for much in their France. His work to reorganise the cavalry had passed largely unnoticed; his new titles, as Governor at Besançon, as a chevalier of Saint-Louis and a peer of the realm, seemed to be mere baubles. As a result, the remembrance of past promotions and glory began to take a fresh grip on his thinking, the spirit of the *Grande Armée* to pump and pulse once more through his bloodstream.

But most of all he had changed sides on account of an ineradicable loyalty to the single man whose will-power was almost always a dominant influence upon his life. It was a loyalty that, originally, had raised him to the heights of his chosen career: only to suffer a regular bombardment of disillusion when Napoleon made mistakes and the Empire's sun at last began to set. Nevertheless his loyalty had a bedrock quality about it; with which the Bourbons and the Allies could never really hope to compete. And in that vital, fascinating face-to-face at Auxerre the changeover was seldom in any doubt. Napoleon had been cynical but persuasive, Michel Ney the subject of his own divided make-up, the nature of which his attempt to bluster over their respective positions could hardly hide. Napoleon took advantage of this. Since when the Marshal had been made to take 'a hundred little cuts'.

Loyalty again played its characteristic role though when he scanned the letter of 11 June: overriding every other consideration and drawing him back into active service for what was to prove the Napoleonic era's grand finale. In effect the summons turned Michel Ney into a kind of French Icarus. Until its arrival he had begun to look upon his fighting days as definitely over. By now he realised that the Emperor's longstanding faith in him was badly shaken— while he himself felt let down and more or less impotent within the cascade of events. Even so, after his initial surprise the upsurge of old attachments was fairly predictable. He *could* have sat on the fence, awaiting the outcome of the fighting in Belgium and then seeking to exploit it to his private advantage. He *could* (as several marshals did) have feigned sickness. Instead he packed his bags and headed north.

Why did Napoleon send for Ney? So late in the scheme of things and after months of proceeding without him?

Well, as usual with Napoleon, there were several reasons underlying his call: only this time all of them based upon practical necessity; none resulting from a change of heart. To begin with he was chronically short of experienced senior commanders. Of those who still adhered to his cause, a number were tied down in holding the southern and eastern frontiers. Davout was doubling as Minister of War and the defender of France's capital. 'I can entrust Paris to no one but you.'[1] (If Ney and Davout had fought side by side at Waterloo as they did at Borodino then the result must almost certainly have been very different. Not even Wellington's skill and proverbial good luck could have withstood them.) The worst blow came when Mortier fell ill—quite genuinely —with attacks of gout and sciatica. It caused the loss of a tough fighting general whose lack of imagination was far outweighed by his being the perfect subordinate.

Another argument for Marshal Ney's recall seemed to be his propaganda value. Napoleon was too adroit a politician to miss such an opportunity. For, 'on the one hand it was a calculated blow against Louis XVIII's prestige to re-employ the former Bourbon Commander-in-Chief; on the other, Ney's preferment might serve to persuade other servants of the Bourbons that their acts of desertion in 1814 could be overlooked in return for new tokens of devoted service to Napoleon's cause.'[2]

In the end, however, what tipped the scales in Ney's favour was his personal popularity with the ordinary French soldier. Until Russia he had been a hero essentially to those who served under him and to a minority who, stationed elsewhere, followed the Imperial campaigns in detail. But since the retreat from Moscow his bravery was a legend throughout the army, and in particular among the conscripts: the Classes of 1814 and 1815, most of whom didn't even know the names of Napoleon's other officers. If the coming campaign proved to be the Empire's Armageddon then there was no one like *Le Rougeaud* for rallying troops to cope with the fearful impact of a repulse.

Ney reached Avesnes on 13 June, having paused along the way just once: to pick up Colonel Heymès (the only close friend and confidant he would have on his staff during the next few days of battle). Alighting from their coach the two men hurried off to report themselves for duty. They found the Emperor 'in the garden of the ex-King's lieutenant . . . before the palace . . . walking up and down, deep in meditation and oblivious to the fine view over the valley.'[3] At the edge of the garden they stood, silent, waiting for him to notice them and show some response. Both knew that after Russia, a string of defeats in Germany and in France, abdication and exile, their former 'invincible' leader was a changed man. Even the heady excitements of a march on Paris, his acclaim by its populace and then one hundred days of the Empire restored had failed to eliminate the processes of age, ill-health and his burden of insecurity. Napoleon's 'faculties had not diminished; his lucidity, intelligence and verve were still intact; but now he took longer to make up his mind, and his energy seemed to flag at times. He brooded and hesitated, expressing his state of mind by silences, depression, and torpor. This caused decisions which no one else dared make to be postponed.'[4]

And so the two waited: wondering, apprehensive. Ney had been aware of these mental and physical changes ever since Borodino, and they had influenced him a very great deal during the abdication crisis. Following upon which he stood accused of being 'shell-shocked' himself; and was referred to as 'knowing less than the last-joined drummer boy' by an Emperor grown highly unpredictable as well as basically unforgiving. But now Ney's loyalty had committed him to the ultimate involvement. How exactly would it be received?

In fact it was received rather well. At least to start with, when Napoleon's behaviour can be taken as reasonably spontaneous. Catching sight of them at last, he evidenced delight—stumping over to surprise Ney by embracing him and then back-slapping Colonel Heymès. He promptly invited the Marshal to dinner, at which he appeared to be all of his old, mercurial self: full of charm, slighting about his European enemies and clearly optimistic that this new campaign would be a success.

The pity was, carried away by so much exuberance, he neglected to tell Ney what his command would be and absolutely nothing about the overall strategy for the campaign; while Ney hesitated to spoil the cordial atmosphere by pressing him. The first real mistake in the run-up to Waterloo had been made. In the coming days Blücher, d'Erlon, Grouchy, Napoleon, Ney, Soult and Wellington would each of them make horrifying mistakes. But Napoleon's were destined to have by far the most serious consequences, compounding as they did Ney's own mistakes at a tactical level and even when the Marshal did get things right fully inhibiting his chances for exploiting the situation. I date the beginnings of catastrophe from what *should* have been and *wasn't* explained to Michel Ney at their so-called reunion meal.

The following morning French Headquarters (and Napoleon with it) moved to Beaumont, just inside the present Belgian border. A sizeable part of the army was already there—together with the stricken Mortier, who wrote in his diary: 'I have violent pains accompanied by chills. M. Percy examined me and warned me that this would continue for forty days. Since I cannot possibly mount a horse, I sent my regrets to the Chief-of-Staff for my inability, in this condition, to share the glory and dangers of the Army . . .'[5] (He could hardly be expected to command the cavalry of the Imperial Guard from a litter!) Meanwhile Ney and Colonel Heymès, being without horses, were left to make their journey to Beaumont in a requisitioned peasant's cart. Rain had transformed the ground into a sea of mud; and not even the cheers of the infantry they passed, nor the grumblers' remarks ('There is *Le Rougeaud* . . . things will pick up now!') could lift their depression. Incredible as it might seem, they continued to have no idea what the Emperor expected of them.

Once in Beaumont Ney lost no time in seeking out Mortier. He was quick to show sympathy for an old comrade in so much pain, but this sick-room visit also had a practical purpose. He arranged to buy Mortier's two magnificent battle-chargers together with all of their attendant harness (still not enough for Waterloo as it turned out). Colonel Heymès also managed to procure a horse, and in this way, at 3.30 in the afternoon of 15 June they finally caught up with Napoleon, now twenty-six kilometres to the north-east at Charleroi.

They found the Emperor sitting outside an inn just beyond Charleroi (the Bellevue) being cheered by soldiers of the Young Guard who marched by in columns. Michel Ney dismounted and hurried over to him. 'Bonjour, Ney,' was the sharp, formal greeting he received. Then, without more ado: 'I want you to take command of the I and II Corps. I am also giving you the Light Cavalry of the Imperial Guard—*but don't use it without my orders.* Tomorrow you will be joined by Kellermann's *cuirassiers.* Go and drive the enemy back along the Brussels road!'[6]

In effect Napoleon had given him nominal and tactical command of the entire Left of the French Army, in all around fifty thousand men. Its front formations were I Corps (21,000 men, including nearly 2,000 cavalry and 46 guns under General Drouet, better known as Count d'Erlon) and II Corps (25,000 men, including another 2,000 cavalry and 46 guns under General Count Reille). Detached from the Reserve to support these were the

Light Cavalry of the Guard under Lefebvre-Desnoëttes, another 2,000 men and horses, plus Kellermann's III Cavalry Corps, a further 3,700. Altogether a considerable strength and incorporating some of the newly reconstituted *Grande Armée's* best fighting men.

Why therefore—and this remains one of the biggest imponderables of the whole Waterloo campaign—did Napoleon avoid telling Ney he was to command such a formidable and important part of France's army until the moment of his much-delayed arrival? (An arrival, incidentally, which need not have been delayed if the Emperor had helped him with horses back in Avesnes.) Can it be that, in spite of a diplomatic thawing-out towards Ney, he was still reluctant to trust him with the third of the three most important subordinate positions in his Army of the North? (Soult, Chief-of-Staff, and Grouchy, as commander on the right-wing, were both appointed in good time.) Was he hoping against hope for a last-minute recovery by Mortier, having called up Michel Ney merely as a standby and so making his final appointment an act of hasty improvisation? We shall never know; but what is certain is that the delay placed the Marshal at an enormous disadvantage. He had no time for proper reconnaissance, nor to become acquainted with his new staff and organisation. Worse still, there was no chance to supervise the formation of his initial front. Reille and the II Corps were well beyond the River Sambre and already in contact with Wellington's outposts.

What appears even more inexcusable though is Napoleon's failure to enlighten Ney regarding his overall strategy and intentions. He could have done this at Avesnes; he should, surely, have given up a few minutes to explain it with the aid of his maps at Charleroi. Instead of which he tossed out the vague, open-ended order 'Go and drive the enemy back along the Brussels road!' and then left Ney to make an individual interpretation of it.

We must pause here to examine the Emperor's vital strategy in more detail. Otherwise it will become increasingly difficult to understand Marshal Ney's mishandling of the action which preceded the battle we now call Waterloo at the crossroads of Le Quatre Bras (meaning the four arms or branches).

As in the best of earlier Napoleonic strategies, this final one combined flair with feasibility and interwove practical politics with a military solution. Since the Allies had turned down his requests for a peace which would leave him as Head of State, but contained inside France's pre-Revolutionary borders, the Emperor felt forced to take the offensive before their mobilisation was complete. He was beset by enemies inside the country as well as surrounding it. The standing army guaranteed his position, but there were many notable defectors and the middle class and the workers had shown themselves prepared to sacrifice him if it meant peace. Even the formerly-docile Chamber of Deputies was showing its teeth. The originator of an Act to depose him in 1814 had just been elected its President. However, very much in Napoleon's favour was the situation that not all the Allies were yet ready to tackle him in the field. Only Wellington's 90,000 Anglo-Dutch, strung out between Brussels and the sea, and Marshal Blücher's 105,000 Prussians, concentrated to the east near Namur, were in any state of preparedness for battle. Intelligence reports stated that they were working up towards an invasion of France by mid-July.

Napoleon estimated he could have at his disposal in the north of France over 120,000 men. His decision, therefore, was to strike into Belgium in June. Apart from catching them unawares, he would strike between Wellington and the Prussians, divide them, then defeat each in turn. Once this was achieved a pro-French revolution in Belgium looked virtually certain, followed by a collapse of the remaining Allies and the muting of his critics in Paris.

His decision made, everything progressed rapidly (and, one must add, quite brilliantly) towards a delivery of the intended knock-out blows. Despite Berthier's defection Napoleon succeeded in the:

> concentration of five corps, the Imperial Guard and the cavalry reserve into a zone thirty kilometres square from a dispersal area of more than two hundred miles. In terms of manpower this involved moving 89,000 infantry, 22,000 cavalry, 11,000 gunners and engineers and 366 guns from locations as far distant as Metz, Lille and Paris. The advancing corps were divided into three powerful columns ready for the advance over the Sambre scheduled for the morning of 15 June.[7]

Furthermore the concentration was accomplished with the greatest secrecy. 'A security band was imposed on the frontier area on 7 June. As operational forces were moved from frontier positions, their places were discreetly filled by National Guardsmen; civilian traffic was carefully controlled, the mails suspended and fishing boats ordered to keep in port.'[8] The Emperor stressed the importance of these security precautions to Marshal Soult in a despatch of 7 June:

> Give the most positive orders for all means of communication to be closed along the fronts of the North, Rhine and Moselle (forces): not a coach or *diligence* must be allowed past. Demand the greatest surveillance to prevent even a single letter getting through if that is possible. See the ministers of police and finance and get them to instruct their agents to intercept absolutely all communications.[9]

Such measures were helped considerably by a fair amount of muddle on the Allied side. Although Wellington and Blücher enjoyed a numerical superiority over the French their liaison work was deplorable. On a purely personal basis, relations between the two commanders were excellent, but Gneisenau, the Prussian No. 2, had an almost pathological dislike of the British general and:

> sweet or sour, the Prussians were in no mood to share either headquarters or lines of communication with Wellington. Gneisenau had chosen Namur for Blücher's headquarters forty-eight miles from Wellington at Brussels. While the British communications ran back westwards and northwards to Ostend and Antwerp, Blücher's lines moved in the opposite direction eastwards through Liège and Aachen into Germany. Between the two armies there was thus a joint or hinge represented by the area on either side of the great paved *chaussée* running due north from Charleroi to Brussels. Napoleon had his eye on the hinge.[10]

On the night of 13/14 June 'advanced Prussian patrols under General Zieten had seen the twinkling lights of innumerable camp-fires in the direction of Beaumont, a few miles from the frontier on the French side of the River Sambre.'[11] This was their first realisation that Napoleon had seized the initiative, and a Prussian general-alert followed. But no one from their headquarters bothered to tell Wellington, who consequently visited the Duke and Duchess of Richmond's house on the evening of 14 June still believing that the French would try to turn his right flank in a sweep towards the coast.

Belgium was full of French sympathisers. Napoleon received reports of the Prussian alert almost as soon as it was ordered. Clearly now he would have to deal with this sector first—and so his plan became to smash Blücher with Grouchy's forces and the Imperial Guard before the Prussians had any chance for a manoeuvre. According to the overall strategy this meant Ney's left-wing would be relied upon to prevent Wellington making a supporting link-up with Blücher. Hence the importance of seizing that vital crossroads at Quatre Bras. Once the French held it in strength Wellington had no direct route to the Prussian positions, and it was even assumed that Ney could detach some of his units for an attack on Blücher's flank. 'Whichever side held Quatre Bras would hamper the enemy's movements, and on 15 June it was not likely that there were more than a few isolated detachments of Allied troops between Charleroi and the crossroads.'[12]

The trouble was—Napoleon being Napoleon—he wanted to keep all his options open. 'In war one must remain flexible!' Perhaps he quite genuinely believed (as some have claimed) that the Prussian alert might be a prelude to their retreat; in which case the entire French army could be swung left against Wellington. But the real damage was done by his failing to explain any of this to Ney. Which in itself was not unusual. Throughout his campaigning life the Emperor had a reputation for peppering his senior officers with cryptic, unexplained orders. Before though there had always been Berthier to dot the i's and cross the t's; to clarify the Emperor's intentions in model despatches. Soult, while he had Napoleon's confidence, proved fairly lost when it came to interpreting his master's voice and if anything tended to increase the confusion. As a result, Ney hurried off to take up his command with 'Go and drive the enemy back along the Brussels road!' still ringing in his ears and imagining that he had a totally different role to play. Instead of making a swift dash for the crossroads and then holding on to it like grim death (a part he was by temperament ideally suited for) he thought himself to be the vanguard of a general advance upon Brussels—with Grouchy left behind to contain Blücher. His initial moves were slow and prudent, therefore: just when a typical Ney 'spurt' could have given him Quatre Bras with scarcely a fight.

'Napoleon has *humbugged* me, by God!' would be Wellington's comment after discovering the full extent of the French army's advances. What he felt about Gneisenau and the Prussian staff went unrecorded.

His first indication that things were not going according to plan and that the French had seized the initiative came just after 3.00 p.m. on 15 June, when 'a Prussian officer covered with dirt and sweat galloped into Brussels with a much-delayed dispatch sent by General Zieten'.[13] It stated that Zieten's 1st Corps had been attacked and their outposts driven in at Thuin.

Now Thuin is south-west and *down* the Sambre from Charleroi, where the despatch had been written at breakfast-time. He still had no idea, consequently, that Napoleon had since occupied Charleroi and was ordering his left-wing up the main road towards Brussels. He continued to expect a flanking move via Mons to the sea. But at least he knew 'something was up' and immediately issued orders for the entire Anglo-Dutch army to be ready to march out at a moment's notice. That same evening, with a display of remarkable *sang-froid*, he attended the Duchess of Richmond's now famous ball.

One hour after Wellington read Zieten's dispatch Michel Ney joined the men placed under his command. There was no opportunity to call a meeting of principal officers, for they were already on the move. Foy, Kellermann, Lefebvre-Desnoëttes and, of course, d'Erlon, he had fought with both in Spain and Russia. Also he had met Piré in Spain. But Reille, the Emperor's former *aide-de-camp*, and Bachelu he knew only by repute. At the same time the detachment to him of Prince Jerome, Napoleon's irresponsible younger brother and now the ex-King of Westphalia, must have caused even a commander with Ney's iron nerve at least a twinge of anxiety.

However, the most important thing was to acquaint himself with the situation up ahead: where Reille's II Corps had begun to climb the bluffs above the little Piéton river. Leaving Colonel Heymès behind to improvise an HQ staff the Marshal rode on to join it. Only just in time—for almost immediately they came upon a strong Prussian outpost at the village of Gosselies. Ney ordered Reille to attack at once and the Prussians soon broke. He noticed they retreated eastwards towards Fleurus, a clue that the main Prussian forces were still a long way from Wellington's. Halting Reille he detached a single division under General Girard to shadow them. (Girard would die in the coming battle against Blücher at Ligny.) Then he told Lefebvre-Desnoëttes to probe northwards again. 'With caution though!' (Napoleon's specific '. . . *but don't use the Guard cavalry without my orders*' had clearly taken effect.)

It was a fine, warm afternoon still; if rather humid and with the hint of thunder to come. Hectare upon hectare of wheat and rye stood tall and yellow under the sun. Lefebvre-Desnoëttes (who complained the Emperor didn't use his Horse Guard nearly enough) was eager to go. He looked a magnificent figure at the head of 2,000 men in green and scarlet. But this was no mere 'parade ground hero'. He thirsted for revenge over the British! Apparently they had treated him badly after his capture at Benavente—which in turn decided him to break parole and escape back to France. As a leader of horse he had Ney's full confidence. 'I know I can rely upon you not to do anything idiotic,' the Marshal added. (No doubt a belated sideswipe at Murat's tactics.) He need hardly have worried. The Light Cavalry of the Guard followed a gigantic sabre, not a golden-tipped wand.

They clattered off along the road for almost four kilometres before suddenly the leading squadron (of Polish lancers) was 'greeted by musketry' at Frasnes. Several companies dismounted. Others began flanking the village, only to discover an enemy retreating in orderly fashion to the crossroads 'identified as Le Quatre Bras'. Lefebvre-Desnoëttes followed at a safe distance; but then he came under heavy fire from what was apparently a much larger force (with artillery) dug in on either side of the main road.

After seeking first-hand information from each of his officers he sent back to Ney this report:

> *Monseigneur.* Arriving at Frasnes . . . we found it occupied by 1,500 Nassau infantry. When they saw that we were manoeuvring to turn their flank they made a sortie from the village where we did in fact surround them. General Colbert advanced up the road to within a musket shot of Quatre Bras; however the terrain was difficult and the enemy, with the woods at their back, kept up a lively fire with eight guns, so we could not get at them. The troops at Frasnes were serving under Lord Wellington. They had not been beaten at Gosselies . . . The troops beaten there marched to Fleurus, and none passed this way. We took 15 prisoners and lost 10 men killed and wounded.'[14]

Nassauers! Even his cavalry commander's claim that the enemy's line of retreat was a sortie didn't disguise the value of such information to Ney. So 'Nosey' held Quatre Bras; and seemingly in strength! The Marshal promptly ordered a battalion of infantry to join Lefebvre-Desnoëttes, put the remainder of Reille's Corps back on full alert and then spurred north through Frasnes to reconnoitre for himself.

It was early evening when he reached the French positions: still fairly light, but with a quantity of shadow thrown across Quatre Bras by the outline of Bossu Wood. Also the rye between the French and the crossroads had been allowed to grow unusually tall that year. Ney rode back and forth, impatiently straining to see over it. Meanwhile the cannon-balls kept zooming across, there was concealed musketry from both sides of the road and three lots of farm buildings appeared to be strongly held, one to the west, two to the east. Given all this cover it was impossible to calculate the numbers of enemy ahead. And so the Marshal missed his first chance. Had he but known the truth about the weakness facing him, brittle and hollow like a chocolate Easter-egg, then he would never have backed off and might well, in Lady Longford's dramatic phrase, 'have ridden straight through to Brussels'.[15]

Quite simply, what had occurred was the result of inspired insubordination. While 'Nosey' continued to misinterpret the French advance 'a couple of intelligent Allied officers . . . were prepared to risk Wellington's wrath and disobey the letter of his orders in order to pursue a course of action they felt more justified on account of their completer knowledge of local events.'[16] At 2.00 p.m. that afternoon Constant Rebecque, chief-of-staff to the Prince of Orange, had agreed to a brigade of General de Perponcher's Dutch-Belgian troops under Prince Bernhard of Saxe-Weimar probing south to Quatre Bras and beyond. Their advance battalion had only just entered Frasnes when the alarm went up. Whereupon the remaining Nassauers (4,000 infantry plus their eight gunnery crews) threw themselves down behind the tall rye and kicked up enough racket to sound like a whole army corps. Certainly enough to deceive Marshal Ney, who saw nothing for it but to break off the engagement pending further instructions from 'on high'. Riding back towards

Gosselies he ordered Reille to bivouac for the night and soon there were French campfires alight in a great arc running from Frasnes to the Sambre.

As he did so Rebecque 'took a second important step when at 8.00 p.m. he authorised Brigadier-General Bylandt to march up from Nivelles to support Bernhard. Shortly after this instruction had been sent out, Wellington's after-noon order reached Ist Corps headquarters, ordering Perponcher's division in its entirety to concentrate at Nivelles, some eight miles from Quatre Bras.'[17] To obey it, Rebecque saw clearly, would be disastrous. He showed the order to Perponcher 'without comment, and the latter, on his own initiative, decided to disregard his commander-in-chief's specific order and to continue the occupation and reinforcement of the Quatre Bras position.'[18] In fact these two officers between them had saved not only Wellington's army but also his personal credibility. For he very nearly marched his men like the Grand Old Duke of York (and as d'Erlon was to do with I Corps, disastrously, on the following day).

Michel Ney rode back to Gosselies and conferred with Colonel Heymès. Not a word in from his own commander-in-chief. Tired though he undoubtedly was, the Marshal saw nothing for it but to ride on towards Charleroi—where he arrived at midnight and found the Emperor asleep. Quickly aroused, Napoleon ordered a late supper and, after listening to Ney's report, at long last began to explain the overall strategy. Perhaps his mind was still clouded by sleep. Perhaps he now firmly believed that Blücher would retreat; in spite of the previous day's activity around Fleurus, where the Prussian outposts had shown their preparedness to make a fight of it. Anyway, he said, the big blow must be delivered against 'Villainton'; and if possible at Quatre Bras. Once in control of the crossroads the French could use Grouchy either to finish off Wellington or give chase to Blücher. A final instruction returned to the controversial subject of available cavalry. Ney should 'cover Lefebvre-Desnoëttes' division with d'Erlon's and Reille's horse *in order to spare the Guard*.'[19] (Desnoëttes winced when this message was passed on.)

Well, at least he had a clear picture of what was expected from him; or so he rode away thinking. And he would have the whole French army to support his attack; again so he thought. Upon reaching Gosselies he permitted himself just two hours' sleep, wrapped up in his big velvet cloak.

Wellington also found himself limited to two hours' sleep that night. Upon realising he'd been duped, and after delivering his celebrated *humbugged!* remark, the Duke bent himself to the task of ordering up reinforcements. To be fair he kept his head. In dressing-gown and slippers he issued the necessary movement orders for all British units to converge on Quatre Bras. And he examined the map again; especially the area around Mont-St.-Jean in front of Waterloo. 'I have ordered the army to concentrate at Quatre Bras; but we shall not stop him there, and if so, I must fight him here.'[20] (The Duke had studied the actual ground a year earlier.) While he considered this a multitude of fashionable people dispersed themselves from the Richmonds' ball as a startled rabble, quite a few to pack their bags.

Michel Ney rose at 7.00 a.m. on 16 June and promptly instructed his officers to hurry along the troops who were now requisitioning breakfasts.

Wellington set off for Quatre Bras at 7.00 a.m. too. (He reached the cross-roads at 10.00 a.m.)

Napoleon on the other hand was up and working by 6.00 a.m.: beginning with a lengthy dispatch to Ney which more or less confirmed their discussion in the small hours. The Marshal 'was to hold himself in readiness for an immediate advance towards Brussels (probably that evening)—*once the reserve reached him*'—and once the Emperor's mind '*was finally made up*'.[21] For the time being 'he was to place one division five miles north-west of Quatre Bras, retain six at Quatre Bras itself, and send out one more towards Marbais to serve as a link with Grouchy.'[22]

It took two hours for this despatch to be prepared, signed and on its way. 'But hardly had the *aides* left Imperial Headquarters than a message from Grouchy arrived reporting that his cavalry screen had spotted strong columns of Prussian troops advancing towards Sombreffe from the general direction of Namur.'[23] At which point Napoleon's course was suddenly crystal clear. His strategy had allowed for it; and from the outset it was always the likeliest event. Since Blücher was obligingly coming at him, then *he* instead of 'Villainton' would receive the first French knock-out blow! And so the Emperor, Grouchy's divisions and those of the Guard marched off to fight and win a victory at the brook of Ligny, just beyond Fleurus—*but with never a word concerning their moves being passed on to Ney*. In fact, until it was too late, French HQ now chose to ignore the Left completely.

Over on the western front all remained relatively quiet that morning. Though the Anglo-Dutch were being reinforced it was a slow business; the French still outnumbered them by three to one, plus sixty guns against their eight. As a result Ney missed his second (and last) chance for an easy break-through. He was under strict orders not to take risks; consequently he proceeded with extreme caution, again just when a touch of his old, hussar-like fury might have won the day.

One should add too that the psychological presence of Wellington had begun to influence French thinking. Both Ney and Reille were veterans of the Peninsular War—and both had learned their lessons against Wellington the hard way. The Marshal for his part had sustained several other defeats since then: but these he could put down to circumstances, even to his own mistakes, whereas at Bussaco he had fought well only to be beaten by a general of very obvious talents. The more he thought about it the more convinced he grew that Wellington was using similar tactics at Quatre Bras. In other words, concentrating a large force of infantry out of sight on the reverse slope behind the crossroads. After Reille had endorsed this view Michel Ney came to a decision. There would be no lightning manoeuvres and certainly no two-pronged attack. Instead they must mass and roll right over the enemy from front to rear in one continuous assault. It would mean casualties at first, but if II Corps faltered then d'Erlon's I Corps ought to be in close support to finish the job. At all costs Wellington must be knocked off balance and then prevented from rallying once the battle was joined. Ironically—and such are the vagaries of war—the Duke had departed from Quatre Bras by this time. At 10.30 a.m. he rode off to witness the beginnings of Blücher's 'damnable mauling' at Ligny—and to promise his support 'provided I am not attacked myself'.[24] When he returned to the crossroads at 2.20 p.m. the French steam-roller was already proving its effectiveness.

Ney and Reille had used what was left of the morning to form column: which they did very thoroughly, with General Piré in front, then Bachelu and Foy, and finally Prince Jerome Bonaparte in a position calculated to stem the flow of his impatience. Soon after 11.00 a.m. Ney received Napoleon's written orders and by 11.45 the French were marching and singing their way up the road north of Frasnes. At the same time they opened up with a battery of fourteen guns. One by one the main defence points in front of Quatre Bras began to fall. Bachelu's division fanned out to capture Piraumont Farm, driving back Perponcher's left, and following upon this Foy took Gemion-court. When Foy next moved against the farm of Pierrepoint, over to the west, he found the opposition too strong, but immediately Ney ordered up Prince Jérôme who did rather well in capturing it. Ney was riding in front here and had his first horse killed under him. To the troops' relief he got up without a scratch.

By 2.30 p.m. when Wellington returned to Quatre Bras there remained only Bossu Wood to be cleared. In fact the Duke is reported to have heard Ney, now mounted on the second of Mortier's battle-chargers, urging his men on with the familiar *Grande Armée* jingle: 'The Emperor rewards those who advance!'[25] The fighting at this point became quite desperate. Wellington was helped by a steady stream of reinforcements, including Sir Thomas Picton's 8,000 men, the black Brunswickers, some more Nassauers and General Halkett's brigade. With these he managed to contain the next French assault. But at 3.30 p.m., when attempting to counterattack with the Prince of Orange's cavalry, he suffered a bloody repulse and lost six of his precious cannon. Meanwhile Reille and Piré had nearly fought their way through Bossu Wood. At long last the French Left seemed set fair to take the cross-roads.

At 4.15 p.m. Michel Ney received a further message from HQ (drafted by Soult at 2.00 p.m.). 'His Majesty's intention is that you shall attack whatever force is before you; and after vigorously driving it back, you will turn in our direction, so as to bring about the envelopment of those enemy troops which I have already mentioned to you (i.e. Blücher). If the latter is overthrown first, then His Majesty will manoeuvre in your direction so as to assist your operation in the same way.'[26] Apart from the sporadic boom of artillery over to the south-east this was his first real indication that Napoleon had opted to dispose of Blücher before Wellington, and that he (Ney) was expected to gain possession of the crossroads as an arterial means of swinging against the Prussian flank. However, it was not the moment to argue. If anything he must redouble his efforts!

Reille's men were tiring, while Wellington's numbers had swollen to around 21,000. But Ney had noticed how not all the Allied troops were of equal quality. Picton over to the right was withstanding all the French threw against him, but a contingent of Dutch and Belgians in Bossu Wood seemed close to cracking. The Duke of Brunswick, rushed to support them, was beaten back and mortally wounded. Whereupon both Netherlanders and Brunswickers broke and fled (some of them back to Brussels where they proclaimed a French victory). Clearly now a final push by Reille at this spot would open the way for d'Erlon's 20,000 fresh troops to seize the crossroads. An urgent message was sent to the Count at Frasnes. 'Let him not lose a moment in bringing up I Corps.'[27]

The news which came back was staggering! La Bèdoyère, Napoleon's zealous and fanatical *aide-de-camp*, had halted I Corps just as it was moving off for Quatre Bras and 'in the Emperor's name' ordered it to proceed towards Ligny. Poor d'Erlon, reconnoitring ahead, was subsequently compelled to go after it. As he pleaded later:

> . . . beyond Frasnes I paused among the Generals of the Guard where I was joined by General La Bèdoyère who showed me a pencilled note which he was taking to Marshal Ney and which called on that Marshal to direct my army corps to Ligny. General La Bèdoyère informed me that he had already given the order for the movement, changing the direction of my column, and he told me where I could rejoin it. I followed the route indicated right away and sent my chief-of-staff, General Delcambre, to the Marshal to inform him of my new destination.[28]

Michel Ney 'was beside himself. He could not know that Napoleon only needed d'Erlon on the Prussian flank to fell the staggering giant with a blow from which he would never recover. All Ney knew was that one mad stroke had suddenly halved the threat to Wellington.'[29]

He was still fuming when another Imperial aide, Colonel Forbin-Janson, rode up with a despatch from Soult timed at 3.15 p.m. It contained the following:

> *Marshal.* I wrote to you an hour ago to inform you that at 2.30 p.m. the Emperor would attack the position taken up by the enemy between the villages of Saint-Amand and Brye. At this moment the action is in full swing. His Majesty desires me to tell you that you are to manoeuvre immediately in such a manner as to envelop the enemy's right and fall upon his rear; the fate of France is in your hands. Thus do not hesitate even for a moment to carry out the manoeuvre ordered by the Emperor, and direct your advance on the heights of Brye and Saint-Amand so as to co-operate in a victory that may well turn out to be decisive. The enemy has been caught in the very act of carrying out his concentration with the English.[30]

Unfortunately, these very specific instructions were accompanied by a verbal postscript from Napoleon to 'look sharp and finish off the business at Quatre Bras': which the tactless *aide* delivered first. The Marshal (understandably) blew up! Whereupon Colonel Forbin-Janson grew so flustered under the resultant tongue-lashing that he forgot to hand over the original despatch. When Ney had cooled down he said: 'Tell the Emperor what you have seen. I am opposed by the whole of Wellington's army. I will hold on where I am; but, as d'Erlon has not arrived, I cannot promise any more.'[31] So the Colonel departed still clutching Soult's despatch and Ney would not read it until much later that evening. He resolved though 'to finish off the business at Quatre Bras' as demanded; and with this now uppermost in mind he sent an *aide* of his own galloping after d'Erlon—who after all still belonged to his command —with orders for I Corps to about-turn.

Around the vital crossroads the scales had most definitely tipped in Wellington's favour. Allied casualties were high, but their perimeter was holding and further reinforcements had brought their total strength up to the 25,000 mark. Unless d'Erlon brought back I Corps quite soon then Ney's plans for rolling right across Wellington were finished. While waiting though he tried a most remarkable gambit. In a battle which involved much chaos, confusion and not a few surprises from either side this stands as the most bizarre incident of all. However, it has Ney's personal stamp all over it and was nearly brilliantly successful: contrasting French flair with typical British and Dutch stubbornness.

At 4.15 p.m. the British officers at Quatre Bras were astonished to find a quantity of French lancers in their midst. (Wellington was away evidently, inspecting his outposts, and these lancers had dashed straight up the Brussels road.) They wrought absolute havoc, especially among the 42nd and 44th Regiments, before the British managed to form square and drive them off. But this was only the preliminary. Ney had ordered up General Kellermann and told him curtly: 'A supreme effort is necessary. That mass of hostile infantry must be overthrown. *The fate of France is in your hands.* Take your cavalry and crush them. Ride them down!'[32] Kellermann himself was sceptical, but Ney silenced him with a 'Go on! Get going then!'[33]

Instead of working up his advance from a walk to a gallop by the customary easy stages, therefore, the III Cavalry Corps commander flung his *cuirassiers* forward hell-for-leather from the start. 'I used great speed,' he later reported to Ney, 'so as to prevent my men shirking, or even perceiving the full extent of the danger which awaited them.'[34]

It was 5.00 p.m. The British were at first dismayed, then horrified. Halkett's 69th Regiment was caught off balance, shattered and lost their King's colour. The 33rd Foot were driven headlong towards Bossu Wood where Reille's men caught them for a second time. Drawing rein Kellermann found himself *actually in possession* of the crossroads. For the French, although sitting on winded horses, it was the only exhilarating part of the day's conflict. Even the British had to admire them. 'The enemy's lancers and *cuirassiers* are the finest fellows I ever saw,' stated Colonel Fraser.

'Then came the reaction. Without immediate support, Kellermann was at the mercy of a raking musket fire from the 30th and 73rd of Halkett's brigade, and the usual assortment of delicacies (grape and canister) from a hidden battery.'[35] Moreover an alarmed Wellington himself had returned to the fray; together with the Gordon Highlanders. 'Don't fire till I tell you,' he shouted as Kellermann swivelled to charge them. When they did fire, with only thirty yards between them, the French suffered badly. 'A British ball got Kellermann's charger but the quick-witted general managed to seize two of his troopers' bits and run off the field between them.'[36] With d'Erlon still nowhere in sight Ney lacked fresh infantry to rush up in support and so the moment of near-victory passed.

At 6.30 p.m. Soult's staff officer, Major Baudus, brought the Marshal news 'that Napoleon now attached little importance to his doings at Quatre Bras but much to the arrival of d'Erlon'. He also brought back (hours too late, of course) the despatch of 3.15 p.m. 'It is evident that the Emperor never comprehended what Ney was up against, despite the report relayed by Colonel Forbin-Janson.'[37] One wonders if he ever delivered it! Wellington's strength at Quatre Bras had now reached 36,000 men, plus 70 guns, but Napoleon still insisted it was only 'an advance guard'.[38] With his second horse just shot from under him Michel Ney stalked off to rally his faltering units. It took him until 9.00 p.m. to organise a safe retreat upon Frasnes in the face of severe enemy counterattacks. He had suffered perhaps 4,000 serious casualties, the Allies 5,000, including 2,500 British killed and wounded. When everything was quiet again d'Erlon returned with his I Corps, having never fired a shot all day. He blamed La Bèdoyére for his useless promenades —and with some justification. Le Bèdoyére in turn would state that he acted on orders from Soult, and therefore Napoleon. But Ney, with equal justification, blamed d'Erlon for not using his own initiative to help clinch what should have been a decisive French 'rolling over' Wellington at Quatre Bras.

At ten o'clock in Frasnes Ney ordered up meat, bread and some red wine. All the wounded had been attended to and bivouacs established. Prince Jerome was invited to supper. 'Their table was a plank resting on two empty barrels, and lit by candles pushed into the necks of bottles,' an *aide-de-camp* recalled later.[39]

Then the Marshal got down to writing his definitive report for Soult:

I have attacked the English positions at Quatre Bras with the greatest vigour; but an error of Count d'Erlon's deprived me of a fine victory, for at the very moment when the 5th and 9th Divisions of General Reille's corps had overthrown everything in front of them, the 1st Corps marched off to Saint-Amand to support His Majesty's left; but the fatal thing was that this Corps, having counter-marched to rejoin my wing, gave no useful assistance on either field. Prince Jerome's division fought with great valour. (*Ney always gave praise where it was due.*) His Royal Highness has been slightly wounded. Actually there have been engaged on our side here only three infantry divisions, a brigade of *cuirassiers* and General Pirê's cavalry. (*A reminder that he had not overworked the Guard cavalry!*) The Count of Valmy (Kellermann) delivered a fine charge. All have done their duty—except I Corps. We have lost 2,000 killed and 4,000 wounded. I have called for reports from Generals Reille and d'Erlon, and will forward them to Your Excellency.[40]

Ten days later, however, a letter to Fouché reveals that Ney had changed his mind. With the battle in front of Waterloo also lost, and upon due reflection, he apportioned the blame for Quatre Bras quite differently. . .

We advanced towards the enemy with an enthusiasm difficult to be described. Nothing resisted our impetuosity. The battle became general, and victory was no longer doubtful, when, at the moment that I intended to order up the first corps of infantry, which had been left by me in reserve at Frasnes, I learned that the Emperor had disposed of it without advising me of the circumstance, as well as of the division of Girard of the second corps, on purpose to direct them upon Saint-Amand to strengthen his left-wing, which was vigorously engaged with the Prussians. The shock which this intelligence gave me confounded me. Having no longer under me more than three divisions, instead of the eight upon which I calculated, I was obliged to renounce hopes of victory, and, in spite of all my efforts, in spite of the intrepidity and devotion of my troops, my utmost endeavours could thenceforth only maintain me in my position till the close of day . . .

By what fatality did the Emperor, instead of leading all his forces against Lord Wellington, who would have been attacked unawares, and could not have resisted, consider this attack as secondary? How did the Emperor, after the passage of the Sambre, conceive it possible to fight two battles on the same day? It was to oppose forces double ours, and to do what military men who were witnesses of it can scarcely yet comprehend. Instead of this, had he left a corps of observation to watch the Prussians, and marched with his most powerful masses to support me, the English army would have undoubtedly been destroyed between Quatre Bras and Genappe; and, this position, which separated the two Allied armies, being once in our power, would have opened for the Emperor an opportunity of advancing to the right of the Prussians, and of crushing them in their turn . . .[41]

17 June was the great interlude day separating the battles at Ligny and Quatre Bras from Napoleon's final downfall in the holocaust of Waterloo. But this doesn't mean to say it was a day without activity. Rather it became at times a frantic preface to the additional battle which both sides now realised they would have to fight.

At Quatre Bras, Ney and Wellington had battered one another into the positions of a stalemate. The French had achieved their prime objective by blocking off any chance of a link-up between Wellington and Blücher. On the other hand, in failing to seize the crossroads, Ney dashed the hopes contained in HQ's garbled messages that he might be able to turn in strength against Blücher's flank. Wellington, despite an initial miscalculation, had therefore achieved *his* primary object. By tenaciously holding on at Quatre Bras, and keeping Ney fully occupied there, he prevented Blücher's 'damnable mauling' becoming a rout. However the Duke could in no way relax after this successful defence. For with the Prussians in flight, then the whole French army was now free to strike at him. And Quatre Bras would be a death trap defended for a second time against such superior numbers.

So Wellington retreated to the ridges before Waterloo which he knew were more suitable for his tactics. He had dined during the evening on 16 June in Genappe, at the Hôtel du Roi d'Espagne. But he returned to Quatre Bras in the early hours and heard the full story of the Prussians' defeat at 7.30 a.m. Blücher had been lost for a time: shot off his horse, twice ridden over and then left lying in a hut crammed with wounded. But he'd rejoined his forces in time to countermand Gneisenau's orders for abandoning 'the perfidious' Duke altogether. The Prussians were currently retreating towards Wavre, some ten miles to the east of Waterloo and Mont-St.-Jean. They would regroup there, 'Old Forwards' decided, and move to assist the Anglo-Dutch as soon as possible. 'It was,' Wellington admitted later, 'the decisive moment of the century.'[42]

The Duke began his own retreat at 10.00 a.m. 'Picton was surly to a degree, and later denounced the *Waterloo* position as one of the worst ever chosen.'[43] Other officers who had fought so heroically to defend Quatre Bras agreed with him. But Wellington was adamant, and soon the road back through Genappe was filled with overloaded carts and trudging infantry. The Duke put together a defensive screen with his cavalry and several pieces of artillery, including Congreve rockets. Even so, it was a hazardous business; and might well have proved disastrous. For if Napoleon had not wasted so much of the morning on trivialities then he could have fallen upon this line of retreat: decimating whole regiments before they reached the positions on the ridge of Mont-St.-Jean which Wellington's quartermaster-general, de Lancey, was still marking out for them. As things turned out though the Emperor, at last assuming personal command of the Left, gave the hurrying Anglo-Dutch just the amount of time they needed. And in fact the only real fighting that occurred on the 17th consisted of cavalry skirmishes and some artillery fire.

Napoleon had breakfasted at 7.00 a.m.—still eating and drinking while he dictated a memorandum to Davout which presented the facts of his victory at Ligny. Minutes later an officer arrived from General Pajol to say that the Prussians were streaming away in the direction of Liège (which suggested back home to Germany!) Also about this time Charles de Flahaut arrived carrying Ney's news: the Emperor's first knowledge that Wellington still held Quatre Bras. Pajol's misinformation would result in Napoleon detaching 33,000 men under Grouchy to look for and then shadow the Prussians. But not immediately. Grouchy was keen to get going, but the Emperor silenced him with a brusque 'I will give you orders when I judge it to be convenient!' Which in effect meant after he had calculated how best to employ his remaining 72,000 men against Wellington. Yet even over this be behaved in an inexplicably lazy manner. After sending off some cavalry patrols for an independent report on Quatre Bras he dictated a long letter to Ney, who 'received the impression that Saturday 17 June was to be a day of rest and revictualling—apart from the occupation of Quatre Bras.'[44]

Besides blaming the *Prince de la Moscowa* for d'Erlon's wasted counter-marching the day before, the letter actually states that:

> The Emperor is going to the windmill at Brye. It is not possible that the English army will be able to take action against you. If it were so, the Emperor would march directly upon them by way of Quatre Bras while you launch a frontal attack with your divisions which, at present, must be reunited, and the British would be instantly destroyed. Inform His Majesty of the exact position of the divisions and of all that is happening in front of you . . . His Majesty's intention is that you should take up

position at Quatre Bras, but if, by chance, that cannot take place, report immediately with the details, and the Emperor will go there himself just as I have informed you. If, on the other hand, there is only a rearguard, attack it and take position. The daylight hours of today will be needed to terminate this operation and complete the munitions, assemble the soldiers who have become isolated, and bring back the detachments. Give the consequent orders and make sure that the wounded are tended and sent to the rear.[45]

Soult wrote it down, changed it from first to third person, signed it and no doubt added to its general indecisiveness. Leaving Napoleon free to enjoy himself on a leisurely perambulation around the battlefield at Ligny. Obviously the Emperor believed, correctly, that Wellington would retreat from Quatre Bras—and neglected to do anything about it. Ney, meanwhile, was deceived by the Duke's conspicuous rearguard activities and sent back a message as follows: 'The enemy is presenting several infantry and cavalry columns that seem to want to take the offensive. I will hold out with Count d'Erlon's infantry and General Roussel's cavalry until the last, and I hope that I may even be able to repel the enemy until His Majesty lets me know his resolve. I will take up a position intermediate to Count Reille.'[46]

It was this (arriving at 11.00 a.m.) which finally jerked Napoleon into action. So the English were prepared to fight again at Quatre Bras! Very well he would fall upon them there and settle things! Fresh orders were dashed off to Ney. '*Marshal*, the Emperor has just taken a position before Marbais with an infantry corps and the Imperial Guard. His Majesty has instructed me to inform you that his intention is that you should attack the enemy at Quatre Bras to drive them from their positions and that the corps at Marbais will support your operation. His Majesty is proceeding to Marbais and awaits your reports with impatience.' Signed: Chief-of-Staff, The Duke of Dalmatia.[47] At the same time Napoleon gave Grouchy verbal instructions to pursue the Prussians via Gembloux.

Poor Ney! In the middle of carrying out instructions 'to reunite' his divisions he received orders for another, immediate attack upon the crossroads. And how long would it be before the extra infantry joined him?

'Shortly after one o'clock, a spurring Napoleon reached Marbais. Pushing on immediately for Quatre Bras,' he was there by two and found the attack still not under way.[48] Just then a captured woman canteen-attendant broke the news. Wellington had flown. There was only his cavalry and a few Congreve rockets at Quatre Bras. The Emperor promptly lost his temper! At breakfast-time he hadn't seemed in the least bit perturbed at the idea of a British retreat. Now though, suddenly, it struck him that a golden opportunity was slipping away: and he directed his anger at Ney. *Someone has lost us France!* he raged. The Marshal, equally taken aback at the news, nevertheless pointed out that all was not yet lost. Why not give chase and damage the British rearguard! This presumably had the desired effect, for a much calmer Napoleon turned to d'Erlon and said: 'Go, my dear general. Place yourself at the head of the cavalry and pursue the rearguard.'[49]

'Fate, however, had turned its face away from Napoleon. It is possible—

even probable—that he would have succeeded in catching up with Wellington and forced him to fight there and then, but for an adverse turn in the weather. At this juncture a colossal thunderstorm burst overhead, and within minutes the ground was turned into a quagmire.'[50] Captain Mercer of the horse-artillery thought his own cover-fire had detonated the swollen sky. 'The first gun that was fired seemed to burst the clouds overhead, for its report was instantly followed by an awful clap of thunder, and lightning that almost blinded us.'[51] Anyway, the rain fell in torrents: which for the British was a genuine 'godsend'. Moreover it occurred just as French lancers (Napoleon with them) crested the rise at Quatre Bras and set off up the *chaussée*. Had d'Erlon's riders been able to fan out across the countryside then they might well have encircled the rearguard. But the rain kept them to the metalled Brussels road, and soon this too came to resemble a river. 'Water swirled along the road', inundating everyone; even so 'the lancers of Colbert and Subervie' were very quickly 'prodding the backs of Sir Hussey Vivian's blue-jacketed Hussars'.[52]

The artillery paused to fire at every opportunity and the cavalry (notably the Life Guards) launched brief counter-attacks.

> At Genappe a few English batteries greeted the emergence of the pursuing pack, obliged to cross the Dyle over a bridge only two and half metres wide. There was fighting on leaving the village. Colonel Sourd, commanding the 2nd Lancers, received six sword cuts and lost his right arm. Surgeon Larrey operated on him by the roadside—after which, with clenched teeth and the stump stiched up, the colonel rejoined his regiment.[53]

Also at Genappe there was the spectacle of the Congreve rockets going off in the wrong direction, one swinging round and chasing Captain Mercer along the road. But otherwise the English rearguard, commanded by Lord Uxbridge, operated with exemplary skill. 'Make haste!—Make haste!' the Earl urged them. 'For God's sake gallop, or you will be taken.' All told they suffered less than a hundred casualties, and once through the bottleneck of the village the rain intensified to mask their progress. No wonder Wellington on Mont-St.-Jean could boast that 'British victories are usually preceded by a deluge!'[54] Even he had never seen a wall of water like this though.

Eventually the wet-through pursuers were called off. By 6.30 p.m. Napoleon was much more interested in the lie of the land and possible positions for a battle the following day. Marshal Ney had instructions that a majority of the troops should bivouac, although on such sodden ground this would hardly be pleasant. Most of the cavalry spent the night in their wet clothes, propped against their horses, or nodding off, slumped forward in the saddle. The French infantry were rather more adept at improvising tents with capes and greatcoats. The Emperor was now at a coaching-inn, La Belle Alliance, within sight of the Mont-St.-Jean ridge. 'Anxious to ascertain that he had in fact run Wellington to ground behind Mont-St.-Jean, he ordered Milhaud's horsemen to ride forward up the road, supported by several batteries of horse-artillery, to make Wellington reveal his presence. The trick worked; a roar of shot and shell from 60 Allied cannon put the Emperor's mind at rest on that score . . .'[55] Satisfied that the Anglo-Dutch would stay and fight he then selected the farm of Le Caillou for his own headquarters, roughly a mile and a half south of La Belle Alliance. In the meantime, after a brief respite, the rain started to fall again.

Ney was as miserable as his troops, but everyone realised that another set-piece would be fought the next day, probably with decisive results. The Marshal helped supervise the arrival and bivouacs of Lobau's (VI) Corps and the Imperial Guard, then rode off through the sticky Brabant mud to a cramped lodging at Chantelet, just east of Le Caillou. Someone had roasted a sheep there, so at least his supper was warm. Generals Reille, Bachelu, Foy, Piré and Prince Jerome Bonaparte dined in the greater comfort of the Hôtel du Roi d'Espagne at Genappe. A waiter told them of a conversation over-heard between two of Wellington's *aides*. He'd served them at lunch: when one was explaining to the other how the Prussian army would march from Wavre for a concerted link-up with Wellington. This was important news; and Prince Jerome would pass it on to HQ the following morning.

What these generals (and certainly Michel Ney) did not know was that Napoleon already possessed disturbing intelligence about Prussian troop movements. A returning patrol of Milhaud's horsemen reported spotting a long column of Prussians marching towards Wavre. 'But the Emperor, confident of Grouchy's ability to keep Blücher busily engaged on the 18th, discounted the news and ignored the warning.'[56]

By this time his mind was entirely obsessed with tactics, the shape of the ground, intended positions, the distance between the two armies: every single thing which might affect the battle to come in his own immediate vicinity. He bent over the maps, memorising each landmark, considering how best to make use of the terrain. And he visited La Belle Alliance again. Yes: the enemy's campfires still glowed on the far ridge, braving a wet darkness. After another glance at his maps he decided that Wellington's position was a mistake. 'The enemy commander,' he noted, 'could do nothing more contrary to the interests of his cause and country, to the whole *mode* of this campaign, and even to the most basic rules of war . . . for he has behind him the defiles of the Forest of Soignies, and if he is beaten any retreat would be out of the question.'[57] The English had won several astonishing victories during the Hundred Years War with a forest at their backs; for their troops it signalled a fight to the death. But Napoleon was too much the military innovator, and too egocentric, ever to accept a lesson from mediaeval warfare.

He returned to Le Caillou at 4.00 a.m., tired but quietly confident. Awaiting him at the farm was a dispatch from Grouchy (written at 10.00 p.m. on 17 June). 'It was rather an indecisive document, but in it Grouchy did surmise that the bulk of Blücher's army might in fact be falling back towards Wavre. In that case, the commander of the French right-wing concluded, *I shall follow them so as to prevent them gaining Brussels and to separate them from Wellington.*'[58] Quite independently, therefore, Grouchy had confirmed Milhaud's previous intelligence. But again the Emperor put it from his mind. He considered the Prussians a spent force, and did not even send any reply to Grouchy until ten in the morning. This was, in the respected opinion of David

Chandler, his first great error of 18 June. 'For had he reacted with even reasonable promptitude and caution and ordered Grouchy to head for Walhain, only a single corps of Blücher's army at the very most would have been able to intervene at Waterloo.'[59]

Beaten they might be, and considerably bruised; but the Prussians were far from being spent. In fact their grizzled commander was even now taking energetic measures to go to Wellington's aid. Chewing cloves of garlic, gulping down draughts of undiluted Hollands gin, Blücher continued with the reassembly of his scattered units. At 11.00 p.m. Baron von Müffling brought him confirmation that Wellington was actually in position and ready to fight at Mont-St.-Jean. Could the Anglo-Dutch expect Prussian support there? Blücher wrote back: 'Bülow's (II) Corps will set off marching tomorrow at daybreak in your direction. It will be immediately followed by the (IV) Corps of Pirch. The I and III Corps will hold themselves in readiness to proceed towards you.'[60] Wellington received this despatch at around 2.00 a.m.[61] And the next day Blücher 'repeated these lavish promises—double what Wellington expected—in a dispatch written about 9.30 a.m. to Müffling: ill as he was, he had made up his mind to lead his troops himself against Napoleon's right-wing. Rather than miss the battle . . . he would be tied to his horse.'[62]

Napoleon still had one 'ally' in the Prussian camp though. Gneisenau was more suspicious than ever of Wellington's intentions; 'that master knave' as he called him. He ordered Müffling 'to penetrate the Duke's inmost thoughts and find out whether he really entertains the firm resolution of fighting in his present position . . .'[63] While to Blücher he put it in even harsher terms. 'The English consider only their own interests. If they are beaten, the Prussian Army will run the greatest risks!' 'To prevent this,' the old warhorse replied, 'we will have to help them.' So, as dawn broke, 'the vanguard of the 4th Prussian Corps (Von Bülow) struck camp and made its way by the narrow, steeply-banked sinking lanes, marching raggedly, to Chapelle Saint-Lambert.' On the other hand, 'the order from Gneisenau was: *march* slowly . . .'[64]

It took a great poet and novelist, Victor Hugo, to put into words the dramatic quality about the late-start on the morning of 18 June.

'If it had not rained in the night between the 17th and the 18th of June, 1815,' he writes, 'the fate of Europe would have been different. A few drops of water, more or less, decided the downfall of Napoleon. All that Providence required in order to make Waterloo the end of Austerlitz was a little more rain, and a cloud traversing the sky out of season sufficed to make a world crumble.'[65]

This is to oversimplify it, of course. The mistakes of men as well as the vagaries of the weather would determine the outcome at Waterloo, and in any case the fate of the opposing sides would hang in the balance almost all day. But the author of Les Misérables is factually correct when he states that the battle of Waterloo did not begin 'until half-past eleven o'clock, and that gave

Blücher time to come up. Why? Because the ground was wet. The artillery had to wait until it became a little firmer before they could manoeuvre.'

'Napoleon was an artillery officer, and felt the effects of this,' he continues:

> There was something of the sharpshooter in his genius. To beat in squares, to pulverise regiments, to break lines, to crush and disperse masses—for him everything lay in this, to strike, strike, strike incessantly —and he entrusted this task to the cannonball . . .
>
> On the 18th of June 1815, he relied all the more on his artillery, because he had numbers on his side. Wellington had only one hundred and fifty-nine mouths of fire; Napoleon had two hundred and forty. Suppose the soil dry, and the artillery capable of moving, the action would have begun at six o'clock in the morning. The battle would have been won and ended at two o'clock, three hours before the change of fortune in favour of the Prussians.

Again this is too simple. If the will and the necessity are there, then guns can and have been moved across far more difficult terrain than the rain-soaked ridge facing Mont-St.-Jean. However, given the Emperor's unusual state of mind at Waterloo, and triggered by his intermittent physical fatigue and distress, one begins to understand how the wetness of the ground was allowed to create certain fatal delays. His unguarded optimism alternated with moments of hesitation; he kept changing his mind; he was no longer the victor of Austerlitz. The old Napoleon would have found a way to move the guns and then attacked while Wellington's men were still rubbing the sleep from their eyes. Instead, at Waterloo he showed vexation but permitted the rolling forward of the guns and the tyranny of the French breakfast to follow their parallel, leisurely courses.

Add to this his erratic dictation of orders, his continuing to ignore the Prussian threat until it was too late to deploy adequate forces to stop it, Soult's muddled and misleading staff-work, Prince Jerome's blind wastefulness around Hougoumont, Grouchy's interminable wanderings to the east, Ney's mistake in failing to support his cavalry attacks, Napoleon's observing this mistake but doing nothing about it; and finally his strange, disastrous behaviour with the reserves, especially the Imperial Guard. Then it becomes clear why Wellington was able to make his own, several slips and get away with them. Nevertheless the canny British field-marshal would be hard-pressed throughout—once the action started; while his gifted individual commanders were several times fearful of defeat. The Emperor could have triumphed at Waterloo, and at one point Marshal Ney nearly won the battle for him.

As Sunday's dawn broke it was cloudy and overcast, but at least the rain had stopped. Immediately observers with both armies began to scan the ground out ahead of them, hopeful of noticing some weakness in the enemy's positions. For it was an unusually small area for so large a battle to take place. Less than three miles square and with a dip of only fifteen hundred yards separating the two ridges in some parts. It gave the seasoned soldiers there a feeling amounting almost to claustrophobia. Any attack would be bound to come at them quickly. Any single breakthrough might well prove decisive.

To the north, most of Wellington's forces were drawn up behind the road coming from Ohain and as usual the Duke kept a good proportion of them hidden. The road itself was narrow, overgrown and for a vital one hundred and fifty yards actually sunken. But where it formed a crossroads with the Brussels-Charleroi road there stood an elm-tree which Wellington thought ought to make a suitable command-post. Both east and west of the crossroads were to be defended, although a majority of his troops lay to the west and one division (General Chassé's) held the village of Braine l'Alleud about a mile off on that side. To the east were General Picton's reserve division and two brigades of 6th Division, plus a smaller number of Lord Uxbridge's cavalry: Ponsonby's and Ghigny's brigades.

In addition though the British commander had decided to occupy and defend for as long as possible several advanced positions in what would obviously become the disputed middle ground. These were (to the west) Hougoumont, a wood and small *château* filled with Hanoverians and Nassauers; in the centre the farmhouse and garden of La Haye Sainte (occupied by the King's German Legion); and further to the east the farms of Papelotte and La Haye and the Château of Frichermont. It was plainly his intention to use them for blunting, or better still diverting the French away from his main forces until the battle had been in progress long enough to tire them. Also, he reasoned, their stubborn defence would gain him valuable time while the Prussians advanced to join in.

There were only two oddities (or slips) in this positioning. First, his stationing seventeen thousand men under Prince Frederick ten miles to the west at Hal. (He would have great need of them once the battle developed!) And secondly, his leaving General Bylandt's brigade fully exposed to the enemy's guns on the eastern side. Whether or not this was an oversight on Wellington's part has never been established—but they were left *south* of the Ohain road, near a sand-pit behind La Haye Sainte, and once the French artillery opened up they took a most fearful pounding.

In contrast, the French dispositions were altogether more symmetrical. West of the Charleroi *chaussée* (and therefore facing Hougoumont) the front line was formed by Count Reille's corps, with Piré on his extreme left and the cavalry of Kellermann and Guyot behind. To the east it was formed by d'Erlon's corps, with Milhaud and Lefebvre-Desnoëttes behind. In addition there was a sizeable central reserve which included Lobau's corps, the cavalry of Subervie and Domon, and of course the Imperial Guard: grumbling as usual but nevertheless keen to take part. In all 72,000 men to strike against Wellington's 67,000.

Ney arrived at Le Caillou soon after eight o'clock and found the Emperor breakfasting off crested silver plate (which suggests better fare than the flour and water most of the French were heating up). With him were Soult, Maret, Bertrand, Count Drouot (responsible for the Guard), Reille and Prince Jerome. Napoleon's mood was currently euphoric: '*So*—now we've got them—those English!' He had already postponed his intended attack until 9.00 a.m., but he remained determinedly optimistic. 'We have ninety chances in our favour with less than ten against . . .'[66]

Others were not so sure. Soult expressed the opinion that Grouchy's 33,000 men should be recalled immediately to protect their eastern flank. The Emperor threw him a dark look. 'Just because you have been beaten by Wellington you call him a good general. Well, I tell you: Wellington is a bad general, the English are bad troops and this whole business'—he glanced down at the remains of his meat—'will be like eating them for breakfast.' 'I sincerely hope so, Sire,' his Chief-of-Staff replied glumly.

The Emperor turned to Reille for support, but that experienced veteran had also suffered at Wellington's hands in the Peninsula. 'Well positioned as Wellington knows how,' he answered, 'and if attacked from the front, I consider the British infantry to be impregnable. They have calm tenacity and superior aiming-power. They can only be beaten by manoeuvring. Also . . .' Furious, Napoleon cut him short; and he similarly dismissed Jerome's information picked up from the waiter in Genappe about 'a concerted link-up between the British and the Prussians'. 'Absolute nonsense!' he snapped. 'After a battle such as we fought at Ligny, the joining-up of the British and Prussians is impossible. Grouchy will deal with Blücher and I shall defeat Wellington. If my orders are carried out, then we will all be sleeping in Brussels tonight. Our coming battle will not only save France . . . it will be celebrated in the annals of the world!'[67]

Ney's sole contribution was to mention that in riding to join them he had noticed English troops moving up the slope beyond La Haye Sainte. (Were they Bylandt's?) Anyway, he suspected Wellington might be trying yet another crafty retreat and advised that the French attack ought to go in as soon as possible. Napoleon agreed with him, but then came his second mistake of the day. He listened to General Drouot describing the wet state of the ground, the problems about moving up guns; he hesitated, and he allowed Drouot to change his mind. Their main action must be postponed, he decided: until 1.00 p.m. Meanwhile he ordered 'a well-done shoulder of mutton' for supper and called for his mare, La Marie. He would ride forward, survey the whole area again, inspect the troops and then dictate his general order of attack.

At ten o'clock—during his tour of inspection—he finally answered Grouchy's despatch of the night before. *His Majesty desires,* Soult wrote, *that you will head for Wavre in order to draw near to us, and to place yourself in touch with our operations, and to keep up your communications with us, pushing in front of you those portions of the Prussian army which have taken this direction and which have halted at Wavre; this place you ought to reach as soon as possible.*[68] As a set of instructions it was vague, it was wishy-washy, it would set Grouchy off in the wrong direction (at least that was his excuse) and worst of all it was sent off hours too late.

Ney accompanied the Emperor on his triumphant inspection. 'Never before,' an officer with I Corps noted, 'have we cried *Vive l'Empereur!* with greater enthusiasm. It was delirious! And what made the picture more solemn and stirring was that out in front, not a thousand yards away, we could see the red line of the English army.' Did Napoleon obtain a psychological advantage by displaying his troops so openly at this juncture? Several commentators appear to think so, and Lady Longford gives us a most vivid description of their *portentous display of power and noise.*

Their head-gear alone made them shine like gods, and showed from what a variety of countries and peoples the conqueror had drawn his might. There were Lancers in red shapkas with a brass plate in front bearing the imperial N and crown and a white plume eighteen inches long, Chasseurs in kolbachs with plumes of green and scarlet, Hussars whose shakos carried plumes kaleidoscopically coloured according to their countless regiments, Dragoons with brass casques over tiger-skin turbans, Cuirassiers in steel helmets with copper crests and horse-hair manes, Carabiniers all in dazzling white with tall helmets of a classically antique design, Grenadiers of the Old Guard in massive, plain bearskins towering above powdered queues and ear-rings of gold. Against the dark, menacing background of the Imperial Guards' long blue coats thousands of pennants fluttered and brilliant uniforms flaunted facings of scarlet, purple and yellow, big bright buttons, trimmings of leather and fur, gold and silver fringes, epaulettes, braid, stripes.

Many a young Dutch-Belgian soldier gazed at the nest of beautiful serpents across the valley wishing his country were still on their side. *Nearby a new English recruit stared:* as if in a trance, the muscles on his round, chalk-white face quivering.[69]

However, if Michel Ney was impressed by the troops and their state of morale, then he was growing considerably uneasy over the French delay. He had good reason to be; and had he known what Wellington now knew his unease would have turned to positive alarm. For 'at 10.00 a.m., while his men were watching the French spectacle unfold, Wellington's sharp eyes had detected Prussian advance guards away to the east, on the fringe of Paris Wood.'[70] Obviously advance guards didn't constitute an entire army, and Blücher's several corps wouldn't be able to join him for hours yet. But from the Anglo-Netherlanders' point of view it was the first encouraging sign of the day—and for the French the most ominous.

At around eleven o'clock, having completed his inspection, Napoleon dictated definitive orders to Marshal Soult. He was, David Chandler stresses, 'determined not to waste any time in fancy manoeuvres. Wellington was to be crushed by a series of unsophisticated sledgehammer frontal blows'.[71]

Immediately the army has completed its positioning, and as soon after 1.00 p.m. as possible, the Emperor will give the order to Marshal Ney and the attack will be delivered upon Mont-St.-Jean in order to seize the crossroads there. To this end the twelve-pound batteries of II and VI Corps will mass with that of I Corps. These twenty-four guns will bombard the enemy troops holding Mont-St.-Jean and General Count d'Erlon will begin the attack by first launching the left division, and, when it becomes necessary, supporting it by other divisions of I Corps. II Corps will also advance, keeping strictly abreast of I Corps. The company of engineers belonging to I Corps will hold themselves in readiness to barricade and fortify Mont-St.-Jean once it has been taken.[72]

On the back of this order Ney took the precaution of adding a pencilled note:

'Count d'Erlon must remember that the attack will be delivered first by the left instead of beginning from the right. Please inform General Reille of this change.'[73]

Clearly, if one interprets such orders to the letter, then Marshal Ney was intended to handle the forward battle at its purely tactical level. But in the opening stages this didn't actually happen. The French artillery opened up with an ear-splitting roar at 11.30 a.m.—and although the Allied batteries replied, Wellington was in trouble almost at once. Bylandt's brigade caught it first. Dozens of his men were blown to pieces while others raced for the nearest cover. But the men defending Hougoumont also came under fire; and it was fortunate that Wellington happened to be on hand to bolster morale. Little did he know the initial French blow would fall upon this position.[74]

Prince Jerome's attack upon Hougoumont was intended as a mere diversion or feint. Napoleon had planned it so, had not bothered to include it in his definitive orders and therefore Marshal Ney did not learn about it until the troops were already on their way. Hence the apparent contradiction in his pencilled note. Nevertheless Jérôme refused to accept the role of being 'a mere diversion'. He set off just after 11.30: underneath the French shells and heading straight for the wood south of the *château*. Jerome was hell-bent upon glory. The wood was strongly-defended, because at the last moment Wellington had decided to reinforce the Hanoverians and Nassauers with Macdonnell's (Coldstream) Guards. But Jerome commanded four superb regiments of veterans and in the next two hours he sent more than half of them to their deaths. Splendidly, heroically, they fought their way through the wood, cleared the intervening orchard and reached the *château's* wall, where 'a stream of bullets through every loophole brought them down in heaps'.[75] At this point he should have retreated, ordered up his howitzers, then blown the wall down at no further cost to his infantry. And Reille, his experienced commanding officer, told him as much. But Jerome went over his head to Napoleon, who let him carry on. The pig-headed Prince sent in another costly attack. Also, disastrously, he began to coax fresh troops out of General Foy to his right for these operations.

His own particular troops—it must be emphasised—fought magnificently. The pity is that they didn't have a more imaginative commander. Legros, a giant second-lieutenant of the pioneers and known to his men as *l'Enfonceur* ('Smasher') took an axe and personally broke the northside door of Hougoumont, only to be shot down in the courtyard. Of his assault group only a single drummer-boy was left alive. However, this so-called diversion, thanks to Jerome, was destined to occupy a vital part of the French for the rest of the battle. True, Wellington had to keep sending down more Coldstream and Scots Guards as reinforcements, but the *château* stayed in British hands, Jerome lost the best part of a division and what was worse, Reille's corps was never freed to move over in support of Michel Ney's attacks upon the Allied centre. Already French mistakes were beginning to accumulate. Hougoumont proved itself exactly the kind of blunting instrument that Wellington had intended.

The next mistake though was again entirely Napoleon's. It was now nearly one o'clock and time for Ney's great assault upon the Allied centre to begin.

'But what was happening on the plateau, between the Paris Wood and Couture St.-Germain, towards La Chapelle-Robert? It was possible to make out something glittering. The field-glasses were trained in that direction. Perhaps a troop movement . . . French? Prussian?[76] Colonel Marbot sent back a messenger from Frichermont. 'Instead of Marshal Grouchy it was Bülow's corps which had materialised!'[77] He had captured a Prussian courier —and this man was brought before the Emperor. He was 'quite prepared to talk and at last the truth dawned. Thirty thousand Prussians were on their way to attack the weakly-protected French right!'[78]

Napoleon could have broken off the engagement at this point. At least he ought to have informed Ney of what was likely to develop on the eastern side. Instead he compounded his earlier errors by drafting another indecisive letter to Grouchy and then detaching Lobau's VI Corps and Subervie's and Domot's cavalry to proceed in that direction. Marshal Ney, he decided, would be left to win the battle for the centre without his most obvious reserves.

In the centre it all began just after 1.00 p.m. with another furious cannonade. Eighty-four guns, including the twenty-four 'beautiful daughters' of the Guard artillery, opened up with what promised to be a massive softening up of the enemy's defences—although in reality this never happened. Bylandt's exposed Dutch-Belgians again suffered heavy casualties. (In fact it was more or less the end of them.) And several of Wellington's own guns on the ridge of Mont-St.-Jean were put out of action. But most of his infantry were out of sight on the reverse slope or sheltering in the sunken roadway. Moreover those cannon-balls which did reach them tended to stick or sink in the mud rather than bouncing and doing their maximum damage. Consequently his troops were still in good order when d'Erlon's I Corps advanced upon them.

At 1.30 p.m. Napoleon (now at La Belle Alliance) ordered the main attack to begin and Ney passed on his order to General d'Erlon. The Marshal (mounted, and with drawn-sword) led I Corps himself across the first hundred yards, but after that d'Erlon took over and the French began to have their initial successes of the day. Admittedly they were moving in a somewhat unwieldy formation (thick blocks, 200 men wide and 25 to 30 men deep) which made them easy targets for the Allied gunners. But they were an *élite* corps: brave and steady under fire. They suffered losses and carried on with their attack.

Quiot's brigade (54th and 55th of the Line) stormed, cleared and occupied the garden of La Haye Sainte—although they failed to take the buildings (garrisoned under the inspired command of Major George Baring). Wellington sent down a part of Colonel Ompteda's battalion to assist them, but these were immediately surprised and cut to pieces by Travers' *cuirassiers*: who then also dispersed Kielmansegge's skirmishers and prepared to charge up west of the farm against Wellington's infantry, lying down on the ridge and led by Halkett, Maitland and Byng. Meanwhile over on the right Durutte took Papelotte, then La Haye, and Donzelot and Marcognet were advancing through the gunsmoke to hit Picton. In the next few minutes things became desperate for Wellington. Officers were being shot down all around him. General Vincent (the Austrian attaché) and Count Pozzo di Borgo (the Czar's official observer) were both wounded. Over to his left General Picton would soon be killed. Marcognet's 25th captured the regimental colour of the British 32nd and crested the ridge, proclaiming victory. In front of them, shattered, fleeing and being booed by the Cameron Highlanders were all that remained of Bylandt's Dutch and Belgians. (They took no further part in the battle.) Ney, surveying the action through his telescope, must have felt reasonably pleased. *Until the counterattack came . . .*

In the end Wellington was saved by his cavalry. And yet—such were the bizarre turnabouts of fortune in this battle—it also led to the destruction of his cavalry. Sir Denis Pack's Gordon Highlanders, the Black Watch and Kempt's brigade were holding their own but with a high casualty rate when the big horses of the Scots Greys suddenly crashed through them, broke the French and sent them tumbling back down into the middle ground. Behind the Scots Greys came the Union and Household Brigades (heavy cavalry) under Sir William Ponsonby and Lord Edward Somerset, with Lord Uxbridge—their overall commander—in the front row.

The trouble was their actual success. Having repulsed and broken up d'Erlon's attack on the ridge the English officers found they just couldn't stop their men. On they galloped, down into the dip, then up against the far ridge. Through towards where the French batteries (loaded, ready) and a quantity of rival horsemen (Farine's brigade and Jacquinot's Lancers) were waiting for them. They had crested the ridge before their horses were winded. They spiked or wrecked fifteen guns and captured two eagles. But then vengeance came swiftly. The remaining gunners opened up at point-blank range, the French cavalry cut into their flank and d'Erlon's retreating infantry prevented their escape. In the next quarter of an hour the British losses were 2,500, including Ponsonby (killed by the Lancers)—and Lord Uxbridge would blame himself for the rest of his life. 'I tried to stop them! I tried! I tried!' At the same time Wellington had more bad news from the western side. The Guards Brigade had turned back Travers's *cuirassiers* but then they in turn ran into an awful crossfire and were nearly decimated by Bachelu's division. So far the battle was not going anyone's way. *Except that the Prussians (Gneisenau notwithstanding) were steadily approaching the theatre of operations.*

3.30 p.m. and Napoleon ordered Ney to capture La Haye Sainte 'no matter at what price'. Given this position and the nearby sand-pit there was still a good chance for them to move on and smash Wellington before Blücher's arrival.

The Marshal responded with alacrity. He attacked the farm buildings at the head of two brigades of I Corps, 'the only troops of d'Erlon's which had so far rallied'.[79] Again the French were thrown back, but Ney thought he saw signs of the troops at Wellington's centre falling back. (He did. One cavalry squadron had panicked. Also there were wounded men moving to the rear.) However Ney decided Wellington was sneaking another retreat and that the French must attack immediately. Accordingly he ignored La Haye Sainte for the moment and ordered up Milhaud's *cuirassiers*. 'Prepare to receive cavalry!' Wellington shouted and immediately his infantry formed squares.

Victor Hugo takes up the story . . .

There were three thousand five hundred of them. They formed a front a quarter of a league in extent. They were giant men, on colossal horses. There were six and twenty squadrons of them; and they had behind them as support Lefebvre-Desnoëttes' division, the one hundred and six picked gendarmes, the light cavalry of the Guard, eleven hundred and ninety-seven men, and the lancers of the Guard with eight hundred and eighty lances.

(Ney's intention was a limited pursuit with only the *cuirassiers*. But in the excitement, and without receiving orders, Lefebvre-Desnoëttes and the other cavalry joined in.)

Ney . . . placed himself at their head. The enormous squadrons were set in motion. Then a formidable spectacle was seen.

All their cavalry, with upraised swords, standards and trumpets flung to the breeze, formed in columns by divisions, descended, by a simultaneous movement and like one man, with the precision of a brazen battering-ram which is affecting a breach, the hill of La Belle Alliance, plunged into the terrible depths in which so many men had already fallen, disappeared there in the smoke, then emerging from that shadow, reappeared on the other side of the valley, still compact and in close ranks, mounting at a full trot, through a storm of grape-shot which burst upon them, the terrible muddy slope of the tableland of Mont-St.Jean. They ascended, grave, threatening, imperturbable; in the intervals between the musketry and the artillery, their colossal trampling was heard. Being two divisions, there were two columns of them; Wathier's (*a hero of Friedland*) held the right, Delort (*a colonel at Austerlitz*) commanded the left. It seemed as though two immense adders of steel were to be seen crawling towards the crest of the table-land. It traversed the battle like a prodigy.[80]

Not to the Allies did they seem to be crawling though. For they were now pounding up the slope very quickly indeed. The British gunners fired into the packed masses with grape and canister, but the rest came on and soon the gunners were forced to abandon their emplacements, seeking the comparative havens of Wellington's infantry squares.

There were thirteen of these squares:

Two battalions to the square, in two lines, with seven in the first line, six in the second, the stocks of their guns to their shoulders, taking aim at that which was about to appear . . . calm, mute, motionless. They did not see the *cuirassiers*, and the *cuirassiers* did not see them. They listened to the rise of this flood of men. They heard the swelling noise of three thousand horse, the alternate and symmetrical tramp of their hoofs, the jingling of the harness, the clang of the sabres, and a sort of grand and savage breathing. There ensued a most terrible silence; then, all at once, a long file of uplifted arms, brandishing sabres, appeared above the crest, and casques, trumpets, and standards . . . three thousand heads with grey moustaches, shouting *Vive l'Empereur!* All of this cavalry debouched on the plateau, and it was like the appearance of an earthquake.[81]

Ney's major mistake was not in attacking with cavalry—on the contrary, he had shown initiative and could have succeeded—but in leading the charges himself when he should have remained behind to cobble together some effective infantry support. Several dozen of his horsemen fell into the concealed Ohain road, but a majority made Wellington's ridge, took possession of it and then swirled around the defiant Allied squares, trying to break into them. However their sabres and pistols were no match for the British rifle and bayonet. Their lancers did some damage and inside the squares it was ghastly. Ensign Gronow recalls how 'we were nearly suffocated by the smoke and smell from burned cartridges. It was impossible to move a yard without treading upon a wounded comrade or upon the bodies of the dead, and the loud groans of our wounded and dying were most appalling.'[82] Outwardly though they stayed intact and they continued to shoot at the French from close quarters. If only Ney had brought up some infantry their musketry could have demolished the squares one side at a time.

On the other hand, if his attack failed due to personal bravery and neglect, then how much more blame is attached to Napoleon for not sending up the infantry himself? He was the overall commander, he could see what was happening in front and he had the men to rectify the situation: half of Foy's division was standing idle, so was Bachelu's division, so were the Guard regiments. Instead he did nothing until it was too late. 'It is too soon by one hour,' he murmured to his *aides*. Perhaps; but is that any excuse for not helping his subordinate? He didn't order off Foy until the cavalry had reformed and were charging Wellington's squares for the fifth and last time. 'The French cavalry made some of the boldest charges I ever saw,' Sir Augustus Frazer recalled: 'They sounded the whole extent of our line. Never did cavalry behave so nobly, or was received by infantry so firmly.'[83] They never once hesitated, and at one point Captain Mercer of the Gunners heard his colonel say: 'I fear all is over for us'.[84] But in the end their horses tired and Wellington's musketry began to take its toll. Moreover they carried nothing with which to spike the English guns, so the crews were able to rush forward and fire upon their retreats; and upon the (too late) advance of General Foy. The last French horsemen hurled their lances into the defending squares like javelins and galloped back. It had become, as one trooper who survived said: 'Like a bloody circus, whose box-seats were held by the English.'[85]

In the fourth charge Ney had led a squadron of *carabiniers* against an opposing square and been filled with admiration for the defenders' endurance. In the subsequent retreat yet another horse was killed under him and our last view of him at this stage in the battle is on foot, his face blackened by powder, frustratedly hitting an abandoned English gun with the flat of his sword.

But if he was frustrated, and so far repulsed, then the Marshal was far from spent. In fact his great moment was just about to come. It was late-afternoon now, getting on for six o'clock, and there was a new sense of urgency on the French side. Sporadic gunfire to the east proclaimed what Napoleon had feared but been keeping quiet about for hours: the involvement of the first Prussians. Lobau, the Young Guard, Subervie and Domon were engaged against them around Plancenoit. Obviously Grouchy had missed out and was marching up and down Belgium like a blind man.

The struggle with Wellington needed to be resolved quickly. The Allied squares had triumphed over the French cavalry, but Ney believed the men left up on the ridge were also close to exhaustion. In truth it was even worse. Colin Halkett's square had withstood eleven charges and had scarcely a man left alive who wasn't wounded. The 69th regiment had disappeared entirely. Alten was demoralised, Adam and Cooke in the hands of the surgeons. 'Wellington was everywhere—not speaking to anyone, but, with that nervous idiosyncrasy peculiar to him, feverishly playing with his field-glasses. *Stand fast*, he told the 95th. *What would they say in England if we were beaten?*[86] How nervous was he? Certainly if Blücher didn't reach him soon, then as a result of Michel Ney's next moves an Allied defeat definitely loomed.

Shortly after 6.00 p.m. Ney renewed his assault against the buildings of La Haye Sainte—but this time with an inspired ferocity the like of which he had not demonstrated since the Great Redoubt at Borodino. Given Wellington's tiredness and his thinning ranks, the Marshal judged, La Haye Sainte could still provide the key to a French victory. It must be taken therefore! And with an astounding burst of energy and enthusiasm he collected men of the 13th Light, added a detachment of engineers and set upon the place.

Major Baring's 375 men of the King's German Legion had held out in the farmhouse and barn all day. But they were getting short of ammunition and Wellington had been unable to reinforce them. In fairness too, the previous attacks on the place had seemed tame compared with this one.

Rushing up at the double Ney's men poured in fire from every angle. Snipers picked off a hundred of his infantry, but within minutes the Marshal's thunderous voice could be heard urging others over the walls—while the engineers smashed and split open the main gate with axes. In no time at all the French were up on the barn-roof, firing down into the courtyard. Baring's stalwarts were doomed, although they defended themselves to the last at bayonet-point. After a fierce hand-to-hand combat only forty-one of them escaped to safety through the north door. By 6.20 p.m. La Haye Sainte belonged to France.

Suddenly now things were going right for Ney. And he began to do everything right. His men continued to perform at the double. An attempt by Alten and the Prince of Orange to recapture the farm ended with Colonel Ompteda dead and the Allies hurled back to their starting-point. Next Ney cleared the sand-pit and started setting up artillery all around the farm. The buildings of La Haye Sainte were filled with his sharpshooters. Count d'Erlon's troops had started a new drive up towards the Wavre road.

Once Ney's guns opened up it meant the whole of Wellington's centre was in deadly peril. The French had cover for themselves but were able to blast into the English from less than five hundred yards (and two crews got their pieces to within three hundred). Lord Fitzroy Somerset, riding beside Wellington, had his arm blown off. Mercer's troop, so gallant at Quatre Bras, was reduced to a bloody shambles. The Allied infantry, after defeating 'the

best horsemen in Europe', clearly couldn't stand much more. As for their
cavalry, Lord Uxbridge was reduced to counting 'the effectives' in half
dozens. 'The centre of our line was wide open,' Alten stated later. And
Kennedy, one of Wellington's few *aides* not to be wounded, wrote: 'At no
other period of the action was the result so precarious.'

'Look!' Ney cried jubilantly to his officers: 'Look at the gaps our fire has
made. And see how they waver! I tell you, Wellington is cracking . . .'
Victory, he sensed, lay within his grasp. All it needed now was to send in a
few fresh battalions of infantry and the ridge would be his. Quickly, tense
with excitement, he scribbled out the request to French HQ.

No one has seriously suggested that the last great blunder of Waterloo was
not Napoleon's except the ex-Emperor himself. He sent Ney nothing. Instead
he reinforced Lobau whose defence against the Prussians was already
crumbling.

To Colonel Heymès who arrived with Ney's request he barked: 'Troops?
Where do you expect me to find them? Am I to make them perhaps?'[87] Yet if
the Marshal had had the few battalions he wanted and overrun Wellington's
positions on the ridge there would have been nothing left for Blücher to save.
Napoleon threw away his chance of victory.

The final outcome of the battle is well-known. Blücher broke through in
two places: with Von Bülow's IV Corps against Lobau and Domon near
Plancenoit; but more importantly with the corps of Ziethen and Pirch further
north, which allowed Wellington to shorten his line and reinforce his battered
centre. Ney was forced to call off his artillery barrage, to abandon La Haye
Sainte and eventually the remnants of the *Grande Armée* were driven from
the battlefield.

However, there had to be one more useless, spectacular and gory act before
the curtain fell on this French tragedy. At a quarter to seven, when the fight
for the centre was almost won, Napoleon had refused Ney any reserves. Half
an hour later, when the overall battle was about to be lost, the Emperor sent
the Imperial Guard to its inevitable destruction.

Wellington had been given time to patch together another defence. The
Prussians were driving forward from the east. And into the smoke-filled valley
between them Napoleon at last committed his *Immortals*. Again one turns to
Victor Hugo for the sense of high drama. 'Conscious that they were about to
die, they shouted *Vive l'Empereur!* History records nothing more touching
than that agony bursting forth in acclamations.' Also he provides us with a
further weather report: 'The sky had been overcast all day long. Then
suddenly, at that very moment—it was nearly eight o'clock in the evening—
the clouds on the horizon parted, and allowed the grand and sinister glow of
the setting sun to pass through, between the elms on the Nivelles road.' Ten
years before the veterans of the Guard had watched a similar sun at Austerlitz.

Napoleon led their battalions out in person, each one commanded by a
general: Friant, Michel, Roguet, Harlet, Mallet and Poret de Morvan.
General Cambronne was adjutant of the I *Chasseurs*. They advanced in
columns with two horse-drawn guns between each two battalions. Together
with some contingents led by Reille and d'Erlon they made a total force of
15,000. Wellington had put together an effective defence of over twenty

thousand, all of whom were now lying down out of sight, resting but with their guns loaded.

At La Haye Sainte the Emperor handed the force over to Michel Ney who would lead this supreme effort and share in its humiliation, the Empire's last throw of the dice. Meanwhile La Bédoyère was sent down the line spreading the news that it was Grouchy, not the Prussians, approaching from the east. 'An unworthy strategem,' Ney commented.[88] Nevertheless he accepted his fate. Minutes later he was in the saddle on his fifth horse of the day and the Guard began its advance: formed into squares now and moving in echelons past the blood-stained litter left by the cavalry.

At first all went well. They were without skirmishers or cavalry units and they soon came under a hail of grape-shot. But as men fell they closed ranks and pushed on, looking like a race of giants with their tall bearskins and long blue overcoats. Ney had his last horse shot from under him but continued to lead them on foot: he, Friant and Poret at the head of the 3rd Grenadiers. At the same time their drummers beat out a steady marching rhythm which planted the beginnings of fear in more than one Englishman. The French Imperial Guard had never been known to fail!

Coming up at an angle, climbing over the previous dead and the remains of a hedge, they quickly put to flight four battalions (mainly Brunswickers) led by Sir Colin Halkett. Halkett himself was shot in the mouth and carried from the field. At the same time, over on the right, they knocked out one of the remaining English batteries. 'Bah!' said one of the Guard's grumblers: 'Victory is a trollop we've taken more than once!'[89]

But then came Wellington's masterstroke and a most terrible carnage ensued. Suddenly the command rang out: *Stand up, Guards!* Immediately, at only thirty paces, 1,500 English Guards under Sir Peregrine Maitland and another 1,200 infantry of General Adam's brigade sprang up from in and around the concealed road to fire a stream of lead into the French. General Michel, Majors Cardinal and Angelet, twenty more officers and over two hundred chasseurs crashed down under their first volley. The rest wavered, hesitated and among observers there was a terrible gasp. *The Guard recoils!*

They fell back in the direction of Hougoumont, where another furious fight developed near the orchard. Maitland and Adam's men were shooting into their front, the King's Hanoverians into their flank and also at this moment the first Prussians arrived! The chasseurs were practically annihilated in this last stand. Colonel Malet was killed; so too was Major Agnès. And yet the entire action had not lasted twenty minutes. Three Old Guard battalions opened fire in a vain effort to cover their retreat. The French service squadrons charged. But just then Wellington, mounted on Copenhagen, waved his legendary hat and the whole Allied line advanced. The Guards' retreat turned into a rout. 'With an immense clamour, drums beating, bands playing, pipers piping and flags unfurled, the English, though much diminished and near exhaustion, came on as the remains of the French artillery blasted away, answered by English and Prussian guns. The French scrambled up to La Belle Alliance, pursued by the whole howling pack.'[90]

Almost the last French cannonball to whistle over took off Lord Uxbridge's leg. It had narrowly missed Wellington himself.[91] Nevertheless Ney's near-victory had become the Duke's real one. He could afford to relax and leave the further pursuit to Blücher . . .

There are many reported glimpses of Ney in these closing moments. Mostly of him trying to rally just a dozen or so men while all around him the cry had gone up: *Sauve-qui-peut!* At least one report adds that tears were streaming down his face.[92] Certainly he was bleeding. Whether he had been grazed by a bullet or caught by a sabre isn't known. But he had a cut along the side of his head.

Also he remained on the battlefield after Napoleon had gone . . .

By this time it was growing dark of course; and eye-witnesses often imagine what they saw. Perhaps those questioned by Victor Hugo did—although it caused him to immortalise the romantic part of Ney with words I find hard to resist:

> Perspiring, his eyes aflame, foaming at the mouth, with uniform unbuttoned, one of his *epaulettes* half cut off by a sword-stroke . . . his *plaque* with the great eagle dented by a bullet; bleeding, bemired, magnificent, a broken sword in his hand, he said: *Come* and see how a Marshal of France dies on the battlefield! But in vain; he did not die. He was haggard and angry. At Drouet d'Erlon he hurled this question: *Are you not going to get yourself killed?* In the midst of all that artillery engaged in crushing a handful of men, he shouted: *So, there is nothing for me! How I wish all these English bullets would enter my chest!* Unhappy man, you were reserved for French bullets!

Count d'Erlon's own account is somewhat different, more typical of the realist in Michel Ney and—I believe—very likely true: since the time for further romantic gestures was clearly over. D'Erlon agrees about the sweating, smoke-blackened face, the tattered, besmirched uniform, the epaulette cut in half and the broken sword; even that Ney said 'Come and see how a Marshal of France dies' (*Venez voir mourir un maréchal de France*). Otherwise, he states, the words were more to the point: 'If they catch us, d'Erlon, you and I will be hanged!'[93]

Our final reported glimpse of the Marshal is of him limping down the road south to Avesnes, leaning on the shoulder of a Guard corporal, then gratefully accepting the offer of a horse from a Polish lancer.

[1] Captain A. F. Becke: *Napoleon and Waterloo*, London, 1939.
[2] *The Campaigns of Napoleon.*
[3] *The Anatomy of Glory.*
[4] Ibid.
[5] Ibid.
[6] Ibid.; also *Marshal Ney, A Dual Life.*
[7] *The Campaigns of Napoleon.*
[8] Ibid.
[9] *Lettres Inédites de Napoleon Ier, 1799-1815.*
[10] *Wellington, The Years of the Sword.*
[11] Ibid.
[12] John Naylor: *Waterloo*, London, 1960.

13 *Wellington, The Years of the Sword.*
14 *The Anatomy of Glory.*
15 *Wellington, The Years of the Sword.*
16 *The Campaigns of Napoleon.*
17 Ibid.
18 Ibid.
19 *The Anatomy of Glory.*
20 *Wellington, The Years of the Sword.*
21 *Correspondance de Napoleon Ier, Vol. XXVIII.*
22 Ibid.
23 *The Campaigns of Napoleon.*
24 *Notes of Conversations with the Duke of Wellington.*
25 John Wilson Croker: *The Croker Papers,* ed. Bernard Pool, London, 1967.
26 Joseph Napoléon Ney: *Documents Inédits du Duc d'Elchingen,* Paris, 1833.
27 *Wellington, The Years of the Sword.*
28 *The Hundred Days.*
29 *Wellington, The Years of the Sword.*
30 *Documents Inédits du Duc d'Elchingen.*
31 *Napoleon and Waterloo.*
32 Ibid.
33 Ibid.
34 Ibid.
35 *Wellington, The Years of the Sword.*
36 Ibid.
37 *The Campaigns of Napoleon.*
38 *Napoleon and Waterloo.*
39 *The Hundred Days.*
40 *Napoleon and Waterloo.*
41 *The Hundred Days.*
42 Houssaye: *1815: Waterloo,* trans. A. E. Mann, London, 1900.
43 *Wellington, The Years of the Sword.*
44 Ibid.
45 Commandant Henry Lachouque: *Waterloo.* London, 1975. See also *Documents Inédits du Duc d'Elchingen.*
46 Ibid.
47 Ibid.
48 *The Campaigns of Napoleon.*
49 Jean-Baptiste Drouet, Comte d'Erlon: *Mémoires,* Paris, 1884.
50 *The Campaigns of Napoleon.*
51 General Cavalié Mercer: *Journal of the Waterloo Campaign,* London, 1870.
52 Lachouque: *Waterloo.*
53 Ibid.
54 *Wellington, The Years of the Sword.*
55 *The Campaigns of Napoleon.*
56 Ibid.
57 *Correspondance de Napoleon Ier.*
58 *The Campaigns of Napoleon.*
59 Ibid.
60 Houssaye: *1815: Waterloo.*
61 Ibid.
62 *Wellington, The Years of the Sword.*
63 Ibid.

64 Lachouque: *Waterloo.*
65 *Les Misérables,* Paris, 1862; trans. Isabel F. Hapgood, London, 1955.
66 Houssaye: *1815: Waterloo.*
67 Ibid.
68 *Napoleon and Waterloo.*
69 *Wellington, The Years of the Sword.*
70 Ibid.
71 *The Campaigns of Napoleon.*
72 *Correspondance de Napoleon Ier.*
73 Ibid.
74 Grouchy was eating strawberries-and-cream when he heard the sounds of the French cannonade. General Gérard urged him to march towards the guns, but the Marshal preferred to obey orders and march upon Wavre. Thus he missed the battle of Waterloo altogether, lost his opportunity to sever the Prussians' line of approach and as a result contributed greatly to the eventual Allied victory. However, one should not blame him too much. The basic mistake lay in Napoleon's/Soult's misleading and delayed despatch.
75 *Wellington, The Years of the Sword.*
76 Lachouque: *Waterloo.*
77 Marbot: *Mémoires.*
78 *The Campaigns of Napoleon.*
79 Ibid.
80 *Les Misérables.*
81 Ibid.
82 R. H. Gronow: *Reminiscences and Recollections,* London, 1900.
83 *Letters of Colonel Sir Augustus Frazer,* London, 1859.
84 Mercer: *Journal of the Waterloo Campaign.*
85 *The Anatomy of Glory.*
86 Lachouque: *Waterloo.*
87 Houssaye: *1815: Waterloo.*
88 *The Trial of Marshal Ney.*
89 *The Anatomy of Glory.*
90 Ibid.
91 *Wellington, The Years of the Sword.*
92 *The Anatomy of Glory.*
93 D'Erlon: *Mémoires.*

CHAPTER 13

Aftermath

Trial and execution

The great battle was over, and lost; but Napoleon still deluded himself that he could carry on as France's Head of State.

Michel Ney believed the opposite. If he'd collected a fatal bullet or been sabred down at Waterloo, as so nearly happened a dozen times, then historians of the Empire could have inserted him into their books as *the most loyal Ney* as well *as the bravest of the brave*. Instead of which, whether by a miracle or just miscalculation he survived the fighting, and only the last vestiges of a romantic were left behind on the trampled rye-fields of Belgium. Within the confirmed and disillusioned realist who lived there was not the slightest possibility of further support for the Emperor. Waterloo, following upon Quatre Bras, broke the spell of fifteen years, forcing Ney to see with an awful clarity that the destinies of France and Napoleon must, finally, be separated. To the fervent French patriot the Emperor had lost credibility. He had deliberately used his popularity to lead the *Grande Armée* into a campaign meant to settle the nation's future and then glaringly failed in his duties as its supreme commander. For this, the Marshal concluded, his lost leader and one-time idol should now accept the blame.

As regards his own future, he entertained 'neither hope nor illusions'. The Bourbon Ultras were sure to come 'hot on the heels of the Allied armies'. Then the witch-hunts would start up in earnest.[1]

Napoleon naturally held other views of where the blame lay; or rather of where it must *seem* to lay, for in order to stand any chance of retaining power he needed a scapegoat to carry the full odium of Waterloo.

The Emperor arrived back in Paris on the morning of 21 June (unaware that Ney had reached the capital one day before him). Looking yellow and waxen, instead of going to the Tuileries he went directly to the Elysée where he found Caulaincourt, and from his reported remarks it is perfectly clear he had made up his mind who was to be the scapegoat. 'The army has done wonders. But it was overtaken by panic. Everything has been lost. Ney behaved like a madman—made me butcher all my cavalry. I can do no more. I need two hours rest before I get down to work . . .'[2]

More followed in the same vein when Davout called to see him. By this time Napoleon was in his bath, flapping his heavy arms in agitation and splashing warm water all over the Marshal's uniform. An equivalent shower of invective then descended upon 'Marshal Ney's performance with the army'. Davout didn't mention he had just met Ney and listened to his account of the battles at Quatre Bras and Waterloo. But he did firmly remind his master that 'Marshal Ney has probably forfeited his life in order to serve you,' adding: 'Your Majesty's most urgent need is to prorogue the two Chambers, otherwise with their passionate hostility they will paralyse everything you try to do.'[3] In other words, why waste time trying to pillory one of France's few remaining heroes? The Emperor should concentrate his efforts on winning the utmost support for yet another assumption of dictatorial powers. Lavalette, who came next, offered roughly the same advice—although he was somewhat disconcerted when Napoleon, now towelling in his cabinet, 'came forward to greet me with a frightful epileptic laugh'.[4]

Half an hour later there was a hurriedly-assembled ministerial meeting, at which Davout urged firmness of purpose, with if necessary armed resistance in the interior. Carnot and Caulaincourt expressed the contrary view that all would be well 'if there was a sincere and genuine union between the Emperor and the two Chambers', and on the whole Cambacérès, President of the Chamber of Peers, and Baron Peyrusse, the Treasurer, agreed with them: promising to use their good offices to help bring this about. But then Fouché, devious as ever, struck a different chord. Like Davout he too had had a meeting with Ney and therefore knew the full extent of the military disaster. Also he was playing for very high stakes indeed, and the last thing he wanted was a further period of armed resistance. 'Why be so alarmist?' he asked soothingly. 'After all, Paris is very quiet. Everyone here is calm.'[5]

Napoleon had hardly replied that it was the excessive calm which worried him when one of the longstanding councillors (Regnault) ventured to suggest what each of them had been thinking, namely the possibility of a second abdication: a supreme sacrifice on the Emperor's part for the good of the nation. 'I do not wish it; but it might be required,' he pointed out. Stung, Lucien Bonaparte called out that rather than see this occur he would institute another *18 Brumaire*. However his brother motioned for him to be quiet. *He*, if anyone, was the person to bully ministers—and he now proceeded to do so, in a cold, contemptuous way that left the majority of them filled with trepidation. No hint of conciliation; and no acceptance of Waterloo—other than to heap the technical blame for it on Marshal Ney. Instead (he told them), far from being an end, the battle represented merely the beginning of a vast new struggle which would go on, if necessary, until the last French farmhouse was a pile of rubble.

Abdicate! Is everyone aware of the consequences of my abdication? Around me . . . the army will rally—remove me and it will dissolve. The soldiers do not understand your political niceties. Nobody understands that I am only the pretext for war. The real object of it is France. If they had rejected me at Cannes, I would have understood, but now when I am part of what the enemy is attacking, *I am part of what France must defend.* Surrender me and you surrender France. It is not liberty that seeks to depose me: *but cowardice!*

It was to be his last official pronouncement as Emperor of the French. Even as he was speaking others were saying things elsewhere in Paris designed to put an end to his rule. With their success, as a living person he disappears from Michel Ney's life-story. And yet this speech is a final indication of how far his original greatness as First Consul had deteriorated under the Empire into self-deceit and tyranny. Rather the fiery furnace for the whole of the French nation than his own deposition . . .

Fouché slipped away from the meeting first; and although decidedly the man of the hour, nevertheless he experienced a slight shiver of fear. '*What a devil!*' he muttered. 'He terrified me. I thought it was going to begin all over again. Luckily, there is no new beginning for him . . .'[6]

It was Marshal Ney's visit to him the previous afternoon which had finally convinced the Head of Police that he must accelerate Napoleon's overthrow and help bring about a second restoration of the Bourbons. From a man whose signature was on Louis XVI's death-warrant this might seem to us no mean achievement, but one should not forget that Fouché was almost the equal of Talleyrand in matters of political scheming. Also he was on the spot, whereas 'Old Tally' had followed Louis XVIII to Ghent. Ney for his part regarded Davout and Fouché as the last effective Government ministers in Paris—and as a result the latter now had in his possession the Marshal's written account of Napoleon's errors at Waterloo, together with his verbal assessment that further resistance was impossible since the army no longer existed. In return he offered to give Ney passports to Switzerland, one in his own name, the other made out for 'Monsieur Michel Theodore Neubourg, merchant, travelling with his secretary Talmas, and his servants, Xavier, Serret, Bonnet and Maton'. Clearly Fouché felt some sympathy for the defeated warrior, sitting there with his bandaged head and a man who was certain to top the Bourbons' *wanted*-list. But in the end Michel made no use of these passports. There have been various explanations why not. My own view, even if simple and obvious, is that he didn't know *how* to run. As a person totally lacking in cowardice, running away was alien to his nature.

Fouché's chosen instrument (and some have argued *dupe*) in the Chamber of Deputies was Lafayette, and as Napoleon used up his time haranguing his ministers so the revered Republican had already got to his feet in the Palais Bourbon:

> Gentlemen, if after many years I raise a voice which I believe all old friends of liberty here will recognise, it is because I feel it my duty to draw your attention to the dangers threatening our country which you alone can save. Sinister rumours have been circulating, and now they are unhappily confirmed. The moment has come to rally round the old tricolour standard, the standard of '89, of liberty, equality and public order. It is this cause alone that we must defend, both against foreign pretension and internal threats. Gentlemen, may a veteran of this sacred

cause, who was never drawn into faction, submit preliminary resolutions, of which, I hope, you will see the necessity . . .[7]

The Chamber declared itself to be in permanent session. 'Any attempt to dissolve (it)' the president Lanjuinais announced, was high treason, 'and any person who makes such an attempt is a traitor to his country and will be treated as such . . .' The Ministers for War, Foreign Affairs, the Interior and the Police were invited to attend, after which Lafayette rose again to pinpoint the blame. He left little doubt who was the architect of their misfortunes. 'Have you forgotten where the bones of our sons and brothers whiten? In Africa, on the Tagus, along the Vistula and in the snows of Russia. Two million Frenchmen have been the victims of this one man who wanted to fight all of Europe. Now, therefore, I say: *Enough!*' His speech set the mood for the whole debate, at the end of which the deputies voted by a large majority for the Emperor's removal. Napoleon hesitated to go to the Palais Bourbon himself; perhaps because he was now devoid of an army to back up anything he might say. Thus, powerless, he himself became the scapegoat of his own parliamentarians of 'The Hundred Days'. Before their day-and-night's sitting was over he had been compelled to sign another, irreversible form of abdication: the first link in a chain of removes which took him from Paris to Malmaison, to Rambouillet, then to Rochefort where he surrendered to Captain Maitland of the British naval ship *Bellerophon* and ultimately to exile on St.-Helena. Also, it meant that Fouché had cleverly manipulated Republican sentiments in order to clear the way for the monarchy's immediate return.

As the long debate was taking place so Ney was still mooching about the capital in a state of acute uncertainty. A gloomy, disconsolate figure, he appeared to be suffering from a mixture of battle-fatigue and post-defeat depression. At his last reported meeting with Ida St.-Elme he spoke with bitterness of the squandered opportunities at Waterloo. 'The victory was in our hands. Our soldiers had never fought with more spirit. And to think of being beaten with such men!'[8]

However, by the next day (22 June) he had recovered sufficiently to make a dramatic intervention in the upper house, the Chamber of Peers, when a debate on the future of France began. The peers were sitting in the Luxembourg; and Ney whose wound was now healing had removed his bandages when he took his place among them. At 2.00 p.m. Carnot read out Napoleon's act of abdication. Then he turned to a prepared speech on the state of the nation, during which he referred with optimism to the army. 'I have a communication from the frontier fortress of Rocroi . . . Marshal Grouchy has defeated the Prussians . . . and with sixty thousand men is coming to the defence of Paris.' 'This is not true!' someone cried in a loud voice. Everyone turned, and by this time Ney was on his feet, his face dark with anger. 'The news which the Minister of the Interior has just given you is false—false in every respect!'

He spoke with an energetic vehemence—and in tones similar to those his men were used to hearing over the boom of cannon:

The enemy is victorious at every point. I have witnessed the disorder, for I commanded the army under the Emperor. After the results of those days of disaster, the sixteenth and eighteenth . . . they dare to tell you we ended by beating the enemy on the eighteenth, and that there are sixty thousand men on the frontier. It is false, Sirs! At the very most Marshal Grouchy has rallied twenty-five thousand men, and when they tell us the Prussian army has been destroyed it simply isn't true. The greater part of the Prussian army has not been in action!

Briefly he gave them a run-down on the Waterloo campaign, praising the troops, but also indicating the mistakes in command which had placed the defeat beyond repair. And finally he predicted: 'In six or seven days the enemy will be in the heart of Paris. There is no means of securing the public's safety other than by opening negotiations!'

The peers listened to him in growing stupefaction. 'Until now, no one had suggested to them that the situation was hopeless. They were not ready for stark realities. The faults Ney spoke of were evidently Napoleon's . . .'[9] And as they left the chamber a group of them remonstrated with him—although mildly. Wasn't he now being the alarmist? But the Marshal replied with conviction. 'I have spoken only in the interests of my country, gentlemen. It is certainly not in my own interest to seek a return of the Bourbons. I know only too well that if Louis XVIII comes back I shall be shot . . .'

The debate was concluded on the following day (23 June), when Ney was subjected to a blistering attack from Charles de La Bédoyère, the ardent Bonapartist of aristocratic descent whose military behaviour at Waterloo had been a contributory factor to the defeat. So violent was this outburst ('. . . vile generals who have abandoned him . . . catastrophes caused by some who are sitting among us now . . .') that even Masséna was heard to remark 'Young man, you forget yourself . . . you are not with your regiment now!' Ney, who was sitting only a few feet from the tribune, remained silent though: an expression of utter disdain on his face. He had given them the facts. If they refused to negotiate, then so much the worse for France. For Louis would be back; that much was certain. Fouché was already at work on the final details. ('*Vice* leaning on the arm of *Crime*' is how Chateaubriand later described it when Talleyrand accompanied the ex-regicide's formal submission to the crown.) And with Louis would come retribution, the so-called 'White Terror'. Only by gaining intermediate concessions from the Allies was there any hope of the French being able to moderate the Bourbons' thirst for vengeance.

Still Ney lingered on in Paris; ignoring the dangers which increased for him almost hourly. On the evening of the 23rd he wrote a lengthy protest to Fouché about 'the calumnies being whispered relative to my performance at Waterloo'. He suspected the Bonapartists of repeating Napoleon's own words:

They are saying I have betrayed my country—I who have always served it with such zeal. Where do all these rumours come from which are spreading with such dreadful rapidity? I could easily investigate this matter, but I fear almost as much to lay bare the truth as to remain in ignorance of it. I would only say that there is every indication that I have been basely deceived, and efforts are being made to cast a veil of treason over the mistakes and extravagances of his (*Napoleon's*) campaign—mistakes that have been carefully omitted from the published bulletin.[10]

It was a letter of genuine indignation, and deeply felt, but the Marshal made an error of judgement in sending it to Fouché, who now looked into everything submitted to him for its possible uses. As Head of Police Fouché had given Ney a very real chance of escape. But aside from this his one interest lay in the furthering of his own political ambitions. The letter included an undeniable truth: that Ney had fought at Waterloo with absolute conviction, and for this reason alone Fouché caused it to be published on 27 June, thereby indicating to his new masters how diligently he was gathering evidence against the Bonapartist Marshal.

From Château-Cambrésis Louis XVIII issued a proclamation promising 'to reward the good and apply the law against those who are guilty'.[11] The statement seemed to contain so much menace that Davout, as Minister for War, refused to surrender Paris to the Allied generals until Wellington had given his word to intervene. As a result, when the Convention of Paris was signed at St.-Cloud on 3 July, its Article XII was specifically designed to protect people like Ney who had sided with Napoleon on his return from Elba. 'Persons and private property shall likewise be respected. The inhabitants, and generally all individuals who are in the capital, shall continue to enjoy their rights and liberties without being disturbed or made the subject of enquiries of any kind regarding the functions they occupy or have occupied, and their conduct and political opinions.'[12]

How far the article went towards preserving Ney's life will take up the remainder of this chapter. For even as the Convention was being signed so the more extreme royalists were busily scrutinising copies of the document for potential loopholes. In the meantime though it is perhaps worth mentioning that nothing was done to safeguard the Marshal's goods and chattels. We have Sergeant Wheeler's eye-witness account of the Prussians at play on his country estate. Blücher's men—together with a smaller British contingent— were far to the south-west of Paris in order to shadow those remnants of the French army falling back towards the Loire. As a result 'we halted,' Wheeler recalls, 'near to a *château* belonging to Marshal Ney. The Prussian army passed our camp to go into position on the other side of the Canal; and they did not forget to destroy everything they could as they moved on. Ney's country seat was none the better for their visit. Everything they could lay their hands on was knocked to pieces.'[13] Further destruction and confiscations were to follow.

Despite Aglaé's continually urging him to go Michel did not finally quit Paris until 6 July: two days before Louis XVIII's triumphal re-entry. Again the King's return was accompanied by the rolls and calls of Allied drums and bugles; and the restoration seemed even more like foreign conquest than liberation when the Prussians endeavoured to blow up the Pont d'Iéna. (In the end they merely succeeded in cracking one of the arches—while in the process they blew one of their own sentries into the river.)

By this time Ney's coach was a good part of the way towards Lyon; from where with luck in another day he could reach Geneva. However, although accepting the necessity to leave the capital, he remained reluctant to depart from French soil: half hoping that the more professional Allied generals would insist upon a rigid amnesty for their opposite numbers on the Napoleonic side.[14] He had with him the passports from Fouché still. Also, just before kissing *goodbye* to Aglaé a masonic friend (Monsieur de Pontalba) had provided him with letters of introduction to a leading merchant in New Orleans, a Monsieur de Marigny. 'When you know him,' these letters stated: 'You will see that he is a man of most modest character. If he notices that his presence causes you any embarrassment or involves you in extra expense . . . he will take himself off. Receive him, therefore, with the greatest simplicity.'[15] But when the Marshal got to Lyon on 9 July a friendly commissaire of police told him the Austrians had closed the Swiss frontiers and it was an excellent excuse to stay put. He took the waters at St.-Alban-les-Eaux and then moved through the lovely, undulating countryside to Roanne, where he lingered for several days. In fact, until Aglaé's confidential agent caught up with him.

At Roanne another friendly officer had signed a *feuille de route* describing him as *Major Michel Reiset of the 3rd Hussars, on a mission to Toulouse.* (Down-river from Toulouse was Bordeaux and ships making the regular run to New Orleans.) Apparently he could take his time. But then on 26 July he was handed Aglaé's letter, containing the details of King Louis' two new decrees. The first, signed by Talleyrand, removed Ney and several others from the List of Peers. The second, countersigned by Fouché, was in the form of an *Ordonnance* which stated that 'the generals and officers who betrayed His Majesty before 23 March, or who attacked France and His Majesty's government sword in hand, will be arrested and placed before a Council of War in their respective areas'. Ney's was the first name on the list; Charles de la Bédoyère came second, and the other seventeen predictably included d'Erlon, d'Ornano, Grouchy, Cambronne (now safe in British hands) and Lavalette. Thirty-eight more were singled out for exile. Aglaé implored her husband not to remain in the Lyon area but to take refuge in the isolated *château* of her distant cousin, Madame de Bessonis, near Aurillac (Cantal).

Michel followed her advice and arrived at the *château* on 29 July. For the next few days he imagined himself safe; not realising that his presence there had been spotted.

There are two conflicting versions of how his arrest actually came about. The first is that a royalist sympathiser visited the *château* and noticed the jewelled sabre from Egypt which Napoleon had given Ney as a wedding-present. Since the only other one like it belonged to Joachim Murat, at present somewhere in Corsica, the man drew his own conclusions and went to the police. Far more likely though is the version traceable to the police records of the Department. Suspicions had been aroused that the *château*'s owners were acting strangely when the *préfet*, a Monsieur Locard, suddenly discovered their relationship with the Princess of the Moscowa. Putting two-

and-two together he immediately ordered the local gendarmerie to go and make an investigation.

On the morning of 3 August Ney was shaving in his room when he heard a great commotion down in the courtyard. He went to the window and saw a whole squadron of mounted gendarmes. 'Who are you looking for?' he shouted to them. '*Le maréchal Ney*,' their captain replied. 'Come up here then,' Ney said: 'And I will point him out to you . . .'—not wishing to risk a proscription of his hosts. As they came puffing up the second flight of stairs he met them at the door. He appeared completely calm, but in no way evasive and certainly not afraid. 'Come in. Please sit down. I am Marshal Ney. I will just finish shaving.'[16]

They took him to Aurillac and confined him in a room at the Town Hall, while Monsieur Locard—who evidently hoped his zeal would be rewarded with a decoration—wrote off to Paris. Although his news caused a sensation there it also produced mixed feelings. The Ultras, naturally, were jubilant; but King Louis feared the possible consequences of a show-trial. 'Everything was done to favour his escape,' he complained: revealing that he knew Fouché had issued the Marshal with passports. 'Now he has done us more harm in getting himself arrested than he did on the day he betrayed us!'[17] Accordingly he rejected the advice of those who wanted him to make the greatest possible example of Ney and instead took up Fouché's suggestion that they hand him over to the army. In this way he could be tried by court martial; and with less argument and publicity. Fouché duly informed the new Minister for War of 'His Majesty's decision' and the latter—none other than Marshal Gouvion St.-Cyr—dispatched a Captain Jomard and three other officers to collect the prisoner.

Their journey back certainly proved eventful, and at times became extremely unpleasant (especially for the Marshal). On the other hand, and fortunately for us, it is all very thoroughly described by Captain Jomard in his official report.

The officers arrived in Aurillac on 15 August. In the meantime Michel had been kept in continuous confinement, yet without any attempt being made to interrogate him. They had orders to take him back manacled—but Jomard for one was anxious to spare him unnecessary humiliation, and they agreed to waive this in return for his solemn promise that he wouldn't try to escape. Then they set off in a post-chaise followed by a squadron of mounted gendarmes.

Already they were considerably impressed by the Marshal's dignity and composure; and when the driver stopped to change horses at Riom they received ample proof that his word was also his bond. This was in an area where Bonapartist army units still moved about unhindered, and as they stood about waiting, suddenly the burly General Exelmans appeared! The redoubtable cavalry-leader was himself on the Bourbons' *Ordonnance* and had clearly come to assist Ney's escape. But although the Marshal greeted his old colleague warmly he refused to break his promise and the journey continued.

At Nevers Jomard found he had lost one of his essential travel passes. He spent two hours obtaining an alternative document, during which time Ney in the post-chaise was subjected to both ridicule and abuse from a mixed crowd

252 MILITARY POLITICS FROM BONAPARTE TO THE BOURBONS

of royalists and occupying Württembergers. At La Charité-sur-Loire, their next stop, the hostility grew even worse and Jomard began to fear the Marshal was in imminent danger of being assassinated. 'The scene of Nevers was repeated, but in a more violent and objectionable manner. Several Württemberger officers, in the presence of their Commandant, used words against the Marshal which decency will not permit me to repeat.' When the captain tried to reason with them the royalists threw stones at him, and from this point on, 'whenever we had to make a stop, we always found ourselves surrounded by large crowds of people many of whom accompanied their abuse with menacing gestures'.

Jomard blames the *Commissaire du Roi*, Major Meyronnet, for promoting these scenes. It was Meyronnet's task to precede them and arrange for the change of horses, but—the captain reports—he might have 'shown greater discretion in announcing our arrival at the places of relay'. He expressed equal disapproval of the *Commissaire's* recruiting a party of Cossacks to escort them between Fontainebleau and Villejuif, even though Ney accepted this final insult 'with resignation' and merely commented: 'How sad if the crime of which public opinion accuses me is allowed to sully the little glory I won in battle!'

At the last stop outside Paris Aglaé was waiting for him, and the considerate Jomard allowed them to spend half an hour alone together in a room at the inn: which leaves the one vital gap in his account. Harold Kurtz in his admirable *The Trial of Marshal Ney* deduces that this was probably the moment when Aglaé acquainted her husband with the continuing sequestrations. She wasn't the sort of person to complain of her own hardship, but she must have told him:

> that the Prussian General Thielmann was lodged in their town house . . . and had commandeered all the horses, carriages and even harness from the stables. She might have shown him the *Journal des Débats* of August 5th which said that some hundred Prussians had been installed in their country house at Coudreaux *where they live as they please, while a contribution of 500,000 francs has been imposed on the nearby town of Châteaudun, the Prussian commandant announcing that the inhabitants were to be reimbursed from the imminent sale of Marshal Ney's properties.*[18]

What she undoubtedly did mention though was that Antoinette's husband Monsieur Gamot, the former Bonapartist *préfet* at Auxerre, was putting together a team of lawyers on the Marshal's behalf. He was not, she stressed, 'entirely without influential friends'.

When Ney returned to the post-chaise—Jomard concludes—he was in tears. 'I astonish you?' he said, blowing his nose. 'Well, let me tell you: I am not crying for myself, but for my wife and for my sons.'

Soon they entered the capital itself, and as they did so the captain made some enquiries of people proceeding towards the Plaine de Grenelle. 'Why, we are going to see the execution of Colonel La Bédoyère,' they told him. 'My God, La Bédoyère,' Ney murmured from the darkness of his corner and again lapsed into silence. Grim-faced, Jomard ordered the coachman to drive on: to the Prefecture of Police, where he discovered the *préfet* Decazes had left instructions for the prisoner to be transferred to the Conciergerie and lodged in the room above Marie Antoinette's old basement-cell.

Élie Decazes was a new name to Marshal Ney. (Which is hardly surprising, since in his previous profession as a lawyer he had never risen higher than being private adviser to Napoleon's mother.) However, the Marshal met the full force of his political ambition as early as the next day (20 August) when the *préfet* conducted his preliminary interrogation. Blond, handsome, much younger-looking than his thirty-five years, Decazes was already a great favourite with Louis XVIII and sufficiently sure of himself to be able to write arrogantly to Fouché: 'Several members of his (Ney's) family have applied for permission to see him. I do not think this is advisable before the prisoner has been seen by me. Should Your Excellency disagree with this opinion, I would be glad to learn of your intentions . . . otherwise I hope to begin the interrogations this evening.'[19] Having helped to capture and prosecute La Bédoyère, obviously he hoped to follow this success with another over his new and far more distinguished prisoner.

In fact Ney could, and should, have refused to answer his questions on the grounds that he was supposed to be the prisoner of the army. Instead of which he gave voice to a deeply-held belief that he would return to again and again until at last, fatally, he had his own way: 'I am under no obligation to answer your questions . . . nor can I be judged by a military commission . . . only by the Chamber of Peers!'[20] 'I am ready to answer any reasonable questions,' he went on, 'but first of all I want to know why I am here—perhaps because they included me on a list where they just call me *Ney?*' He also asked for legal counsel, which Decazes refused on the point that he was at present in custody *au secret*. 'In that case,' the Marshal retorted, 'my arrest was arbitrary, and against all the forms established by law!'

After some further wrangling Ney did embark upon a vigorous defence of his conduct at Besançon and Lons-le-Saunier. He described his efforts to organise an effective resistance to Napoleon, quoting important witnesses to his loyalty before the night of 13/14 March. There were the officers sent by the Count of Artois, he pointed out; also his letters to Oudinot and Suchet. 'I addressed my regimental officers . . . in the interests of the King. I hoped to make them do their duty. I said in public that if I encountered any hesitation I would take the rifle of the nearest *grenadier* and set an example!'

'In that case,' Decazes asked him slyly, 'how do you explain the change in your attitude? And how can you justify your behaviour on the 14th? Surely your duties were still the same . . .'

'Yes,' Ney admitted. 'On that terrible night I was swept away as if by a torrent. I know certain people blame me. I blame myself. I did wrong. *But I am no traitor.*'

'Who carried you away?' Decazes wanted to know. 'Was it not your example that carried the others away, the officers and troops under your command?'

But Ney rejected the charge, describing how even Bourmont and Lecourbe had agreed to his reading Napoleon's proclamation to the troops; and that every officer who'd opted for the King had been left free to return to Paris.[21]

'It's not true then,' the *préfet* asked him with sudden finality, 'that you approached His Majesty and demanded a large sum of money in return for your services?'

He had clearly gone too far. 'That's a damnable lie,' the Marshal shouted in a rage. 'I'll kill the man who says it's true! Do you say so, Sir?'

'No, not I, *Monsieur le Maréchal,*' Decazes replied hurriedly, 'I simply heard it suggested, by whom I do not remember.' He stood up. After failing to acquire any incriminating evidence, and having made the prisoner angry he decided it was wiser to withdraw.

Two days later he tried again though: still hoping to worm a confession of premeditated desertion out of Ney. 'If you did not plan to join Bonaparte before your arrival at Lons,' he asked, 'why was it that you changed your mind so abruptly?'

Ney was prepared for this question however:

> Well, it was like a dyke breaking before the flood. It is difficult to explain, but perhaps it was the effect of many assertions by Napoleon's agents. All seemed lost for the King. Nevertheless, I did not formally change until I read the proclamation to the soldiers. I had received no dispatch, no emissary from Napoleon until the night of the 13th. I was not in touch with anybody. I had no idea what had happened previously. No doubt I was wrong to read the proclamation, but I was carried away by events . . .[22]

He admitted to being deceived by Napoleon's claim that there had been a European agreement over his escape from Elba, but that was all. And he steadfastly refused to name the Emperor's emissaries to him at Lons. Even in the present situation honour counted for rather more than survival.

Decazes was clearly not getting anywhere. 'The interrogations established nothing beyond the fact that in the early hours of 14 March Ney had lost his head and now realised it.'[23] Having failed to establish adequate proof of treason the *préfet* therefore decided to let justice take its course. He advised Gouvion St.-Cyr to proceed with the court martial and informed Aglaé and Monsieur Gamot that their appointed lawyers (*Maître* Berryer, his son Antoine and André Dupin) could have access to the prisoner as often as they liked. The truth seems to be that Decazes was intent on one thing only: the downfall of Fouché; and when he realised Ney could be of little use, and that his case promised to be tricky, he quickly lost interest in it. Instead he turned to other, easier methods of blackening his chief's name—and succeeded admirably. 'Reports *unmasking* Fouché as the enemy of the monarchy were written for Decazes by Count Hubert de Brivazac-Beaumont, and then shown to the King.'[24] Coming on top of a formidable whispering campaign against the ex-regicide they led to his abrupt dismissal as Head of Police on 15 September, followed by 'exile' to a minor ambassadorship in Saxony.

This goes a long way towards explaining why Fouché did not reply to Aglaé's letter begging him to intervene on her husband's behalf, and enclosing a copy of the petition she was sending to the King. This petition declared Ney had acted 'without premeditation or perfidy', and she reminded

Louis how the Marshal had urged Napoleon's abdication at Fontainebleau, thereby becoming a leading agent for the return of the Bourbons.

> He expressed with the vehemence that is characteristic of him, his desire to do good service to Your Majesty, and his language, in the midst of his family and among his most intimate friends, was the same as at the Tuileries. The honesty and loyalty of his whole career, his very talkativeness, his strong beliefs—in a word all his good qualities, all his defects, combine in proving that he was sincere. I will say even more, Sire. My husband could not have dissimulated in this respect even if he had had the infamous idea of doing so. His very nature would have made it impossible . . .[25]

She ended by pointing out that he had joined Napoleon only in the last resort and to avoid further bloodshed. 'Please . . .' she implored him, 'spare the father of my young sons, whom Your Majesty offered to protect. The French are rejoicing at the return of their King and I pray that I might be able to rejoice at the sparing of my husband.'

Louis did not reply to the petition; and Fouché was too deeply involved in his own last battle for survival to help a man whose cause he regarded as already lost. Moreover the policeman's fall, when it came, dealt a further blow to Aglaé's hopes. It meant that Talleyrand's was now the sole voice of comparative moderation within earshot of the King. All the rest about Louis' person were Ultras, who never ceased to proclaim 'Ney is a traitor and must die!'

Marshal Gouvion St.-Cyr ('The Owl') regretted ever having taken on the job at the War Ministry. Despite the pressures from above, his efforts to put together a respectable-looking court martial had hit a number of totally unexpected and embarrassing snags.

First of all he endeavoured to have Moncey, the oldest marshal still on active service, as President; supported by Augereau, Masséna, Mortier and three generals. But Augereau and Masséna pleaded they were ill, while the amiable Mortier informed the Minister privately that he could never be a party to the judgement 'if it went against the Prince of the Moscowa'. Moncey flatly refused to sit in any capacity—and indicated he would accept being stripped of his titles rather than change his mind.

The Minister reminded them that under a law passed in the Year V and still operative they were liable not only to demotion but also imprisonment if they declined to serve. This at least had the effect of restoring Augereau and Masséna to better health, but Moncey continued to refuse; whereupon King Louis appealed to him personally. The Marshal's reply was again *no!*—and before very long an alleged facsimile of his letter was in clandestine circulation on the streets of Paris:

> *Sire:* Placed in the cruel dilemma of having to choose between disobeying Your Majesty or being false to my own conscience, I feel that I must explain. I do not enter into the question of deciding whether Marshal Ney is innocent or guilty. But, Sire, if those who direct Your Majesty's

councils thought only of your welfare, they would tell you that the scaffold has never made friends. Do they imagine that death is so terrible for those who have so often braved it? Is it the Allies who require France to immolate her most illustrious citizen?

Allow me to ask Your Majesty, *the letter continues*, where were his accusers while Ney was fighting upon so many fields of battle? Did they follow him, did they accuse him during twenty years of toil and danger? If Russia and the Allies cannot pardon the conqueror of the Moscowa, can France forget the hero of the Beresina? At the crossing of the Beresina, Sire, in the midst of that awful catastrophe, it was Ney who saved the remains of the army. I had in it relatives, friends and finally soldiers, who are the friends of their commanders. Am I therefore to put to death him to whom so many Frenchmen owe their lives, so many families their sons, their husbands, their relatives! Excuse, Sire, the frankness of an old soldier. Nor do I fail to see that I may thus draw down upon myself the hatred of certain courtiers; but if, as I go to the grave, I can say with one of your own illustrious ancestors, *All is lost but honour!* then I shall die content.[26]

If an accurate version of his letter, then it was also one of the most eloquent tributes ever paid to Ney by a professional colleague. And it undoubtedly expressed what many Frenchmen were thinking. But the Ultras were now everywhere in the ascendant. Talleyrand himself was on the point of falling (to be replaced by the Duke of Richelieu), and within weeks Gouvion St.-Cyr's portfolio passed to the Duke of Feltre. Gouvion departed with feelings of relief and except as a voting peer he too had exited from the brief remainder of Michel Ney's life. However his last act at the War Ministry was to offer the presidency of the *Conseil de Guerre* in charge of Ney's case to Marshal Jourdan, the prisoner's former commander with the Army of the Rhine. In the end this would prove to be a strangely significant appointment.

  Jourdan responded with alacrity. An adherent to Napoleon's cause at the beginning of the Hundred Days, in fact he took no part in the Waterloo campaign—and since then had appeared anxious to demonstrate his loyalty to the restored regime. He quickly secured the three additional generals for the *Conseil* (Claparède, Gazin and Vilatte) and drafted as his *rapporteur* (the man in charge of interrogation, calling of witnesses, etc.) General Grundler, a former colleague of Ney's in Russia. Also he impressed upon Grundler the necessity for speed. King Louis and those who surrounded him were growing impatient.

  Grundler swung into action at once. He officially interrogated Ney on 14 and 15 September, then again on 7 October and 4 November; and in between he interviewed potential witnesses, eventually reducing them in number from over a hundred vociferous royalists to a manageable twenty-four. At his meetings with the Marshal in the Conciergerie the atmosphere was far more relaxed than it had been with Decazes, and yet they added nothing to the known evidence:

except when Ney stressed rather more firmly than he had done to Decazes that his principal motive at Lons-le-Saunier had been to avoid civil war.

He (Ney) answered all questions with an accustomed air of wishing to tell the whole truth, but made it clear from the outset that he did not recognise the competence of the Court Martial *to try my case on the basis of the Ordonnance of July 24th.*[27]

Incidentally, his determined attitude on this last point contradicted the advice of his own team of lawyers. Maître Berryer even stated that he believed the King had agreed to a court martial in order to save him. But Ney thought differently. He had good reason to be suspicious of the Bourbons, and also he suspected one or two of the judges. 'Some . . . of those fellows would shoot me down like a rabbit!' he told Dupin.[28] More importantly though he insisted upon his rights and privileges as a Peer of the Realm, which meant that he considered his defence a legal plea and as such should be placed before the highest in the land. Berryer *père* shook his head; he insisted the Marshal's professional equals constituted their best hope of acquittal or at least a light sentence. 'But what good is freedom without honour?' Michel asked him: a question which could easily have served as his own epitaph.

While these heated discussions continued to absorb (and at times divide) the defence so Grundler was proceeding with his assembly of suitable witnesses. Jourdan had told him that any witness called would have to give evidence on two main points: Ney's final interview with King Louis before his departure from Paris, and the subsequent events at Lons-le-Saunier. The General therefore sifted those who were reasonably reliable from the other, obvious hatemongers on the basis of a written questionnaire:

Do you swear to tell the truth and to speak without hatred? What orders did Marshal Ney receive before 14 March? How did he carry them out? Who prevented them from being carried out? What do you know about the proclamation of 14 March? Did the Marshal's example influence his officers and men? What was the political situation in the Marshal's military district? Did the Marshal have the opportunity of opposing effectively the progress of the invasion of Napoleon Bonaparte?[29]

It meant Grundler had to work around the clock, but on 6 November he informed Jourdan that 'the preliminaries are now complete' and as a result the President set the court martial to begin on 9 November.

On the morning of the 9th Parisians awoke with a definite sense of expectancy. There had been a good deal of speculation in the newspapers, and this led to four or five thousand citizens, both pro- and anti-Ney taking an illegal holiday. To the more sophisticated it looked as if the Ultras would have their show trial after all, because not only were the sessions to be held in the larger arena of the Palais de Justice, but they broke the rules for a court martial by being open to selected members of the public. Not that the latter meant much to the ordinary man in the street, since the best places had been reserved for Metternich, Lord and Lady Castlereagh, various minor princes and princesses of Germany and what seemed like half the Czar's officers. Instead, to the average Frenchman, left standing with his hands in his pockets in the Rue Cherche-Midi, and separated from the trial by the capital's police force as well as contingents of the National Guard, the fate of Marshal Ney began to take on an altogether deeper meaning. In the absence of the Emperor himself, to the cabinet-maker from the Rue St.-Antoine, the porter from Les Halles, the student from the Latin Quarter across the river and the dyers and weavers from Les Gobelins, what happened at Ney's trial was regarded not as the passing of an Empire but as the opening of a new struggle by the self-made and patriotic in France against traditional, unpopular rulers buttressed by their foreign allies. It was a struggle which then went underground, but nevertheless continued and finally toppled the main-line Bourbons during the second French Revolution of 1830, replacing them with a 'Citizen King', Louis-Philippe.

However to these same sincere men, left glowering and occasionally shouting abuse at the cordon of gendarmes in the Rue Cherche-Midi, there would have been little real satisfaction even if they had gained admittance that morning: for on the first day the star-attraction failed to appear.

Metternich complained 'What a bore!' and several well-connected ladies yawned into their gloves, but the initial sessions were monopolised by the court's functionaries arguing about legalities and stating their own adopted positions. There was a last-minute attempt by Masséna to wriggle out of sitting on the grounds that his rows in Spain and Portugal with Ney might introduce a degree of prejudice; but Jourdan refused to accept this and in the end all four marshals and the three generals entered the court-room together at half past ten o'clock. After the formal convening a distinguished *avocat*, Delacroix-Frainville, led for the defence with an argument that no marshal of France had ever been subjected to a court martial, and eventually this was the cue for André Dupin to question the competence of the seven officers to pass judgement upon one of their colleagues. Although Ney was a marshal, he stated, he was also a peer of France, and as such he must be judged by his peers. 'This Court Martial not only can,' he added, 'but *must* declare itself to be incompetent.'[30] Jourdan flushed darkly at the suggestion but was unsure of himself when it came to so fine a legal point. He therefore declared the case adjourned until the following day.

Ney made his entrance into court the next afternoon, and as he entered the room it is interesting to note that the guards snapped to attention. He was dressed in the full blue-and-gold uniform of a general of France and wore the red Grand Cordon of the *Légion d'Honneur*. Calmly, he walked forward and bowed respectfully to the judges. Then he took his seat in a large armchair facing them.

Jourdan, who overnight had sought legal advice, now warned the court-room that anyone 'who shows a lack of respect for this Court of Justice . . . will be arrested'. And he proceeded to a formal identification of the Marshal. In reply to his questions Ney read the following prepared statement: 'My name is Michel Ney. I was born at Saarlouis on 10 January 1769. I am the Duke of Elchingen, Prince of the Moscowa and Marshal of France, a *Chevalier* of the Order of St.-Louis, I wear the Grand Cordon of the Legion of Honour, also the decorations of the Iron Crown of Italy and the Order of Christ.' Jourdan then made a sign for him to be seated again and requested the defence to continue with their presentation.

Berryer *père*, an able if somewhat pedantic lawyer, spoke in all for two hours. He emptied the public-section of its remaining VIPs, but slowly and inexorably he did begin to make a definite impact upon the judges, especially when he quoted Article 33 of the constitutional charter which stated that the Upper House 'took note of the crime of high treason and attempts on the security of the state'. As a result, he concluded: 'A man whom twenty-five years of glory have placed in the first rank of Frenchmen . . . has every right to make himself heard and develop his defence before that same Upper Chamber.' Only one course lay before the judges, this implied: to declare the proceedings illegal and allow Ney to transfer his case to an assembly of the Peers.[31]

Oddly enough he was supported in this view by General Grundler, who spoke next. The General had been called upon essentially to acquaint the court with the results of his own investigations, but these included the fact that (a), the case had no precedent in French history, and also (b), Ney was definitely a *Peer of the Realm* at the time when he was accused of committing the offence. Moreover, the General admitted, it had been difficult to acquire genuinely concurring evidence on *any* of the points in his questionnaire.

For the prosecution Monsieur de Joinville tried to pooh-pooh the idea of a second trial. 'The Marshal,' he stated, 'is a peer created by Napoleon, and therefore not entitled to the ancient privileges which membership of the Upper Chamber normally affords.' In putting it this way he blundered badly of course. Each of the four marshal-judges also owed his promotion to Napoleon—even though they had since made individual compacts with the monarchy. Jourdan looked glum; and announced that as President he wished to adjourn for further discussion with his colleagues.

The writing was now clearly on the wall. Either proceed to try Marshal Ney on the basis of some extremely controversial witnesses or quit the stage themselves.

This was at four o'clock. At a quarter past five the judges returned and Jourdan read out their decision: 'Having deliberated on the question of whether the court is competent to judge Marshal Ney who is accused of high treason, it declares itself to be incompetent by a majority of five to two.'[32]

Ney was temporarily elated, and he complimented Berryer *père* on his overall handling of the defence. 'Ah, Monsieur . . . you have rendered me a very great service.'[33] He had now to his own way of thinking achieved a double victory. Firstly, in preserving his honour as a soldier from condemnation by his military equals; and then in securing the right to defend himself against the charge of treason before the highest in the land. However, in both respects he was being extraordinarily naive. For one thing his sense of honour blinded him to the possibility of the court martial giving a very light sentence or even acquitting him altogether. As Davout said, no soldier of them would condemn such a man, 'Not even Ragusa'.[34] Secondly, the Marshal was forgetting that by denying the Ultras retribution on this occasion he would simply cause them to intensify their efforts to gather a majority against him in the Chamber of Peers.

Perhaps the man who really stands condemned though is Marshal Jourdan. Having jumped at the chance to find favour in the Bourbons' eyes by accepting the presidency of the court martial, it was then inexcusable for him not to see the case through to its logical conclusion. Despite defence pleas to the contrary the court martial *was* competent to try Ney—and therefore in a position to save his life. This was certainly the attitude of the two marshals who voted against dissolution. Mortier had already indicated that he would never condemn Ney, and Augereau later commented: 'We were cowards. We ought to have insisted upon our right in order to save him from himself.'[35] Nevertheless, and for purely selfish reasons, Jourdan led the other officers to vote against their carrying on. To show leniency towards Ney would have meant incurring the royal displeasure. But how could he do otherwise when Grundler's rag-bag of evidence promised so little. Consequently, at the first signs (from Grundler) that exoneration might become necessary the President of the Court, like Pontius Pilate, took the earliest opportunity to wash his hands of the entire affair.

Back inside the Conciergerie Michel was again kept closely confined. He had been practising on the flute, but now the instrument was taken away from him because the authorities feared he might use it to signal plans for an escape.[36] On the other hand he was allowed regular visitors. Aglaé and his four sons came almost daily, usually bringing a *quiche* or some smoked sausage and cream cheese with the good pink wine of Toul to supplement his prison diet. Other visitors included Monsieur Gamot and Antoinette, Colonel Heymès and his *notaire*, Monsieur Batardy. He was permitted books; also snuff and tobacco. But his main complaint was the lack of exercise, for the yard outside was too small to walk about much and upon a man previously as vigorous as Ney this had the effect of making him liverish.

At the same time the Marshal's relatives and friends were continually in the offices of his lawyers preparing their final bid to save his life. Aglaé in addition wrote to the foreign ambassadors, to the Prince Regent, even to the Duke of Wellington; but without any notable success. Sir Charles Stewart, Great Britain's Ambassador in Paris, replied that 'His Royal Highness the Prince Regent cannot interfere in a matter of internal French policy', while the Duke 'regretfully declined' to become involved for the same reason. Obviously everything must now be won or lost in the Chamber itself, for although Michael Bruce did his best to whip up a pro-Ney campaign in the London press, in the end not one of the Allied Powers intervened on the Marshal's behalf.[37]

The Ultras, enraged by the court martial's dissolution, soon showed they were not to be deterred. As early as the following day the Duke of Richelieu appeared before the Upper House and 'in language reminiscent of the Terror' moved that the Peers would have full responsibility to decide the case of Marshal Ney. The Duke of Broglie recalled that the Chief Minister spoke 'like a madman' and certainly with vindictiveness.[38] 'It is not only in the name of the King,' Richelieu shouted, 'that we shall carry out the office . . . but in the name of a France left angry for too long . . . and also in the name of Europe that the cabinet implores and requires you to pass judgement on Ney. We have stated,' he added, 'that the Chamber owes the world retribution. And it must be prompt, for it is essential that the indignation expressed by all be contained. An extended impunity will only breed new evils, greater than those from which we are seeking to escape!'[39]

But the Ultras did not have it all their own way. 'The Chamber of Peers did not accept Richelieu's proposition that, in a case of an indictment for high treason, they could follow their customary procedural methods.' As a result the *procureur général*, Bellart, was forced to draw up a new *Ordonnance* which he presented to the Chamber on Monday 13 November. Under it 'the forms of a legal and public trial as normally observed in French tribunals in criminal cases were now to be the basis of the proceedings. The Chamber heard with satisfaction that not the ministers, but Bellart as *procureur général* would take charge of the prosecution. There was to be the normal process of instruction, of interrogation of the prisoner and of witnesses on the Chamber's behalf, at the end of which the indictment was to be drawn up.'[40] In other words, '. . . from this day onward the Upper Chamber was transformed into a tribunal, and the Government's original plan of rushing the trial through like any political project was given up.'[41]

The trial proper opened on 21 November in the Palais du Luxembourg. Because this meant crossing the river from the Conciergerie, with the dangers of a rescue attempt and the possibility of crowd intervention, it had been decided to confine Ney in an improvised cell at the palace itself, near the end of the Archives-section. From here he made his entry into the Assembly Hall escorted by four gendarmes, wearing the undress uniform of a French general and on this occasion with both the Legion of Honour and the Order of St.-Louis upon his chest.

Altogether a hundred and sixty-one peers had gathered in the Chamber. Talleyrand was the most notable absentee, while Marshal Augereau excused himself on the grounds that he had sat in judgement upon Ney already. There were no ladies present, but Metternich and other foreign dignitaries again occupied the best public seats and the whole of Richelieu's cabinet attended. Chancellor Dambray opened the proceedings by addressing his fellow-peers in the Galerie de Rubens. 'I am sure,' he advised them, 'that this Chamber will prove impartial . . . that its members will be interested in the truth only . . . and that they will lay aside all prejudice so as to allow the prisoner the most ample latitude for his defence.'[42] He then indicated that they should file away to take their places in the hall, and upon the last of them being seated the accused was introduced. Several people present noted that he appeared a trifle pale, but no doubt this too can be attributed to his lack of exercise and fresh air.

Leading for the crown, Bellart read out a long prepared statement (countersigned by every member of the Government) which specifically accused Ney of treachery against the King, desertion to the enemy, waging war against the nation and promoting bloodshed among the citizens. Furthermore, it charged him with declaring for Napoleon 'when his troops were still willing to follow him against the usurper . . . and action which led directly to a collapse of the royal forces.'[43] On all of these charges, the *procureur général* declared emphatically, not only was Marshal Ney guilty, but under the *Code pénal* and in parts of the law of 21 Brumaire in the Year V, the burden of his guilt clearly necessitated the passing of the death-sentence. (The former defending *avocat* of the Republican General Moreau had certainly travelled a long way since then . . . backwards! Or is this just another example of the legal chameleon?)

Dambray next asked the accused if he had any evidence to submit in his defence, adding that the Chamber, ever mindful of the great deeds he had performed for France, would give him a sympathetic hearing. Ney stood up, bowed to the court and read from a paper prepared by his counsel directed against several items in the court's procedure. It was the cue for Bellart and Berryer to start another technical hassle; Berryer's argument being that the Chamber would have to pass a new law in order to try a person accused of what had been described in the prosecution's statement as a criminal matter. Bellart went scarlet with indignation, but Dupin pointed out that 'the law according to which Ney could be judged did not yet exist on the statute books',[44] and with this the Chamber retired to debate the point. One hundred and forty-six of the peers voted against the defence counsel's submission, and Dambray therefore announced they must continue to present their evidence and adjourned the case until 23 November. Although a fine point it was a definite indication that Ney had miscalculated in trusting his life to the peers: that in fact he was now confronted by a large majority of Ultras in the Chamber.

Even so, the court reopened to another legal wrangle. In order to speed things up Bellart made the concession that Ney had not betrayed the King before 14 March, but Dupin would not agree to this. 'It is not enough that the prosecution agrees the point,' he insisted, 'it is a matter which has to be solemnly proven.' And he requested a further adjournment in order to re-examine certain witnesses. Dambray felt obliged to grant the request (until 4 December), but this time he warned the defence that such tactics, far from helping the accused, served only to exasperate the court.[45]

No doubt they kept his words in mind, because on Monday 4 December they raised no new objections. Ney read another prepared statement under Article XII of the Convention of Paris and confirmation of its amnesty in the peace treaty signed on 20 November. Then the first witnesses were called.

General de Bourmont stated that when Ney showed him Napoleon's proclamation on the morning of 14 March, both he and Lecourbe had argued strongly against joining the Emperor. (Lecourbe had died a month before; so apart from Ney there was nobody to contradict him on this point.)

'Why then,' the defence demanded to know, 'did you go with *Monsieur le maréchal* to the parade ground?'

Bourmont hedged. 'I went there to see what would happen. How our troops would react to the Marshal's proposal. Because he was thoroughly resolved to take them over to the Emperor's side. Only half an hour after reading the proclamation, he was wearing the Grand Eagle of the *Legion d'Honneur* with the image of the usurper!'

Now Ney was on his feet—and pointing an accusing finger at Bourmont. 'This witness,' he thundered, 'has had eight months in which to prepare his testimony, and he has used his time well. In making his declaration at Lille he must have imagined that I would be treated like La Bédoyère and that we should never meet again face-to-face. I am not an orator, and I only regret that General Lecourbe is dead, but I shall invoke his support before a higher tribunal, before God who hears us, who will judge you and me, General de Bourmont!' He then gave his own detailed account of the events at Lons, and ended with a further accusation against the witness: 'I asked for light and advice from men I believed had enough affection and energy to tell me I was wrong. Instead of which you drew me on and pushed me over the precipice! It was General de Bourmont who assembled the troops to hear the proclamation. He had two hours for reflection. If he considered my behaviour criminal, why did he not arrest me? I was alone. I had not a man with me . . . not even a saddle-horse on which to escape.'[46]

Dambray was impressed by the force of the Marshal's words. 'Who gave the order for the troops to parade?' he asked Bourmont.

'I did,' the General admitted, 'but following a verbal order from the Marshal . . .'

Again Ney jumped to his feet: '*He assembled them after he had read the proclamation!*'; and Bourmont agreed (somewhat lamely) that 'it was at about eleven o'clock'.

'Why was it,' Dambray again asked him, 'that although you disapproved of the Marshal's behaviour, still you followed him to the parade ground, knowing what he meant to do there?'

'I wanted to see if there were any signs of opposition among the troops to the course the Marshal was preparing to take . . .'

'But did you help to produce any opposition?'

'I had no time. I could do nothing unless I killed the Marshal . . .'

Ney registered both disgust and contempt at this. 'You would have done me a great service, Monsieur. It was your duty to kill me!' The witness was plainly deteriorating under the strain. The Marshal made only one other significant interruption against him and that came after Berryer *père* had taken up the cross-examination.

'*Monsieur le Général.* May I ask if it was also out of curiosity that you attended the party on the evening of 14 March at Lons?'

'I did so to prevent the Marshal suspecting me,' Bourmont replied. 'And to avoid arrest . . .'

'I arrested no one!' Ney called out. 'I left everyone free. Neither you nor anyone else objected. You held an independent command. You could have had me arrested . . . but instead you came to dine with me. You remember. You were there. Why were you in such a happy mood?' Also while on his feet he took the opportunity to clear up another point: namely, the accusation that within half an hour of reading Napoleon's proclamation to the troops he (Ney) had pinned the Grand Eagle of the *Légion d'Honneur* on his chest. 'In fact,' the Marshal stated: 'Until I returned to Paris, I continued to wear the King's decoration.' And he had a good witness to support this, because sitting beside his defence team was Monsieur Cailsoue, a well-known jeweller, who could testify that the Marshal's Grand Eagle was in his Paris safe on 14 March and had remained there until 25 March.

Bourmont clearly had nothing further to offer the prosecution and Bellart decided to move on. The Peers next heard the Secretary of the Tribunal read the testimony of the dead Lecourbe, which hardly incriminated Ney and ended with some remarks in his favour. 'It may have been,' Lecourbe had told the examining magistrate, 'that some officers and even parts of the infantry could have resisted for a while, but the moment they were brought into contact with troops loyal to Bonaparte, they, too, would have been swept away.' 'Who were opposed to Ney and in what way did they show it?' the magistrate had asked him. 'There was no opposition,' Lecourbe replied, 'or if there was I saw none of it.'[47]

The *préfets* of the Jura and of Bourg both testified that they had seen Ney wearing the *Légion d'Honneur* on 14 March; (not knowing the defence had Monsieur Cailsoue to refute their claims). And General Grivel of the National Guard stated Ney had ordered him off the square at Lons immediately after reading the proclamation and had uttered violently anti-Bourbon remarks. But Count Philippe de Ségur restored the balance by his simple, moving testimonial. 'I saw Marshal Ney before he left Paris to take up his command. Everything the Marshal said on that occasion breathed honour and fidelity . . . and was typical of a great soldier who had brought glory to French arms.'

Dambray closed the day's proceedings after summoning Marshal Oudinot to identify the two letters Ney wrote to him on 12/13 March describing how he intended to repel the invader. The normally undemonstrative Berryer *père* gave his son a reassuring smile. The session seemed to have gone extremely well for them, and their own big guns were still to come.

The next morning began quietly, although by midday Monsieur Cailsoue had finally put paid to the crown's case about the decorations worn at Lons-le-Saunier. The jeweller produced his account-books to show where Ney's Grand Eagle had been on 14 March and an embarrassed Bellart was forced to withdraw the charge.

But the proceedings finally erupted after lunch when Berryer *père* called Marshal Davout and began to question him on what he had intended in negotiating the Convention of Paris. The *procureur général* jerked to his feet and yelled at Chancellor Dambray: 'The Act itself is the only valid evidence. What have the later opinions of the Prince of Eckmühl to do with it?'

'Surely,' Berryer broke in, 'the point at issue is the interpretation of the Act. What did the Minister for War hold that the article provided, and what was his understanding of it before he allowed it to be signed by the envoys of the French Government? That, Monsieur, is the question before the Chamber.' The argument raged back and forth between the two lawyers until at length Dambray ruled for the prosecution. 'The Act must stand for itself,' he announced: 'The views of those who were partners to its ratification are not to be admitted to the record.'[48]

At this Ney himself rose and spoke directly to the Chancellor. 'I have counted,' he said, 'on the protection of the Convention. Otherwise, do you not see that I would have preferred to die sword in hand? I was arrested in direct violation of the Convention. I only remained in France because I had every faith in it.'[49] The implication was unmistakable. That the Bourbons had deliberately reneged on the signed agreement and consequently this whole *soi-disant* trial was a sham: no more than legal cover for a private vendetta against the man who now symbolised Bonapartist popularity.

Dambray could hardly fail to realise the significance of Ney's words, and hurriedly made a sign for Berryer *père* to call his next witnesses, Bignon, Bondy and Guilleminot: the French signatories to the Convention. General Guilleminot confirmed his role as Davout's Chief-of-Staff when signing the document. 'I had orders to insist on an amnesty of persons, whatever had been their opinions, their positions, or their conduct. I had orders to break off negotiations otherwise, but the point was conceded without difficulty. The army was ready to attack . . . Article XII made it lay down its arms.'

But really this and what followed proved to be something of an anti-climax —even the lengthy summing up by Bellart and Berryer's equally long reply on the following day. Ney had touched every single member of the Chamber on a very sore spot: in effect that each of them was now serving a regime which was dishonest; a king and his followers who had broken promises made in a signed treaty. And yet these same authorities were now pursuing a distinguished French officer for allegedly breaking his promises to them. The large number of Ultras in the Chamber showed by their hardening expressions what they thought of this surprise countercharge. But a sizeable minority of the Peers were left deeply troubled by Ney's words. To these, the fact that the King had broken faith with the nation over the very first document to which he had given the royal assent augured badly for the future.

'What did the accused do on the morning of the 14th?' Bellart demanded of them. 'Truth speaks here for the prosecution. Before the army . . . he read out his inflammatory proclamation. It was he, the Commander-in-Chief, who took the troops over to the side of the usurper whom he had earlier promised to bring back in an iron-cage . . .

'History offers no parallel to this story of treason. The prosecution has shown as false the idea that the Marshal was irresistibly carried away. He, like a number of other brave officers, could have fled from crime and dishonour . . .

'Twenty-five years of political unrest have made us indulgent,' he added, his voice rising to its inevitable crescendo. 'They have weakened our principles of morality. Is the attempt now to be made to apply that same degraded morality to the case of Marshal Ney? He, surely, is not among those who could plead ignorance as their excuse . . .

'I now leave it to the counsel for his defence to distort whatever facts they can in order to reduce the guilt and save some bits from the wreckage of his honour. I have finished, and leave it to your conscience to examine everything contained in the indictment.'[50]

By comparison Berryer *père* was more restrained. But in nearly four hours of addressing the Peers he not only reviewed the glorious episodes in Ney's career, but painstakingly dissected and contradicted as false every point and witness in the prosecution's case. The basis of his defence though remained the confused state of the nation at the time of Napoleon's return and the intensity of Ney's patriotism.

> Peers of France, remember how all were declaring for the invader, or so it seemed to the Marshal. He was an honest soldier and a great one. He had no political ambitions. He had observed many changes of government and in each seen the will of the people at work. Behind these forms of government lay our country, our nation, France, and to him this

was all-important. The Marshal's only considerations were how his actions might affect France. It was the constant object before his eyes . . . and he pursued it with a religious devotion. This incontestable truth, demonstrated by so many splendid exploits, ought to sweep away all doubts about his character. Once more we must attribute the act with which Marshal Ney is charged entirely to his ardent desire to avoid the shedding of French blood by other Frenchmen.[51]

Strangely enough, Ney reaffirmed his patriotism—and more or less instinctively—by what proved to be his final intervention at the trial. Near the end of his summing up Berryer *père* returned, quite deliberately, to the Convention of Paris. It brought an immediate interruption from Bellart, who waved at him a new Government decree to the effect that the Convention was not relevant in Ney's case and the Peers should confine their debate to the facts of the indictment.

Defence counsel had been informed of this decree (signed by Richelieu and the whole Cabinet) only during their lunch-break. Briefly, because obviously prepared in haste, it authorised Dambray to forbid, 'as a safeguard to National dignity . . .', the invoking of a Convention 'concluded by the agents of a party in open revolt against the legitimate Sovereign'. Berryer recognised now there was no hope of gaining an acquittal. The document swept away any last pretensions that Ney's trial had been intended as a fair one. But in accepting defeat the lawyer was determined to make these facts widely known—hence his provoking the interruption from Bellart. In this way he forced Dambray to have the decree read aloud; also to confess how at a secret session that morning most of the Peers had agreed to it.

'In that case,' André Dupin replied, rising: 'I have an observation to make, not involving the Convention but the second treaty of Paris, signed only a fortnight ago. I understand that under the terms of this treaty we have given the defendant's native city of Saarlouis to Prussia. He now belongs to a country which is no longer subject to the King of France, and is therefore protected by international law. We cannot call him a Frenchman.'[52]

Half the Chamber seemed about to fall off their chairs with astonishment; following which there was a good deal of abuse hurled at Dupin by the Ultras. *Another lawyer's trick*, the Duke of Uzès shouted. But as the din died away it was Ney who was seen to be on his feet and doing all the talking, putting an end to further debate. '*I am French!*' he shouted. 'I must beg to disagree with my eminent counsel. But I am French, *and I will die French!*'

Dambray had to gavel a burst of applause from the public-section of the hall, but Ney continued to hold the floor. 'So far,' he said, 'my defence has appeared to be free. I now perceive that it is to be fettered. I thank my generous defenders for what they have done and are still prepared to do; but I beg them to desist altogether rather than defend me imperfectly. I am accused in defiance of the faith in treaties and I am forbidden to appeal to them. I do like General Moreau therefore, I appeal away from you to Europe, to posterity, to God!'[53]

There was more shouting after this, until suddenly the voice of Ney could be heard over it for the last time. A voice ringing out with authority and command, and the Peers grew silent: like a bunch of recruits meeting their first officer. '*I forbid my counsel to speak again unless they are allowed to speak freely!*'

Bellart was the first to recover and took this as his cue for advising Dambray that the prosecution had no more to add. They should, he urged, proceed to judgement without delay. 'Have you anything further to tell us?' Dambray asked the Marshal. '*Nothing whatsoever!*' Ney replied. The Chancellor therefore ordered the hall to be cleared of everyone except the sitting Peers. He gathered up his papers. The next few hours threatened to be more exhausting than the trial itself. But then he had not overheard Ney's final remarks to the Berryers and Dupin: 'Of course it was all decided beforehand . . . even I can see that now!'[54]

Just how much had been decided was revealed later that night when the Peers arrived at their verdict. Ultra persuasion and intimidation were obviously the key-factors. Against which the defence evidence relied upon its appeal to the individual conscience.

Dambray had announced to the Chamber that each member would be required to vote on three questions of fact. *First:* Did Marshal Ney receive certain emissaries on the night of 13/14 March? *Second:* Did Marshal Ney read a proclamation in the square at Lons-le-Saunier, inviting the troops to rebellion and defection? *And third:* Has the Marshal been guilty of an attempt against the safety of the State?[55]

In point of fact the answer to the first question had never become a matter for dispute, but it was now the excuse for those peers who felt most troubled to register their dissent. After a proportion had voted 'yes' the name of the Marquis d'Aligre was called. 'I vote *No!*' he shouted. 'I cannot in all conscience condemn the accused, since you have refused to let him speak about the Convention of Paris in relation to his case.' Lanjuinais agreed with him, while Count Nicolai took it a stage further: 'I do not like the way this trial has been conducted,' he said after voting '*no*'. When the votes were tallied a hundred and eleven had said 'yes' and there were forty-seven who said '*no*'.

Over the second question only three peers dissented: d'Aligre, Lanjuinais and Nicolai. And with the third it developed into a Ultra triumph. One hundred and fifty-seven voted 'yes'. D'Aligre and Richebourg said 'yes' but demanded clemency, while Lanjuinais voted 'yes'—subject to the Convention. The one unqualified negative came from the young Duke of Broglie. 'There is no crime unless there is criminal intent,' he flung at them. 'And no treason without premeditation or a design to betray. I vote *No!*'[56]

Dambray banged his gavel again to restore order. In view of the overwhelming result it was now just a question of the sentence. The Chancellor announced that if the majority voted for death, a second ballot would be taken in case certain peers wished 'to appeal to His Majesty for mercy'.

At the first ballot thirteen voted for deportation, one hundred and forty-two for execution by firing-squad and one—a Count Lynch—for death by the guillotine. On the second ballot seventeen voted for deportation (including Broglie, Gouvion St.-Cyr and General Colaud); five abstained and requested a petition for clemency to be addressed to the King; and one hundred and thirty-seven voted for the military-style execution. Among the latter were five Marshals of France: old Kellermann of Valmy, Pérignon, Sérurier, Marmont and Victor. The last two names come as no surprise, but one would have thought the other three—especially Kellermann—might have voted differently.

By this time it was 7 December, a Thursday. After leaving the hall Michel had been escorted to his improvised cell, and there he still remained—asleep. The Berryers and Dupin had visited him in the course of the evening and found him 'tranquil, dining with an excellent appetite and in perfect peacefulness. In each corner of the room there was a gendarme, who . . . were members of the *garde du corps* (the royal bodyguard) in disguise.' Dupin suggested that if the sentence was death they should appeal for mercy. But Ney did not think so. He appeared more concerned about seeing his family. At length he shook hands with the two younger lawyers and embraced Berryer *père*. His last words to them were: '*Adieu*, my dear defenders. This is perhaps our final meeting, but we shall see each other again up above.'[57] Obviously he had resigned himself to his fate.

It was three o'clock in the morning when Monsieur Cauchy, Secretary to the Chamber of Peers, woke Michel in his place of imprisonment. 'I apologise for disturbing you, *Monsieur le maréchal*,' he stammered, 'but it is my painful duty to . . .' and he held up a piece of paper.

'Well, Monsieur Cauchy,' Ney replied, rubbing his eyes, 'that can't be helped. Everyone must try to do his duty. What is it?'

The little archivist cleared his throat and began to read. 'To Michel Ney, Marshal of France, Duke of Elchingen, Prince of the Moscowa . . .'

'Come to the point, Monsieur. Never mind the formalities.'

Cauchy read on until he reached a part describing Ney's crime 'against the succession to the throne'. The prisoner could not repress a smile. 'That refers to a law designed for the benefit of Napoleon's family,' he pointed out.

Over the death-sentence itself he also came up with a grim jest. 'Why such stuffy language, Cauchy? Isn't it better to say: *must bite the dust!* So much easier for an old soldier to understand . . .'

Cauchy told him finally the execution would take place that same morning, at nine o'clock. And the Marshal answered: 'Whenever they like. I am ready.'

Colonel de Montigny, the officer in charge at the Luxembourg, explained that they would be joined shortly by Count Léon de Rochechouart, the Military Commandant of Paris who now had the job of arranging the execution. In the meantime, Montigny asked him, was there anything the Marshal wished to be done for him? Michel requested permission to see his notary, then his wife and children—to which the officer replied that they would be sent for at once.

Cauchy added, 'The Abbé de Pierre, parish priest of St.-Sulpice, is also waiting to see you . . . to offer the consolation of religion.'

'I shall send for him later, perhaps . . .'

Again Cauchy apologised. 'Perhaps the Abbé is not acceptable to *Monsieur le maréchal*. In which case I shall be glad to call in any other priest of his choosing . . .'

For the first time Ney's voice betrayed a degree of impatience. 'You are worrying me, Monsieur Cauchy, with your priests. The Abbé de Pierre is no doubt as good as any other, but I intend to appear before God as I have appeared before men. I am not afraid.'

Rochechouart had just arrived and recalls that one of the soldiers in the room argued about this with the Marshal. '*Monsieur le maréchal,* forgive me, but you are wrong. I am not as brave as you, although I am as old in the service. I have fought under you, and I have never gone into battle more fearlessly than when I first of all recommended my soul to God!' Ney was moved by the man's sincerity. 'Very well, *mon brave,*' he answered, 'perhaps you are right. I will see the Abbé after I have seen my family.'[58]

Montigny produced Monsieur Batardy in double-quick time. Their meeting did not take long. The Marshal had prepared detailed instructions for him and it was simply a question of going over certain points and then saying *good-bye.* Among the items which the notary took away was a letter addressed to Monsieur Monnier, his sister Marguerite's husband, in which he stated:

> My trial is finished. I am condemned to die. (*But then he had added thoughtfully*) Break this very gently to my dear father, who is himself on the verge of the grave. Before another twenty-four hours are gone I shall appear before God, full of deep regrets over not being able to serve my country for much longer. God will know, as I have openly claimed, that I have no other feelings of remorse. Embrace my sister for me, and give your children my love. I hope they will love my own children, in spite of this terrible tragedy which has fallen upon them. Farewell. I embrace you with all the feelings of a devoted brother. The Marshal, Prince of the Moscowa. *Ney.*

Afterwards Michel bathed, shaved and lay down on his bed to doze before Aglaé arrived with his sons at six o'clock.

When they entered the room Aglaé screamed and fainted. The strain upon her had been considerable. His sons, now aged between twelve and three, broke into fits of weeping. Michel remained the only calm person present. When Aglaé recovered he spoke to her with great affection and tenderness on several domestic matters, and instructed his sons to protect her at all times. 'Honour the name *Ney,*' he told them, 'because I am leaving it to you without stain, and after the hates and passions of these last few days are over I believe the world will also come to remember it with honour.'[59]

Recovering somewhat, Aglaé announced that she was setting off for the Tuileries. Marshal Marmont, she said, had promised to go with her to see the King.

Michel avoided mentioning what he already knew. That Marmont had twice in the same night voted for his death. Nevertheless he encouraged her to go, deciding that at least she would leave him with feelings of hope. He kissed her without reserve; then in swift succession each of his four sons. Afterwards Antoinette Gamot took the boys to her home while Aglaé set off for the palace . . .

Ney next indicated to Rochechouart that he was prepared to see the Abbé de Pierre. At this point the guards withdrew, and although the two men remained in conversation for an hour we have no record of what passed between them. For Rochechouart, otherwise our accurate authority for these last scenes in Ney's life, had a problem of his own to resolve.

His firing-squad was already assembled: four sergeants, four corporals and four ordinary riflemen. The order to shoot would be given by an officer he could fully trust, Major de St.-Bias. Also he had taken the necessary precautions for transporting the prisoner 'without danger of escape' to the place of execution: by tradition for military men the Plaine de Grenelle where La Bédoyère had been shot. But then suddenly there was a panicky message from the War Ministry. From the Chamber of Peers' passing sentence to the recruitment of the firing-squad everything so far had appeared to be proceeding both with speed and in secret. *And yet now, somehow, the news had got out and was spreading like wildfire!*

'A large crowd,' the message stated, 'has begun to converge on the Plaine de Grenelle. It is decidedly anti-Royalist and in a very hostile mood. Half-pay officers, veterans of the *Grande Armée* . . . and young boys carrying clubs. Some have driven out to the parade-ground in cabs . . . others are arriving on foot.' There was nothing else for it but to choose another place, and Rochechouart issued instructions for all those involved to reassemble just a few hundred yards from the Luxembourg: against the wall leading up from the Avenue de l'Observatoire. 'We will still use the coach though,' he added: 'And I want two lieutenants to travel inside with the Marshal.'[60] At the very moment he was saying this the Princess of the Moscowa took her seat in an ante-room at the Tuileries. Alone—there being no sign of Marmont—she listened while a junior *aide* said: 'His Majesty is at breakfast . . . I must not trouble his digestion.'[61] 'But my husband will be shot unless the King . . .' Already the officer was walking away, his ears closed, and when Antoinette found her there a few minutes later she was on her knees praying.

Michel dressed himself in black knee-breeches, black silk stockings, a white cravat and a long, dark blue frock-coat. Under no circumstances would he subject himself to the indignity of appearing in uniform and having his decorations and epaulettes torn off. 'And another thing,' he informed Rochechouart: '*I*, not you, and certainly not one of your officers, will give the order to fire!' Then he walked down and outside to where the coach stood waiting. The Abbé de Pierre accompanied him, only to be seized with a fit of trembling. 'Ah, Monsieur le Curé,' Ney murmured sympathetically: 'I do understand.' He buttoned his coat up to the neck. It was drizzling with rain. 'This is a wretched day,' he added, glancing at the black swollen sky. But on the point of getting into the coach he noticeably brightened. '*Please:* get in first,' he motioned to the Abbé, smiling, 'remember I shall be going before you presently!' Rochechouart ordered his mounted escort to fall in on either side of the vehicle.

At the place of execution Michel handed the Abbé his gold snuff-box together with a purse; 'Give the box to Madame Ney. And this small amount of money use for the poor of St.-Sulpice.' The Abbé, in tears, embraced him and fell on his knees in prayer. As he did so the Marshal was already striding over to a place near the wall, spurning the offer of a blindfold. 'I've fought a hundred battles for France,' he said to St.-Bias, 'do you seriously think I fear bullets?'

From the wall he took four paces towards the execution party, faced them squarely and held up his black, flat-topped hat. 'Soldiers, when I give the command to fire, fire straight at the heart. Wait until the command. It will be my last to you. I protest against my condemnation. I have fought a hundred battles for France,' he repeated, 'and not one against her.' 'It was all done,' Rochechouart remembered, 'with an attitude I can never forget: noble, calm and dignified beyond reproach.'[62]

Soldiers, Ney called, striking the hat down over his heart: *Fire!* and the united volley of musketry rang out.

As the smoke cleared so the officers hastened to examine the body. 'Thank God there was no need to administer the *coup de grâce*,' Rochechouart reported: the Marshal had died instantly.[63] He had fallen forward upon his face. They therefore turned the body over, laid it on a stretcher and left it on public view for fifteen minutes, ironically in accordance with a Revolutionary Law of 1793. Then four men placed the stretcher on a gun-carriage and escorted it two hundred yards to the Hospice de la Maternité. A Sister of Mercy bathed away the blood while others removed the cravat and coat. Later on that day 'the remains of Marshal Ney were visited by a number of important individuals: peers, generals, officers, ambassadors . . .'[64]

At the Tuileries Aglaé still sat and waited. At last, at half past nine, the Duke of Duras appeared. 'Madame,' he said shortly: 'I fear the audience you have been hoping for with His Majesty would now be . . . well, without object.' The Princess started to rise, horrified. 'You mean he has already been . . .' Then she fainted away.

In itself it seems but a minor detail to add; and certainly at the time it would have done little to relieve Aglaé's distress. Yet as Antoinette Gamot consoled Ney's widow on their cab-journey back to tell the Marshal's sons, so Count de Rochechouart was wondering how to explain away an unpleasant discovery at the Carrefour de l'Observatoire. One of his officers returning from the Hospice had confirmed what the Paris Commandant already suspected: that only eleven bullets struck the victim before he fell. And further examination revealed a damaged brick at the top of the wall. It could not have been a bad shot; for the firing-squad—whose names were a closely-kept secret—had been picked on their individual abilities as marksmen. No, the only explanation is that whoever disobeyed orders was clearly expressing a single soldier's regard for honour and patriotism—and his contempt for the Bourbons who had tried so hard to sully them. This anonymous act has been described as showing 'extremely good taste': and so it did. But in effect it was also the first act on behalf of an unusual, but always remarkable nation to restore Michel Ney to his due place in French history.

---

[1]   The Bourbon proscriptions were far more extensive than is generally realised. The new government promised that no one would be arrested for his opinions, but their police acted differently. Ney and La Bédoyère are the best-known victims because both were shot after the equivalent of show-trials. However Generals Barrois, Lallemand, Arrighi, Colbert, d'Ornano, Merlin de Douai and Hulin all suffered imprisonment, and within three months 12,000 officers, NCOs and ordinary rankers were exiled from Paris, with a further 700 kept under police surveillance.

Lefebvre-Desnoëttes escaped the firing-squad only by shaving off his luxuriant cavalry-moustache and posing as a commercial traveller. Subsequently he made his way to America. Delaborde went into hiding in France; but Chartrand and Mouton-Duvernet were caught and executed, while Drouot and Cambronne suffered imprisonment by the British and the French in turn. Lavalette was the amazing one though. While under sentence of death he escaped from the Conciergerie after one of his wife's visits dressed in her top-clothes. He was then smuggled out of France by Michael Bruce, Sir Robert Wilson and John Hely-Hutchinson (later the Earl of Donoughmore), an action for which the Bourbons tried and briefly incarcerated all three Britishers the following Spring.

2   *Les mémoires du Général de Caulaincourt.*
3   L. J. Gabriel de Chenier: *Histoire de la vie politique, militaire et administrative du Maréchal Davout*, Paris, 1866.
4   *Mémoires et Souveniers du Comte Lavalette.*
5   Louis Madelin: *Fouché*, Paris, 1905.
6   Houssaye: *1815. La seconde abdication—La terreur blanche*, Paris, 1905.
7   *The Hundred Days.*
8   *The Bravest of the Brave.*
9   *The Hundred Days.*
10 *The Bravest of the Brave.*
11 Houssaye: *1815.*
12 Ibid.
13 *The Letters of Private Wheeler*, ed. Capt. B. H. Liddell Hart, London, 1951. (Wheeler was promoted to Sergeant following the battle of Waterloo.)
14 Considering Ney's humane treatment of British officers and men in Spain and Portugal this did not seem an unreasonable hope.
15 *The Bravest of the Brave.*
16 Ibid.
17 Marmont: *Mémoires.*
18 *The Trial of Marshal Ney.*
19 Ernest Daudet: *Louis XVIII et le Duc Decazes*, Paris, 1899.
20 Ibid.
21 Ibid.
22 Ibid.
23 *The Trial of Marshal Ney.*
24 Ibid.
25 *The Bravest of the Brave.*
26 Ibid.
27 *The Trial of Marshal Ney.*
28 André Dupin: *Mémoires*, Paris, 1855-61.
29 *The Trial of Marshal Ney.*
30 Dupin: *Mémoires.*
31 Nicholas Berryer: *Souvenirs de 1774-1838*, Paris, 1839.
32 Henri Welschinger: *Le maréchal Ney, 1815*, Paris, 1893.
33 Berryer: *Souvenirs.*
34 *Histoire du Maréchal Davout.*
35 *Napoleon's Marshals.*
36 *The Bravest of the Brave.*
37 Why Wellington did not intervene on Ney's behalf remains an enigma and is a permanent blot on his reputation. The Duke was never a vindictive man, he had a professional soldier's healthy regard for his talented opponents and, of course, he possessed the necessary qualifications to intervene: being 'Commander-in-Chief of

the Troops of the Allies in France for the protection of the King'. Aglaé was not the only one who appealed to him. His own subordinate officers were clearly unhappy at Ney's being put on trial, and it is even claimed that the Duke of Broglie sought Wellington's help as a fellow-Mason. Elizabeth Longford offers two possible explanations for his non-intervention: both based upon his own letters. The first suggests that he was insulted by Louis XVIII and therefore stayed away from the Tuileries during the vital weeks when he might have demanded Ney's life. *My belief is,* he wrote to Lord Alvanley, *that they had offended me on purpose to drive me away, that I might not interfere to prevent Ney's death. For not long after it was accomplished, the King sent the Count of Artois to me to express great regret at my absence, and hopes that I . . . had not been offended in any way.* If true, then his wounded pride served the Bourbons' objectives most effectively. This letter is far more specific and accusative than the one he wrote to Miss Angela Burdett-Coutts from Walmer Castle in 1849. Admittedly the latter was composed thirty-five years after the event, but if one believes it then his attitude to Ney's case seems at best ostrich-like: *I might or I might not have had great influence on the King! I did not interfere in any way! I did not consider it my duty to interfere! There was clearly no claim upon me as the General Officer who had signed the treaty or convention of St. Cloud.* Anyway, apart from 'declining' by letter, he also told Aglaé face-to-face 'that his was not an independent voice'—and this she undoubtedly passed on to her sons. Napoleon Joseph Ney in growing up seriously considered provoking Wellington to fight a duel, while many years later Colonel Edgar Ney (the 4th son) is reported to have told the 2nd Duke of Wellington how bitter the family felt over the 1st Duke's failure to help.

38 Duke of Broglie: *Personal Recollections*, London, 1887.
39 Ibid.
40 *The Trial of Marshal Ney.*
41 Ibid.
42 *Le maréchal Ney, 1815.*
43 Ibid.
44 Dupin: *Mémoires.*
45 *Le maréchal Ney, 1815.*
46 Ibid.
47 Ibid.
48 Ibid.
49 Ibid.
50 Ibid.
51 Ibid.
52 Dupin: *Mémoires.*
53 *Le maréchal Ney, 1815.* Ney's sudden reference to Moreau can be misleading. Presumably he had in mind his former commander's claim that he was being tried for purely political reasons and eventually the truth would become known. But one must remember that Moreau definitely did indulge in plots against Napoleon even if he was innocent of any involvement in Cadoudal's assassination attempts.
54 Berryer, *Souvenirs.*
55 *Le maréchal Ney, 1815.*
56 Ibid.
57 Berryer: *Souvenirs.*
58 Comte de Rochechouart: *Souvenirs sur la Révolution, l'Empire et la Restauration,* Paris, 1899.
59 Ibid.
60 Ibid.

[61] *The Trial of Marshal Ney.*
[62] Rochechouart: *Souvenirs.*
[63] Ibid.
[64] Harold Kurtz, based on a police report. The body was quickly interred in a grave without a headstone at Père Lachaise in Paris.

ENDPIECE

Figures in a Timescape

Time, although it steals from each of our lives, can, in its forward movement, also act as a restorer . . .

Quarrels die down; people choose what they like to remember, or regret what they have lost. But then—usually much later—certain facts (previously 'overlooked') are apt to turn up and equip a new assessment. Once this happens the earlier *men of fashion* just seem to crumble away. So that a person of quality, who stood somewhat apart from them while he lived, can come through in reputation and impress us for the better.

Among Napoleon's other marshals, following the defeat at Waterloo there were hurried moves either to serve the restored Bourbons or accept an honourable retirement.

Not by all though. 'Murat died before a firing squad as well; court martialled after trying to regain his original kingdom of Naples. While Brune (in battle the Marshal with a charmed life) was shot and stabbed to death by a Royalist mob at Avignon.

Berthier's passing proved to be more mysterious. Even today there is argument about whether he was murdered or committed suicide: the latter out of remorse for ignoring Napoleon's call upon the return from Elba. He had escorted Louis XVIII to Ghent, but then went to Bavaria and at Bamberg was found dead in the street. 'Some say . . . murdered by six masked men, others that, seeing Allied troops on the march towards France, he hurled himself from a window . . .'[1]

Augereau and Mortier were both dismissed from the army for refusing to try Ney in 1815; and Moncey, doyen of the marshalate, went to prison for three months—although later both he and Mortier were reinstated.

Soult and Grouchy, who campaigned with Ney in Belgium, preferred exile to the certainty of a trial dominated by Bourbon vindictiveness. The former went to Dusseldorf; Grouchy escaped to America. But in 1819 they returned to France under a general amnesty. And in 1830 Soult was made Minister for War.

Victor died without anyone shedding tears in 1841. As for Marmont, arch-betrayer of the Emperor and a leader of those prepared to vote for Ney's death, his turncoat triumphs were of strictly limited duration. Initially loaded with honours by the Bourbons (a peer of France, 450,000 francs in gratuities, Major-General of the Royal Guard and Minister of State), he mistakenly sided with them again during the revolution of 1830 and as a result spent the last twenty-two years of his life in rootless exile. A modern Cain, he devoted himself to some overlong *Mémoires*: the model for all vain and rightly-ignored officers. (Only Talleyrand and perhaps Fouché had matched him for shallow egotism during the course of the Empire. But at least they were recognised as cunning men. Marmont owed Napoleon much more.)

Another who lived on and contributed to Ney's rehabilitation was Stendhal. As Henri Beyle he yawned his way through an impecunious middle-age as France's consul at Civita Vecchia. But as 'Stendhal' he was writing novels that in French literature have no equal until the best work of Flaubert and Proust. He remained an ardent Bonapartist; and his judgements upon the last Bourbons are devastating: the flabby autocracy (Louis XVIII, Charles X) transposed to venal opportunism under a self-styled bourgeois, Louis Philippe. In *The Charterhouse of Parma* Ney actually appears under his own name. The chapter called *War*, which so influenced Tolstoy and afterwards Hemingway, portrays his strength and resolve as exemplary in a battle that is otherwise all fragments and chaos.

Stendhal also appreciated the subtle powers of time. 'I run a chance,' he wrote, 'of being read in 1900 by the souls whom I love . . . '[2] He didn't die until 1842. And one wonders if he attended the first performance of Hector Berlioz' Requiem in 1837. For this monumental mass (composed for enlarged symphony orchestra including eight sets of timpani, choirs of both adults and children, and four brass bands) is really the companion to his prose. Requiem can be taken as a misnomer. What is displayed so finely is the passion, the great building and the optimism of Napoleon's France. It proclaims that then life meant more than just hard work and an improved taste to the bread.

However the writer who did rather more to restore Ney's good name was Philippe de Ségur. His two-volume history of the expedition by the French into Russia appeared in 1825, and suddenly a whole new generation of readers was acquainted with the facts concerning the Marshal's heroism there. Even the Duchess of Angoulême admitted to being moved by what Ségur had written. 'My God, why did we not know of such heroism earlier?' she exclaimed to her ladies-in-waiting. 'We ought to have saved the Marshal's life!'

Time at long last worked officially for the restoration of Ney in 1853 with the unveiling of Rude's statue. (Napoleon's remains were already back from St.-Helena and at Les Invalides: quite soon, ironically, to be covered by a slab of red porphyry purchased from Russia.) The Council of Paris positioned the Marshal's figure not only at the scene of his execution, but where it could be noticed and admired by families returning from their Sunday-afternoon perambulations in the Luxembourg gardens. By this date, of course, the Bourbons themselves had passed away. Frenchmen now regarded the First Empire as a 'golden age' in their history, and before very long, in the ordinary citizen's view, Michel Ney was returned to the band of national heroes. No matter that he lost at Waterloo. So did Napoleon; so did France. But the fall had been a glorious affair! From here on all criticisms of Ney would belong to that special province of military men—or be made by one or two novelists.

No less a sculptor than Auguste Rodin called Rude's statute 'the most beautiful in Paris'.

Certainly it is one to fire the imagination. Upwards, from spurred heels to raised right-arm with drawn sabre, its whole stance has been moulded to express defiance and courage. A powerful physique is tightly buttoned inside the short-tailed 'undress' uniform he wore when mounted; the hat is set at a jaunty angle. And his features, by an out-thrust chin and fierce glare project single-mindedness, a devotion to the good fight that today we would call 'true grit'. This is the Ney who charged at Elchingen, never thinking of his own safety. It is the seven o'clock-saviour of the *Grande-Armée* at Eylau and the real victor of Guttstadt and Friedland. It is even, everything considered, the man who finally captured La Haye Sainte but went down to defeat along with the Imperial Guard against Wellington's last resistance on the same evening. In other words: Ney the intrepid, the romantic, the 'bravest of the brave'.

However: the most beautiful statue in Paris offers a limited viewpoint. For what writing this book has taught me is that Michel Ney never was quite the simple, straightforward warrior of popular legend. Rude could hardly hope to show the man as he was in Spain and Portugal: disgusted with Masséna's corruption and strategy. Nor his humiliation later at the Camp of Boulogne. And how can a single bronze describe the tortuous relationship with Napoleon? The love-hate feelings that underwent their severest test at Borodino, when Ney gambled with his life once more to hold the Great Redoubt and Napoleon threw away the victory . . .

Some of the Marshal's complexities only revealed themselves as events wore on. His unexpected hard-headedness during the abdication scenes of 1814, for instance. Also his fatal hesitation at Quatre Bras: when he believed the Iron Duke was waiting for him again.

But the most difficult questions of all still surround his loyalty. The twisted knot of loyalties that took him back to Napoleon's side for 'The Hundred Days'. Does any other satisfactory explanation exist for the way he kept charging at Waterloo, using up his last men and horses in an effort to save the Empire?

Now another November has passed, and I work on these lines sitting in the upper-room of a friend's house in Ardèche. Outside it has been dry and sharp as the white wine of the region. Day after day of empty sunlight falling through a mistral-swept, quite intense cold. In the corner an oil-stove flickers fitfully. Meanwhile I am looking at a second statue of Ney.

It is a porcelain figure, less than a foot high, placed near my writing table. It has him muffled in a fur-lined greatcoat, triangular hat jammed down over his ears—and in his hands a rifle with fixed-bayonet, the only one of Napoleon's marshals ever to be represented thus. The face is pale with cold, but set firm, as if determined to meet the worst with equanimity. This is Ney on foot, fighting alongside his veterans to defend the retreat from Moscow. 'The last Frenchman to quit Russian soil . . .'[3]

To me, this is a Ney who grew up alongside the person Rude saw as being naturally and incurably romantic. The dashing *sabreur* turned into a skilled commander as well as a realist. Tragically though, the majority of his youthful ideals had been either chipped away or abruptly shattered in two decades of warfare. So that the man who came back alive from Russia's snows was practical, no longer trapped into any grand gestures or dreams—but also chastened.

What was he left with? He felt tired of campaigning. He wanted to take an early retirement. But one ideal remained intact. That of loyalty. In Ney's case to a mingling of his adopted France with personal honour, immediate family, the men he commanded and ultimately—despite their previous quarrels—Napoleon.

And it was loyalty which brought the twin aspects of his character, the romantic and the realist or the romance and the real, together again on the field at Waterloo.

Mount Felix, France, Iberia and Central Europe, 1973-1981

[1]   *Napoleon's Marshals.*
[2]   *Letters.*
[3]   *Napoleon's Marshals.*

Index

279